Topological Insulators

Volume 6

Marvin Cohen
Editor Emeritus

Contemporary Concepts of Condensed Matter Science

E. Burstein, A. H. MacDonald and P. J. Stiles

Topological Insulators

Volume 6

Marcel Franz
University of British Columbia
Vancouver, Canada

Laurens Molenkamp
University of Würzburg
Würzburg, Germany

ELSEVIER

AMSTERDAM • BOSTON • HEIDELBERG • LONDON
NEW YORK • OXFORD • PARIS • SAN DIEGO
SAN FRANCISCO • SINGAPORE • SYDNEY • TOKYO

Elsevier
The Boulevard, Langford Lane, Kidlington, Oxford, OX5 1GB, UK
Radarweg 29, PO Box 211, 1000 AE Amsterdam, The Netherlands
225 Wyman Street, Waltham, MA 02451, USA
525 B Street, Suite 1800, San Diego, CA 92101-4495, USA

First edition 2013

Library of Congress Cataloging-in-Publication Data
A catalog record for this book is available from the Library of Congress

British Library Cataloguing-in-Publication Data
A catalogue record for this book is available from the British Library

ISBN: 978-0-444-63314-9
ISSN: 1572-0934

For information on all Elsevier publications
visit our website at store.elsevier.com

Printed in the United States of America
Transferred to Digital Printing, 2013

13 14 15 16 17   9 8 7 6 5 4 3 2 1

Working together
to grow libraries in
developing countries

www.elsevier.com • www.bookaid.org

Contents

Part I
Theoretical Foundations
1. Topological Band Theory and the $\mathbb{Z}_2$ Invariant
C.L. Kane

2. Theory of Three-Dimensional Topological Insulators

Joel E. Moore

3. Models and Materials for Topological Insulators

Chaoxing Liu and Shoucheng Zhang

4. Field-Theory Foundations of Topological Insulators

Xiao-Liang Qi

Part II

Experimental Discoveries
5. Quantum Spin Hall State in HgTe

C. Brüne, H. Buhmann and L.W. Molenkamp

6. Topological Surface States: A New Type of 2D Electron Systems

M. Zahid Hasan, Su-Yang Xu, David Hsieh, L. Andrew Wray and Yuqi Xia

Contributors

L. Andrew Wray Joseph Henry Laboratories, Department of Physics, Princeton University, Princeton, NJ 08544, USA
Advanced Light Source, Lawrence Berkeley National Laboratory, Berkeley, California, 94305, USA
SLAC National Accelerator Laboratory, Stanford, CA 94305, USA

Haim Beindenkopf Joseph Henry Laboratories and Department of Physics, Princeton University, Princeton, NJ 08544, USA

C. Brüne Physikalisches Institut, Experimentelle Physik 3, Universität Würzburg, Würzburg, Germany

H. Buhmann Physikalisches Institut, Experimentelle Physik 3, Universität Würzburg, Würzburg, Germany

Xi Chen State Key Laboratory of Low-Dimensional Quantum Physics, Department of Physics, Tsinghua University, Beijing 100084, China

Ke He Beijing National Laboratory for Condensed Matter Physics, Institute of Physics, The Chinese Academy of Sciences, Beijing 100190, China

David Hsieh Joseph Henry Laboratories, Department of Physics, Princeton University, Princeton, NJ 08544, USA
Department of Physics, California Institute of Technology, Pasadena, CA 91125, USA

C.L. Kane Department of Physics and Astronomy, University of Pennsylvania, Philadelphia, PA 19104, USA

Chaoxing Liu Department of Physics, The Pennsylvania State University, University Park, Pennsylvania 16802-6300, USA

Xucun Ma Beijing National Laboratory for Condensed Matter Physics, Institute of Physics, The Chinese Academy of Sciences, Beijing 100190, China

L.W. Molenkamp Physikalisches Institut, Experimentelle Physik 3, Universität Würzburg, Würzburg, Germany

Joel E. Moore Department of Physics, University of California, Berkeley, CA 94720, USA
Materials Sciences Division, Lawrence Berkeley National Laboratory, Berkeley, CA 94720, USA

Alberto F. Morpurgo University of Geneva, Switzerland

Naoto Nagaosa Department of Applied Physics, University of Tokyo, Tokyo 113-8656, Japan
Cross Correlated Materials Research Group (CMRG) and Correlated Electron Research Group (CERG), ASI, RIKEN, Wako 351-0198, Japan

Jeroen B. Oostinga University of Würzburg, Germany

Xiaoliang Qi Department of Physics, Stanford University, Stanford, CA 94305, USA

Pedram Roushan Joseph Henry Laboratories and Department of Physics, Princeton University, Princeton, NJ 08544, USA

Ari M. Turner Institute for Theoretical Physics, University of Amsterdam, Science Park 904, P.O. Box 94485, 1090 GL Amsterdam, The Netherlands

Ashvin Vishwanath Department of Physics, University of California, Berkeley, CA 94720, USA

Yuqi Xia Joseph Henry Laboratories, Department of Physics, Princeton University, Princeton, NJ 08544, USA

Su-Yang Xu Joseph Henry Laboratories, Department of Physics, Princeton University, Princeton, NJ 08544, USA
    Advanced Light Source, Lawrence Berkeley National Laboratory, Berkeley, California, 94305, USA

Qi-Kun Xue State Key Laboratory of Low-Dimensional Quantum Physics, Department of Physics, Tsinghua University, Beijing 100084, China
    Beijing National Laboratory for Condensed Matter Physics, Institute of Physics, The Chinese Academy of Sciences, Beijing 100190, China

Ali Yazdani Joseph Henry Laboratories and Department of Physics, Princeton University, Princeton, NJ 08544, USA

M. Zahid Hasan Joseph Henry Laboratories, Department of Physics, Princeton University, Princeton, NJ 08544, USA
    Princeton Institute for the Science and Technology of Materials, School of Engineering and Applied Science, Princeton University, Princeton, NJ 08544, USA

Shoucheng Zhang Department of Physics, Stanford University, Stanford, CA 94305, USA

Contemporary Concepts Of Condensed Matter Science

Contemporary Concepts of Condensed Matter Science is dedicated to clear expositions of the *concepts* underlying experimental, theoretical, and computational developments, new phenomena and probes at the advancing frontiers of the rapidly evolving subfields of condensed matter science. The term "condensed matter science" is central, because the boundaries between condensed matter physics, condensed matter chemistry, material science, and biomolecular science are disappearing.

The overall goal of each volume in the series is to provide the reader with an intuitively clear discussion of the underlying concepts and insights that are the "driving force" for the high-profile major developments of the subfield, while providing only the amount of theoretical, experimental, and computational details, data, and results that would be needed for the reader to gain a conceptual understanding of the subject. This will provide an opportunity for those in other areas of research, as well as those in the same area, to have access to the concepts underlying the major developments at the advancing frontiers of the subfield.

Each volume (∼250 printed pages) is to have a preface written by the volume editor(s), which includes an overview of the highlights of the theoretical and experimental advances of the subfield and their underlying concepts. It also provides an outline of the sections on key topics, selected by the volume editor(s), and authored by key scientists recruited by the volume editor(s), that highlight the most significant developments of the subfield.

Each section of a given volume will be devoted to a major development of the subfield. The sections will be self-contained—it should not be necessary to go to other sources to follow the presentation or the underlying science. The list of references will include the titles of the publications. The section will also provide a list of publications where the reader can find more detailed information about the subject. The volume editor(s) will remain in touch with the section-authors to insure that the level and presentation of the material is in line with the objective of the series.

The volumes in this series emphasize clear writing whose goal is to describe and elucidate the concepts that are the driving force for the exciting developments of the field. The overall goal is "reader-comprehension" rather than to be "comprehensive," and the goal of each Section is to cover "key conceptual aspects" of the subject, rather than be an "in-depth review." The model is a well-presented colloquium (not a seminar!) which invites the audience to "come think with the speaker" and which avoids in-depth experimental, theoretical, and computational details.

The audience for the volumes will have wide-ranging backgrounds and disparate interests (academic, industrial, and administrative). The "unique" approach of focusing on the underlying concepts should appeal to the entire community of condensed matter scientists, including graduate students and postdoctoral fellows, as well as to individuals, not in the condensed matter science community, who seek an understanding of the exciting advances in the field.

The volumes are particularly suited to graduate students and postdoctoral researchers who wish to broaden their perspectives, or who plan to enter a field covered by a volume in the series. They also provide an excellent way for researchers entering interdisciplinary efforts to obtain an understanding of concepts outside their specialized focus, but central to their new endeavors.

Quantum mechanics of electrons moving in the spatially periodic potential of the ionic lattice—the venerable band theory of solids—forms one of the key pillars of modern physics. It underlies our understanding of all crystalline solids, explains why some are insulators and some metals, and by extension enables much of the technological progress in the information age. Developed originally by Felix Bloch in 1928 the band theory has been long thought fully understood and devoid of surprises. Yet, starting in 2004, our understanding of crystalline solids underwent an unexpected revolution when theorists began to explore the effects of topology on the physics of band insulators. It turns out that the inclusion of ideas from topology and the homotopy theory brings new key elements into this old field: whereas previously it has been thought that all band insulators are essentially equivalent, the topological band theory predicts two fundamentally distinct classes of band insulators in two spatial dimensions and 16 classes in three dimensions. These "topological insulators" exhibit a host of remarkable physical properties including topologically protected gapless surface states, unusual electromagnetic responses, and emergent exotic quasiparticles, previously thought impossible in such systems. After nearly 80 years of existence the band theory of solids predicted a new state of quantum matter.

Within a short period of time this new state of quantum matter has been discovered experimentally both in two-dimensional thin film structures and in three-dimensional crystals and alloys. Theoretical and experimental studies of these materials are ongoing with the goal of attaining fundamental physical understanding and exploiting them in future practical applications. It now appears that topological insulators are quite common among the narrow-bandgap semiconductors with heavy elements; today there are dozens of substances predicted to exhibit this behavior of which many have been experimentally confirmed. Aside from the canonical examples of HgTe quantum wells and Bi_2Se_3, Bi_2Te_3 crystals, topologically non-trivial behavior has been predicted and observed to occur in a host of materials, including perovskites, pyrochlores, chalcogenides, Heusler alloys, and heavy-fermion compounds. There exists even a naturally occurring topological insulator, a mineral called Kawazulite (after Kawazu mine in central Japan), with a chemical formula $Bi_{2.07}(Te_{1.95} Se_{0.97}S_{0.03})$, close to the archetypal topological insulators Bi_2Te_3 and Bi_2Se_3.

The discoveries outlined above of course did not happen in a vacuum but were enabled and foreshadowed by a body of intervening theoretical and experimental work. The early seminal results of Tamm and Shockley in 1930s that elucidated the role of the electron surface states in ordinary semiconductors, paved the way for the discovery of the transistor and led to the advent of the modern semiconductor technology. Theoretical work in the 1980s on the quantum Hall fluids by Laughlin, Wen, Thouless, and Haldane uncovered the important role of topology in solids while parallel developments in the context of superfluids were pioneered by Volovik and others. An important connection between the band inversion and the surface states also had been noted by Volkov and Pankratov, in the 1980s. On the experimental front, studies of HgTe, Bi_2Te_3, and Bi_2Se_3 have been ongoing since the 1950s and the existence of the surface states in the latter two can in fact be inferred from this early work. Interestingly, Bi_2Te_3 has been widely used over the past decades as a highly performing thermoelectric material in practical applications, including commercially available refrigerators.

Even with all these ingredients in place it still took a major leap forward to predict and discover topological insulators. The key to this effort was gaining a thorough understanding of the role of the time-reversal symmetry in electron systems, culminating in the mathematical formulation of the Z_2 topological invariant. Subsequent experimental discoveries relied in turn on state-of-the-art growth facilities and an arsenal of probes developed earlier to study materials ranging from high-T_c cuprates to graphene.

This book chronicles the work done worldwide that has led to the recent discoveries of new topological states of quantum matter. Its 11 chapters, authored by leading researchers in the field, are designed to provide the reader with a comprehensive overview of the subject. Chapters 1–4 lay down the theoretical framework by explaining the physical background and the new ideas that were essential in formulating the theory of topological insulators, constructing the relevant models, and predicting their physical properties. Chapters 5–8 describe the experimental discoveries of topological insulators and the subsequent effort at their detailed characterization. The remaining chapters, 9–11, are forward-looking. They deal with the most recent experimental studies, theoretical work on the new related states of quantum matter, and visions for future technological applications enabled by them.

Volume editors
Marcel Franz and Laurens W. Molenkamp
April 2013

Part I

Theoretical Foundations

Topological Band Theory and the $\mathbb{Z}_2$ Invariant

C.L. Kane

Department of Physics and Astronomy, University of Pennsylvania, Philadelphia, PA 19104, USA

Chapter Outline Head

Topological Insulators. http://dx.doi.org/10.1016/B978-0-444-63314-9.00001-9

1 INTRODUCTION

A central goal in condensed matter physics is to characterize phases of matter. Some phases, such as magnets and superconductors, can be understood in terms of the symmetries that they spontaneously break. In recent decades, it has become apparent that there can exist a more subtle kind of order in the pattern of entanglement in a quantum ground state. The concept of topological order was introduced to describe the quantum Hall effect [1,2]. The quantum Hall state does not break any symmetries, but it has fundamental properties (such as the quantized Hall conductivity, and the number of conducting edge modes) that are insensitive to smooth changes in materials parameters and cannot change unless the system passes through a quantum phase transition. These properties can be understood as consequences of the topological structure of the quantum state.

While the topological characterization of the quantum Hall effect is an old story, interest in topological order has been rekindled by the discovery of topological insulators [3–13]. A topological insulator, like an ordinary insulator, has a bulk energy gap separating the highest occupied electronic band from the lowest empty band. The surface (or edge in two dimensions) of a topological insulator, however, necessarily has gapless electronic states that are protected by time reversal symmetry. Like the integer quantum Hall state, which has unique gapless chiral edge states [14], the surface (or edge) states of a topological insulator are topologically protected and exhibit a conducting state with properties that are unlike any other known 1D or 2D electronic systems.

The concept of topological order [2] is often used to characterize fractional quantum Hall states [15], which require an inherently many body approach to understand [16]. However, topological considerations also apply to the simpler integer quantum Hall states [1], for which an adequate description can be formulated in terms of single particle quantum mechanics. In this regard, topological insulators are similar to the integer quantum Hall effect. Due to the presence of a single particle energy gap, electron-electron interactions do not modify the state in an essential way. The phenomenology of topological insulators can be understood in the framework of the band theory of solids [17]. It is remarkable that after more than 80 years, there are still treasures to be uncovered within band theory.

In this chapter we will provide a pedagogical introduction to the foundations of topological band theory and explain how these ideas can be used to characterize the integer quantum Hall effect and topological insulators.

2 TOPOLOGY AND BAND THEORY

We begin by reviewing the key elements of topology and band theory. We will introduce the notion of topological equivalence and explain its role in band theory, and we will describe the deep connection between the bulk topology

(a)   **(b)**

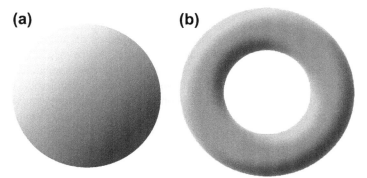

FIGURE 1 The surfaces of a sphere ($g = 0$) and a doughnut ($g = 1$) are distinguished topologically by their genus g.

and protected boundary modes. In Section 2.3. We will discuss the Berry phase, which is a key conceptual tool for the analysis of topological phenomena.

2.1 Topology

Topology is a branch of mathematics concerned with geometrical properties objects that are insensitive to smooth deformations. This is most easily illustrated by the simple example of closed two-dimensional surfaces in three dimensions (see Fig. 1). A sphere can be smoothly deformed into many different shapes, such as the surface of a disk or a bowl. But a sphere cannot be smoothly deformed into the surface of a doughnut. A sphere and a doughnut are distinguished by an integer topological invariant called the genus, g, which is essentially the number of holes. Since an integer cannot change smoothly, surfaces with different genus cannot be deformed into one another, and are said to be topologically distinct. Surfaces that can be deformed into one another are topologically equivalent. Determining the topological invariants that characterize a given object is an interesting math problem. For surfaces, a beautiful theorem, known as the Gauss-Bonnet theorem, states that the integral of the Gaussian curvature, K over a surface defines an integer topological invariant called the Euler characteristic [18],

$$\chi = \frac{1}{2\pi} \int_S K dA. \tag{1}$$

It can easily be checked that $\chi = 2$ for a sphere of radius R, where $K = 1/R^2$. More generally, the Euler characteristic is quantized and related to the genus by $\chi = 2 - 2g$. The topological invariants that we will be concerned with in this chapter are similar, though they will characterize more abstract objects.

2.2 Band Theory

How can topology be used to characterize phases of matter? Here we will explain the topological classification of insulators. An insulator is a material that has an energy gap for electronic excitations, which separates the ground state from all excited states. This allows for a notion of topological equivalence based on the principle of adiabatic continuity. Insulators are equivalent if they can be changed into one another by slowly changing the Hamiltonian, such that the system always remains in the ground state. Such a process is possible if there is an energy gap E_G, which sets a scale for how slow the adiabatic process must be. Thus, insulators are topologically equivalent if there exists an adiabatic path connecting them along which the energy gap remains finite. It follows that connecting topologically inequivalent insulators necessarily involves a phase transition, in which the energy gap vanishes.

The topological classification of general gapped many body states is a formidable problem that has not been completely solved. A tremendous simplification occurs if we consider a subclass of states that can be described by the band theory of solids. Such band insulators can be effectively described in the independent electron approximation, where the many body ground state is represented as a Slater determinant of single particle states. This does not mean that electron interactions are being ignored. The existence of an energy gap means that the many body state remains topologically equivalent when finite strength interactions are turned on. We thus assume that the state in question can be adiabatically connected to noninteracting electrons, and topologically classify the band structures. It is then important to address whether the topological distinctions found within band theory persist when interactions are added. We shall see that this is indeed the case.

A second key assumption we will make in this chapter is that the material is crystalline, which allows us to take advantage of translation symmetry. This assumption can be relaxed, and we will touch on the issue of disorder later in this chapter. Translation symmetry allows the single particle states to be labeled by their crystal momentum $\mathbf{k}$. According to Bloch's theorem, they may be written $|\psi(\mathbf{k})\rangle = e^{i\mathbf{k}\cdot\mathbf{r}}|u(\mathbf{k})\rangle$, where $|u(\mathbf{k})\rangle$ is a cell periodic eigenstate of the Bloch Hamiltonian,

$$H(\mathbf{k}) = e^{i\mathbf{k}\cdot\mathbf{r}} H e^{-i\mathbf{k}\cdot\mathbf{r}}. \qquad (2)$$

$H(\mathbf{k})$, or equivalently its eigenvalues $E_n(\mathbf{k})$ and eigenvectors $|u_n(\mathbf{k})\rangle$, defines the band structure. An insulating band structure has an energy gap separating the highest occupied band from the lowest empty band. Lattice translation symmetry implies $H(\mathbf{k}+\mathbf{G}) = H(\mathbf{k})$ for reciprocal lattice vectors $\mathbf{G}$. The crystal momentum is therefore defined in the periodic Brillouin zone, with $\mathbf{k} \equiv \mathbf{k}+\mathbf{G}$, which has the topology of a torus T^d in d dimensions. Thus, an insulating band structure can be viewed as a mapping from the Brillouin zone torus to the space of Bloch Hamiltonians with an energy gap.

2.3 Topological Band Theory and the Bulk-Boundary Correspondence

One of the objects of topological band theory is to classify topologically distinct Hamiltonians $H(\mathbf{k})$. By doing so, we are classifying distinct electronic phases. The most important consequence of this occurs when there is a spatial interface between two topologically distinct phases. Imagine an interface where a crystal slowly interpolates as a function of distance y between a two topologically distinct phases. Somewhere along the way the energy gap has to go to zero, because otherwise the two phases would be equivalent. There will therefore be low energy electronic states bound to the region where the energy gap passes through zero.

A second object of topological band theory is thus to characterize those gapless states. We will see that they too can be classified topologically, and that there is a deep principle, which we will refer to as the bulk-boundary correspondence, which relates the boundary topological invariants to the difference in the bulk topological invariants. This interplay between topology and gapless modes is a ubiquitous phenomenon in physics, and has appeared in many contexts [19, 21–24].

2.4 Berry Phase, and the Chern Invariant

A key role in topological band theory is played by the Berry phase [25]. The Berry phase arises because of the intrinsic phase ambiguity of a quantum mechanical wavefunction. The Bloch states are invariant under the transformation

$$|u(\mathbf{k})\rangle \rightarrow e^{i\phi(\mathbf{k})}|u(\mathbf{k})\rangle. \tag{3}$$

This transformation is reminiscent of an electromagnetic gauge transformation, and invites the definition of the Berry connection,

$$\mathbf{A} = -i\langle u(\mathbf{k})|\nabla_{\mathbf{k}}|u(\mathbf{k})\rangle. \tag{4}$$

$\mathbf{A}$ is similar to the electromagnetic vector potential. Under (3) it transforms as $\mathbf{A} \rightarrow \mathbf{A} + \nabla_{\mathbf{k}}\phi(\mathbf{k})$. Though $\mathbf{A}$ is not gauge invariant, the analog of magnetic flux is. For any closed loop C in $\mathbf{k}$ space, we may define the Berry phase,

$$\gamma_C = \oint_C \mathbf{A} \cdot d\mathbf{k} = \int_S \mathcal{F} d^2\mathbf{k}, \tag{5}$$

where $\mathcal{F} = \nabla \times \mathbf{A}$ defines the Berry curvature. For notational simplicity, we will assume here that $\mathbf{k}$ is two dimensional. The generalization to higher dimensions is straightforward.

The Berry phase has many applications in physics, and describes the phase acquired under an adiabatic cycle. In the present context, it will be useful for classifying loops in momentum space. In the following section we will attach physical meaning to it.

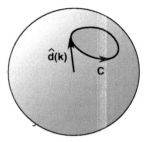

FIGURE 2 The Berry phase in a two band theory is given by half the solid angle swept out by $\hat{\mathbf{d}}(\mathbf{k})$.

It is useful to understand the Berry phase for the simplest two level Hamiltonian, which may be expressed in terms of Pauli matrices $\vec{\sigma}$ as

$$H(\mathbf{k}) = \mathbf{d}(\mathbf{k}) \cdot \vec{\sigma} = \begin{pmatrix} d_z & d_x - id_y \\ d_x + id_y & -d_z \end{pmatrix}. \tag{6}$$

This Hamiltonian has eigenvalues $\pm|\mathbf{d}|$. We ignore a term proportional to the identity because that does not affect the eigenvectors, which depend only on the unit vector $\hat{\mathbf{d}} = \mathbf{d}/|\mathbf{d}|$. $\hat{\mathbf{d}}$ can be viewed as a point on a sphere S^2.

A classic result, shown by Berry [25], is that for a loop C the phase associated with the ground state, obtained from (4) and (5) is (see Fig. 2)

$$\gamma_C = \frac{1}{2} \left(\text{Solid angle swept out by } \hat{\mathbf{d}}(\mathbf{k}) \right). \tag{7}$$

In particular, when C corresponds to a 2π rotation of $\hat{\mathbf{d}}$ in a plane, the Berry phase is π. The Berry curvature is given by the solid angle per unit area in $\mathbf{k}$ space, which is simply half the solid angle element for the mapping $\hat{\mathbf{d}}(\mathbf{k})$,

$$\mathcal{F} = \frac{1}{2} \epsilon_{ij} \hat{\mathbf{d}} \cdot (\partial_i \hat{\mathbf{d}} \times \partial_j \hat{\mathbf{d}}). \tag{8}$$

An important consequence of (8) is that the Berry curvature integrated over a closed 2D space (such as a 2D Brillouin zone T^2) is a multiple of 2π that is equal to the number of times $\hat{\mathbf{d}}(\mathbf{k})$ wraps around the sphere as a function of $\mathbf{k}$. This defines a topological invariant called the Chern number [18], which for a closed surface S may be expressed as

$$n = \frac{1}{2\pi} \int_S \mathcal{F} d^2 \mathbf{k}. \tag{9}$$

The quantization of the Chern number is more general than the two band model, and follows from the fact that for a loop C on a closed surface, the "inside" of C used in (5) is arbitrary, so that the surface integral over the inside and the outside must agree with one another up to a multiple of 2π. It follows that the Berry curvature integrated over the entire surface must be $2\pi n$. This quantization is also closely related to the quantization of the Dirac magnetic monopole. Note the similarity with (1). $\mathcal{F}$ can be viewed as a curvature, similar to the Gaussian curvature K.

In the following sections we will discuss the physical meaning and consequences of this and other topological invariants.

3 ILLUSTRATIVE EXAMPLE: POLARIZATION AND TOPOLOGY IN ONE DIMENSION

In this section we will consider the simplest setting for topological band theory, which is one dimension. This will allow us to introduce several key concepts in their simplest form, including the electric polarization, the Chern number, and topologically protected boundary states. We will introduce the Su, Schrieffer Heeger (SSH) [20] model, which provides a simple and solvable theory that illustrates these ideas.

3.1 Polarization as a Berry Phase

In elementary electrostatics, the electric polarization $\mathbf{P}$ is defined as the dipole moment per unit volume. Polarization leads to bound charges in the bulk $\rho_b = -\nabla \cdot \mathbf{P}$ and on the surface $\sigma_b = \mathbf{P} \cdot \hat{n}$. In one dimension, the polarization P is related to the end charge,

$$Q_{\text{end}} = P. \tag{10}$$

In this section, we show how to determine the polarization from a 1D band structure. The problem is not trivial because a band structure is generally defined on a system with periodic boundary conditions, so Q_{end} is inaccessible. The solution, which has emerged in the theory of ferroelectricity [26–29], provides a beautiful application of Berry's phase. The 1D polarization is the Berry phase of the occupied Bloch wavefunctions around the 1D Brillouin zone,

$$P = \frac{e}{2\pi} \oint_{BZ} A(k)dk. \tag{11}$$

The integral is over the 1D Brillouin zone, which is equivalent to a circle S^1. Detailed derivations of (11) can be found in the literature. Here we will motivate the result using physical arguments.

The first piece of circumstantial evidence is that both the polarization and the Berry phase share a similar intrinsic ambiguity. Q_{end} in (10) is not completely determined because integer charges can be added or removed from the ends without changing the bulk. Thus, $Q_{\text{end}} = P$ mod e. P in (11) has a similar

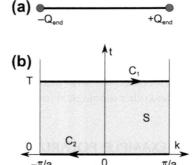

FIGURE 3 (a) The polarization in 1D determines the end charge modulo e. (b) The change in the polarization is given by the difference of Berry phases on loops C_1 and C_2.

ambiguity because under a gauge transformation $|u(k)\rangle \to e^{i\phi(k)}|u(k)\rangle$ with $\phi(\pi/a) - \phi(-\pi/a) = 2\pi n$, $P \to P + ne$. At first sight, this appears to violate (5), which implies γ_C is gauge invariant. However, it must be noted that the 1D Brillouin zone defines a special kind of loop that is not the boundary of an interior, so that the reasoning in (5) does not apply. On such nontrivial loops, γ_C changes by $2\pi n$ under "large" gauge transformations, in which the phase winds by $2\pi n$ around the hole.

Though the polarization and the Berry phase are both ambiguous up to an integer, *changes* in either quantity are well defined and gauge invariant. Imagine that the band structure is a function of a control parameter t. Then the change in the polarization between $t = 0$ and $t = T$ is

$$\Delta P = P_{t=T} - P_{t=0} = \frac{e}{2\pi}\left[\oint_{C_1} - \oint_{C_2}\right] \mathbf{A} \cdot dk$$

$$= \frac{e}{2\pi}\int_S \mathcal{F}\, dk\, dt. \tag{12}$$

In this case, as shown in Fig. 3, $C_1 - C_2$ is the boundary of an interior, so Stokes' theorem can be used to express the integral in terms of the gauge invariant Berry curvature $\mathcal{F}$.

A second piece of circumstantial evidence is that the polarization is something like the expectation value of er, while the Berry phase is something like the expectation value of $2\pi i \nabla_k$, so if you are willing to believe $r \sim i\nabla_k$ the equality follows. This is not convincing, though, because neither $\langle r \rangle$ nor $\langle i\nabla_k \rangle$ are defined for extended Bloch states. A somewhat more rigorous version of this argument is to introduce a basis of localized Wannier states associated with lattice sites R [28],

$$|\phi(R)\rangle = \oint_{BZ} \frac{dk}{2\pi} e^{-ik(R-r)}|u(k)\rangle. \tag{13}$$

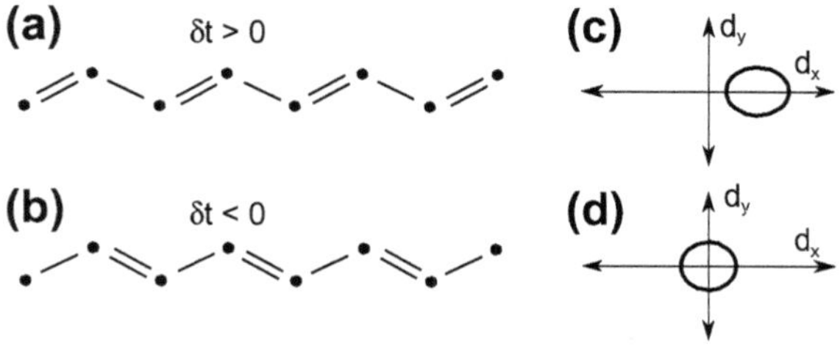

(a) δt > 0

(b) δt < 0

(c) d_y, d_x

(d) d_y, d_x

FIGURE 4 (a,b) The two distinct groundstates of the SSH model. Figures (c) and (d) show the unit vector $\hat{\mathbf{d}}(k)$.

The Wannier states depend on the gauge choice for $|u(k)\rangle$, but for a sufficiently smooth gauge they are localized so that $\langle r \rangle$ is well defined. The polarization can then be written

$$P = e\langle \phi(R)|r - R|\phi(R)\rangle = \frac{ie}{2\pi}\oint dk \langle u(k)|\nabla_k|u(k)\rangle. \tag{14}$$

3.2 Su, Schrieffer, Heeger Model

In this section we consider an important model that illustrates the analysis of the preceding sections. It will also provide a setting for introducing domain wall states. The Su Schrieffer Heeger (SSH) [20] model was introduced as a model of the conducting polymer polyacetylene, which at half filling undergoes a Peierls instability to a dimerized state. What makes this interesting is that, as shown in Fig. 4, there are two different dimerized states. We will see that there is a sense in which these two states are topologically distinct (and a sense in which they are not). Importantly, interfaces between the two states are associated with zero energy boundary states. The SSH model provides the simplest two band model for describing these topological phenomena.

To model polyacetylene, SSH introduced a 1D tight binding model

$$H = \sum_i (t + \delta t)c_{Ai}^\dagger c_{Bi} + (t - \delta t)c_{Ai+1}^\dagger c_{Bi} + h.c. \tag{15}$$

Here we have arbitrarily defined a unit cell with two atoms, labeled A and B. The dimerization is characterized by δt, which leads to an energy gap. The two dimerization patterns are distinguished by the sign of δt. For simplicity, we consider spinless electrons here (even though the real SSH model includes spin). The filling is one electron per unit cell.

To analyze (15) we Fourier transform and write

$$H = \sum_k H_{ab}(k)c_{ak}^\dagger c_{bk}, \tag{16}$$

where

$$H(k) = \mathbf{d}(k) \cdot \vec{\sigma} \tag{17}$$

and

$$d_x(k) = (t + \delta t) + (t - \delta t)\cos ka,$$
$$d_y(k) = (t - \delta t)\sin ka, \qquad\qquad (18)$$
$$d_z(k) = 0.$$

Viewing the two band $H(k)$ in terms of $\mathbf{d}$, it is important to note that $d_z = 0$. It follows that $H(k)$ possesses a "chiral" symmetry defined by the operator $\Pi = \sigma^z$, which *anticommutes* with the Hamiltonian: $\{H(k), \Pi\} = 0$. This chiral symmetry leads to a particle-hole symmetric spectrum because any eigenstate $|u_E\rangle$ with energy E has a partner $|u_{-E}\rangle = \Pi|u_E\rangle$ with energy $-E$. This symmetry is not intrinsic, though, and will be violated in real polyacetylene (for example by second neighbor hopping). Nonetheless, it is useful to consider its effects.

Consider the polarization, which can be expressed in terms of the Berry phase using (4) and (11). For $\delta t > 0$, $d_x(k) > 0$ for all k so $\hat{\mathbf{d}}(k)$ sweeps out no solid angle, and $P = 0$. For $\delta t < 0$, however, $d_x(k \sim \pi/a) < 0$, so that $\hat{\mathbf{d}}(k)$ rotates by 2π leading to Berry phase π and $P = e/2$. The polarization can be understood easily in the strong coupling limit, $|\delta t| = t$, in which electrons reside in localized states on the strong bonds. It is clear that passing from the $\delta t = +t$ state to the $\delta t = -t$ state involves moving each electron over by half a unit cell, resulting in a polarization $e/2$.

In the presence of the chiral symmetry, the polarization must be a multiple of $e/2$. On the other hand, if the symmetry constraint is relaxed, then $\mathbf{d}$ can tip out of the xy plane, and the polarization can vary continuously. Thus, in general, there is no topology in 1D: all 1D insulating band structures are topologically equivalent. But imposing chiral symmetry leads to topologically distinct states that are distinguished by their quantized polarization. To get from the $\delta t > 0$ state to the $\delta t < 0$ state without violating the chiral symmetry requires a point where $\mathbf{d}$ vanishes, signifying a quantum phase transition. This is an example of the general principle that enhanced symmetry can lead to richer topological structure.

Real polyacetylene does not have chiral symmetry, but it does have the spatial symmetry of inversion about the center of a bond. This is expressed by $H(-k) = \sigma^x H(k)\sigma^x$, and leads to a quantized polarization, even in the absence of chiral symmetry. The SSH model also resembles the Bogoliubov-de Gennes Hamiltonian for a one-dimensional topological superconductor. In that case there is an intrinsic particle-hole symmetry, expressed by $H(-k) = -\sigma^z H(k)^*\sigma^z$, which also leads to distinct topological phases.

3.3 Domain Wall States and the Jackiw Rebbi Model

The interface between the two ground states of polyacetylene gives rise to a soliton state with a polarization charge $\pm e/2$ on the boundary. The electronic structure in the presence of such a domain wall has a zero energy midgap state.

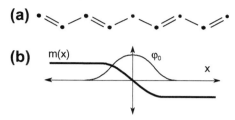

FIGURE 5 (a) A domain wall in the SSH model. (b) The midgap state is associated with a change in sign of the mass m.

The $\pm e/2$ states arise when this "zero mode" is empty/occupied. The existence of the zero mode is easily understood in the strong coupling limit $|\delta t| = t$, since there is an unpaired site on the boundary, as shown in Fig. 5(a). This state is protected in the sense that it is impossible to get rid of it without closing the bulk energy gap. Its existence can be traced to the fact that it is on an interface between topologically distinct states. Such topological zero modes were first found in a 1D field theory by Jackiw and Rebbi [19], who presented a simple exact solution for the zero mode. The SSH model provides a physical realization of the Jackiw Rebbi model.

Here we will present the Jackiw Rebbi solution, starting from the SSH model (15). The chiral symmetry $\{\Pi, H(k)\} = 0$ implies that eigenstates come in pairs at $\pm E$. It is possible, however, for a state at $E = 0$ to be its own partner, $|u_0\rangle = \Pi|u_0\rangle$. If this is the case, then this zero mode is topologically protected, because to move it away from $E = 0$ would require another state to appear out of nowhere.

To explicitly construct this zero mode, it is helpful to develop a low energy continuum theory for (15). We consider the limit $\delta t \ll t$ and focus on the low energy states near $k = \pi/a$. We thus let $k = \pi/a + q$ and expand for small q. In real space we then let $q \to -i\partial_x$. This results in a low energy Hamiltonian of the form

$$H = -iv_F\sigma^x\partial_x + m\sigma^y, \tag{19}$$

where $v_F = ta$ and $m = 2\delta t$. This Hamiltonian has the form of a massive $1 + 1$D Dirac Hamiltonian, with spectrum $E(q) = \pm\sqrt{(v_Fq)^2 + m^2}$.

To describe the zero mode we allow m to vary spatially with a kink such that $m(x \to +\infty) < 0$ and $m(x \to -\infty) > 0$, as shown in Fig. 5(b). A zero energy solution $H|u\rangle = 0$ can easily be constructed by multiplying on the left by $i\sigma^x$ and considering eigenstates $|z\pm\rangle$ of σ^z with eigenvalue ± 1. Integrating the resulting first-order equation leads to a single normalizable solution,

$$\psi_0(x) = e^{-\int_0^x m(x')/v_F}|z+\rangle. \tag{20}$$

This zero mode is topological in that it does not depend on the precise form of $m(x')$. It only depends on the sign change. It is guaranteed to be at zero energy if

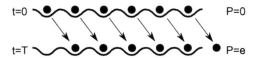

FIGURE 6   In a Thouless charge pump the polarization changes by e in each cycle.

there is chiral or particle-hole symmetry. The antikink with $\pm m(x \to \pm\infty) > 0$ is similar, but involves $|z-\rangle$.

3.4 Thouless Charge Pump, and the Chern Invariant

We have seen that without extra symmetries, such as chiral symmetry, particle-hole symmetry, or inversion symmetry there are no topological band structures in one dimension. In this section we describe another topological phenomenon that occurs in 1D in the absence of symmetries.

Consider a one-dimensional insulating Hamiltonian that changes with time adiabatically in a cyclic manner, so that $H(k, t) = H(k, t + T)$. At every time t, the system has a polarization P, that is well defined up to an integer charge e. As t changes, the change in P is completely defined. After one full cycle, the Hamiltonian returns to its original value. However, since the polarization is only defined modulo e it is possible that the polarization changes by

$$\Delta P = ne \qquad (21)$$

for integer n (see Fig. 6). Such a system defines a topological charge pump [30,31], in which n electrons are transported across the system in every cycle, despite the fact that the energy gap remains finite.

In Section 3.1, we showed that the change in the P in one cycle, given by 12, is related to the Berry curvature integrated for $-\pi/a < k < \pi/a$ and $0 < t < T$. Since $H(t) = H(t + T)$, t is defined on a circle. Thus, the domain of integration of the surface integral in (12) is the 2-torus, T^2 defined by k and t. It follows that the integer n in (21) is the Chern number, defined in (9),

$$n = \frac{1}{2\pi} \int_{T^2} \mathbf{F} \, dk \, dt. \qquad (22)$$

We thus conclude that the cyclic families of 1D insulators defined by $H(k, t)$ are classified by the Chern number, and that topological invariant characterizes the quantized charge pumped per cycle.

4 INTEGER QUANTUM HALL EFFECT

The integer quantum Hall effect occurs when a two-dimensional electron gas is placed in a strong perpendicular magnetic field [32]. The quantization of the

FIGURE 7 Laughlin's argument shows that when flux $\phi_0 = h/e$ is threaded down the cylinder the polarization changes by e.

electrons' circular orbits leads to quantized Landau levels. If n Landau levels are filled and the rest are empty, then an energy gap separates the occupied and empty states just as in an insulator. Unlike an insulator, though, an electric field causes the cyclotron orbits to drift, leading to a Hall current characterized by the quantized Hall conductivity, $\sigma_{xy} = ne^2/h$.

Landau levels can be viewed as a band structure. Since the generators of translations do not commute with one another in a magnetic field, electronic states cannot be labeled with momentum. However, if a unit cell with area hc/eB enclosing a flux quantum is defined, then lattice translations do commute, so Bloch's theorem allows states to be labeled by the 2D crystal momentum **k**. In the absence of a periodic potential, the energy levels are simply the **k** independent Landau levels. In the presence of a periodic potential with the same lattice periodicity, the energy levels will disperse with **k**. This leads to a band structure that looks identical to that of an ordinary insulator. What is the difference between the quantum Hall state and the ordinary insulator? They are distinguished by topology.

4.1 Laughlin Argument, and the TKNN Invariant

In an important 1982 paper, Thouless, Kohmoto, Nightingale, and den Nijs (TKNN) [1] showed that the integer in the integer quantized Hall conductivity is precisely the Chern number. Their calculation was a straightforward application of linear response theory, which showed that the Kubo formula for σ_{xy} is identical to (9). Rather than repeat their analysis here, we will give a slightly different physical motivation for this result, which relates it to the quantized charge pump discussed in Section 3.4.

The key insight is provided by Laughlin's argument for the integer quantum Hall effect [33]. Suppose we have an integer quantum Hall state on a cylinder and we adiabatically turn up the magnetic flux Φ threading the cylinder from 0 to the flux quantum $\phi_0 = h/e$ (see Fig. 7). The changing flux induces a Faraday electric field $d\Phi/dt$ going around the cylinder, which in turn generates a Hall current $I = \sigma_{xy}d\Phi/dt$ going down the cylinder. At the end, a net charge $\sigma_{xy}h/e$ has been transported from one end to the other. When $\Phi = \phi_0$, the vector potential can be eliminated by a gauge transformation, so that the Hamiltonian has returned to its original form at $\Phi = 0$. It follows that the charge transferred must be an integer number of electrons $Q = ne$, from which the quantization $\sigma_{xy} = ne^2/h$ follows.

Viewed as a 1D system, the cylinder with threaded magnetic flux is precisely a Thouless charge pump with $t = \Phi$. The Chern number characterizing the pump can be evaluated by summing over all of the occupied one-dimensional subbands of the cylinder with radius R. These are indexed by a discrete azimuthal momentum $k_y^m(\Phi) = (m + \Phi/\phi_0)/R$. From (22), we find

$$n = \sum_m \frac{1}{2\pi} \int_0^{\phi_0} d\Phi \int dk_x \mathbf{F}(k_x, k_y^m(\Phi)). \tag{23}$$

Changing variables from Φ to k_y^m, it can be checked that the sum of integrals becomes a single integral over the 2D Brillouin zone $S = T^2$, given by (9).

4.2 Haldane Model

An example of the quantum Hall effect in a band theory is provided by a simple model of graphene in a periodic magnetic field introduced by Haldane [34]. This model is important because it provides a simple 2 band description of the quantum Hall effect. It also provides a stepping stone to the 2D quantum spin Hall insulator.

Graphene is a 2D form of carbon that is a material of high current interest due to experimental advances [35–38]. What makes graphene interesting electronically is the fact that the conduction band and valence band touch each other at two distinct points in the Brillouin zone. Near those points the electronic dispersion is linear, and resembles the dispersion of massless relativistic particles, which are described by the Dirac equation [39,40].

The simplest theory of graphene is a tight binding model that takes into account the p_z orbitals of each atom on a 2D honeycomb lattice.

$$H_0 = -t \sum_{<ij>} c_i^\dagger c_j. \tag{24}$$

Since there are two atoms per unit cell, this leads to a two band model (ignoring spin) that can be expressed in the form of (16, 17) with

$$d_x(\mathbf{k}) = -t \sum_{p=1}^{3} \cos \mathbf{k} \cdot \mathbf{a}_p,$$

$$d_y(\mathbf{k}) = -t \sum_{p=1}^{3} \sin \mathbf{k} \cdot \mathbf{a}_p, \tag{25}$$

$$d_z(\mathbf{k}) = 0.$$

Here $\mathbf{a}_1 = a(0,1)$ and $\mathbf{a}_{2,3} = a(\pm\sqrt{3}/2, -1/2)$ are the three nearest neighbor vectors pointing from the A sublattice to the B sublattice. The combination of inversion ($\mathcal{P}$) and time reversal ($\mathcal{T}$) symmetry requires $d_z(\mathbf{k}) = 0$. $\mathcal{P}$ takes $d_z(\mathbf{k})$ to $-d_z(-\mathbf{k})$, while $\mathcal{T}$ takes $d_z(\mathbf{k})$ to $+d_z(-\mathbf{k})$. An important consequence of this is that there can exist point zeros of $\mathbf{d}(\mathbf{k})$. These occur at the two distinct

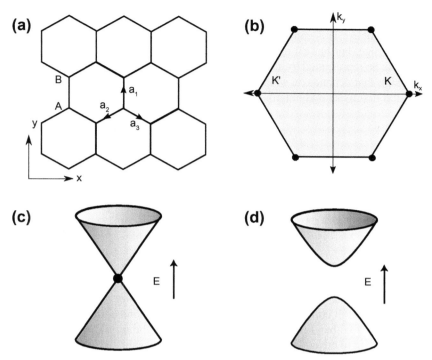

FIGURE 8 (a) Graphene's honeycomb lattice. (b) Graphene's Brillouin zone with two distinct corners **K** and **K**′. (c) Massless Dirac spectrum. (d) Massive Dirac spectrum.

corners $\mathbf{K} = (4\pi/3\sqrt{3}a, 0)$ and $\mathbf{K}' = -\mathbf{K}$ of the hexagonal Brillouin zone. For small $\mathbf{q} \equiv \mathbf{k} - \mathbf{K}$, $\mathbf{d}(\mathbf{q}) = v_F \mathbf{q} \cdot \vec{\sigma}$, with $\hbar v_F = 3ta/2$. Taking $\mathbf{q} \rightarrow -i\nabla$, the continuum theory has the form of a 2D massless Dirac Hamiltonian,

$$\mathcal{H} = -i\hbar v_F (\sigma^x \tau^z \partial_x + \sigma^y \partial_y), \qquad (26)$$

which has a linear dispersion $E(\mathbf{q}) = \pm\hbar v_F |\mathbf{q}|$ shown in Fig. 8(c). Here we have introduced $\tau^z = \pm 1$ to represent states near $\mathbf{K}/\mathbf{K}'$. The degeneracy at $\mathbf{q} = 0$ is protected by $\mathcal{P}$ and $\mathcal{T}$ symmetry. By breaking these symmetries the degeneracy can be lifted. For instance, $\mathcal{P}$ symmetry is violated if the two atoms in the unit cell are inequivalent. This leads to a nonzero $\mathbf{k}$ independent

$$d_z^{\mathrm{CDW}}(\mathbf{k}) = \lambda_{\mathrm{CDW}}. \qquad (27)$$

If λ_{CDW} is small, then the continuum theory acquires a mass term,

$$\Delta\mathcal{H}^{\mathrm{CDW}} = m_{\mathrm{CDW}}\sigma^z, \qquad (28)$$

with $m_{\mathrm{CDW}} = \lambda_{\mathrm{CDW}}$. The electronic dispersion $E(\mathbf{q}) = \pm\sqrt{|\hbar v_F \mathbf{q}|^2 + m_{\mathrm{CDW}}^2}$ then exhibits an energy gap $2|m_{\mathrm{CDW}}|$. This state describes an ordinary insulator.

Haldane imagined lifting the degeneracy by breaking time reversal symmetry. This can be done by applying a magnetic field that is zero on the average, but has all of the spatial symmetries of the honeycomb lattice. This can be represented by an imaginary second neighbor hopping term, which has a sign that depends on whether the electron makes a left or right turn going from the first to second neighbor. This leads to

$$d_z^H(\mathbf{k}) = \lambda_H \sum_{p<p'=1}^{3} s_{pp'} \sin \mathbf{k} \cdot (\mathbf{a}_p - \mathbf{a}_{p'}), \qquad (29)$$

with $s_{pp'} = \pm 1$ when $p' = p \pm 1$ mod 3. This also introduces a mass to the Dirac points. Since $d_z(-\mathbf{k}) = -d_z(\mathbf{k})$, the masses at $\mathbf{K}$ and $\mathbf{K}'$ have *opposite* sign, so that in the continuum theory,

$$\Delta \mathcal{H}^H = m_H \sigma^z \tau^z \qquad (30)$$

with $m_H = 3\sqrt{3}\lambda_H$. Haldane showed that this gapped state is not an ordinary insulator, but rather has a quantized Hall conductivity $\sigma_{xy} = e^2/h$.

The Hall conductivity can be understood by considering the Chern number for the two band model in terms of (8). When the $m_{\text{CDW}} = m_H = 0$, $\hat{\mathbf{d}}(\mathbf{k})$ is confined to the equator $d_z = 0$, with a unit (and opposite) winding around each of the Dirac points where $|\mathbf{d}| = 0$. For $m_{\text{CDW}} > m_H$, $|\mathbf{d}|$ is nonzero everywhere, and visits the north pole near both $\mathbf{K}$ and $\mathbf{K}'$. The net solid angle subtended is thus zero, and $\sigma_{xy} = 0$. For $m_H > m_{\text{CDW}}$ the masses at $\mathbf{K}$ and $\mathbf{K}'$ have opposite sign, so that $\hat{\mathbf{d}}(\mathbf{k})$ visits both the north and the south pole, and wraps the sphere once. Thus $\sigma_{xy} = e^2/h$. For $m_{\text{CDW}} = m_H$, $d_z(\mathbf{K}) = 0$, so that the system is at a gapless quantum critical point characterized by a single 2D massless Dirac fermion.

4.3 Chiral Edge States, and the Bulk Boundary Correspondence

A fundamental consequence of the topological classification of gapped band structures is the existence of gapless conducting states at interfaces where the topological invariant changes. Such edge states are well known at the interface between the integer quantum Hall state and vacuum [14]. They may be understood in terms of the semiclassical skipping orbits that electrons undergo as their cyclotron orbits bounce off the edge (Fig. 9(a)). Importantly, the electronic states responsible for this motion are *chiral* in the sense that they propagate in one direction only along the edge. These states are insensitive to disorder because there are no states available for backscattering—a fact that underlies the perfectly quantized electronic transport in the quantum Hall effect. The existence of such "one way" edge states is deeply related to the topology of the bulk quantum Hall state.

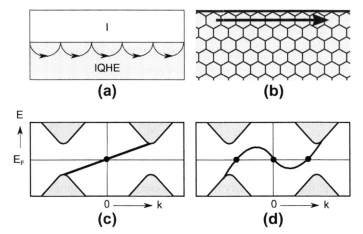

FIGURE 9 (a) Edge states as skipping cyclotron orbits. (b) Edge states in Haldane's model on a semi-infinite plane. (c) The chiral edge states connect the valence band near **K** and **K'**. (d) By changing the Hamiltonian near the edge, the details of the edge states change, but $N_R - N_L = 1$ remains fixed.

A simple theory of the chiral edge states can be developed using the two band Dirac model (28). Consider an interface where the mass m at one of the Dirac points changes sign as a function of y. We thus let $m \to m(y)$, where $m(y) > 0$ gives the insulator for $y > 0$ and $m(y) < 0$ gives the quantum Hall state for $y < 0$. Assume $m' > 0$ is fixed. Translation symmetry in the x direction allows us to consider plane wave states $\psi_{q_x}(x, y) = e^{iq_x x}\phi(y)$. For each q_x, the problem is identical to the Jackiw Rebbi problem discussed in Section 3.3. The zero energy mode $\phi_0(y)$ has precisely the form of (20), and leads to a band with dispersion

$$E(q_x) = \hbar v_F q_x. \tag{31}$$

This band of states intersects the Fermi energy E_F with a positive group velocity $dE/dq_x = \hbar v_F$ and defines a right moving chiral edge mode.

The chiral edge states can also be seen explicitly by solving the Haldane model in a semi-infinite geometry with an edge at $y = 0$ (Fig. 9(b)). Figure 9(c) shows the energy levels as a function of the momentum k_x along the edge. The solid regions show the bulk conduction and valence bands, which form continuum states and show the energy gap near **K** and **K'**. A single band, describing states bound to the edge, connects the valence band at **K'** to the conduction band at **K** with a positive group velocity.

By changing the Hamiltonian near the surface the precise dispersion of the edge states can be modified. For instance, $E(q_x)$ could develop a kink so that the edge states intersect the Fermi energy three times—twice with a positive group velocity and once with a negative group velocity, as in Fig. 9(d). The difference $N_R - N_L$ between the number of right moving and left moving modes, however, cannot change, and is an integer topological invariant characterizing the interface. The value of $N_R - N_L$ is determined by the topological structure of the bulk states. This is summarized by the *bulk-boundary correspondence*:

$$N_R - N_L = \Delta n, \tag{32}$$

where Δn is the difference in the Chern number across the interface.

5 $\mathbb{Z}_2$ TOPOLOGICAL INSULATORS

Since the Hall conductivity (and hence the TKNN invariant) is odd under time reversal, the topologically nontrivial states described in the preceding section can only occur when time reversal symmetry is broken either by an external magnetic field or by magnetic order. However, the spin orbit interaction allows a *different* topological class of insulating band structures when time reversal symmetry is unbroken [4]. In this section we will introduce the $\mathbb{Z}_2$ topological insulators and show that they exhibit protected boundary states.

The possibility of a time reversal invariant 2D topological insulator was first noticed in a model of graphene with spin orbit interactions [3]. We will begin by introducing that model of a quantum spin Hall insulator, from which it is straightforward to establish the existence of edge states, and to see why they are protected. We will then go onto consider the quantum spin Hall insulator in more generality and show that it is protected by a $\mathbb{Z}_2$ topological invariant. We will discuss how the $\mathbb{Z}_2$ invariant can be computed, as well as its consequences. Finally, we will briefly introduce the three-dimensional topological insulators, which will be discussed in detail in the following chapter.

5.1 Quantum Spin Hall Insulator in Graphene

In Section 4.2 we argued that the degeneracy at the Dirac point in graphene is protected by inversion and time reversal symmetry. However, that argument ignored the spin of the electrons. The spin orbit interaction allows a new mass term in (26) that leads to a topological insulating state [3].

That such a mass term is possible can be easily understood by considering the symmetries of the possible mass terms. Without spin, it is clear that the only possible terms in (26) that can open a gap are $\lambda_{\mathrm{CDW}}\sigma^z$ (which violates inversion) and $\lambda_H \sigma^z \tau^z$ (which violates time reversal). The spin degree of freedom allows a new mass term of the form

$$\delta\mathcal{H}^{\mathrm{SO}} = \lambda_{\mathrm{SO}}\sigma^z \tau^z s^z, \tag{33}$$

where $s^z = \pm 1$ represents the spin degree of freedom. This term respects all of the symmetries of graphene, so it must be present. However, since the spin orbit interaction in carbon is weak, this term is small, and has not yet been observed. Nonetheless, it is of conceptual value to consider its effects.

Taken separately, the Hamiltonians for the $s_z = \pm 1$ spins violate time reversal symmetry and are equivalent to Haldane's model for spinless electrons, which gives quantized Hall conductivity $\pm e^2/h$. An applied electric field thus leads to Hall currents for the $s_z = \pm 1$ spins that cancel each other, but generate a net spin current $\mathbf{J}_s = (\hbar/2e)(\mathbf{J}_\uparrow - \mathbf{J}_\downarrow)$ characterized by a quantized spin Hall conductivity $\sigma_{xy}^s = e/2\pi$. This is the origin of the name "quantum spin Hall effect." However, it must be emphasized that this quantized spin Hall conductivity is an artifact of an oversimplified model in which the spin s^z is conserved. In reality, spin is not conserved, and this quantization will break down in the presence of s^z nonconserving interactions.

Since it is two copies of a quantum Hall state, the quantum spin Hall state must have gapless edge states. Unlike the quantized Hall conductivity, these edge states remain robust even when spin is not conserved. These edge states have the special "spin filtered" property that up and down spins propagate in opposite directions. They were later dubbed "helical," in analogy with the correlation between spin and momentum of a particle known as helicity [41]. They form a unique 1D conductor that is essentially half of an ordinary 1D conductor. Ordinary conductors, which have up and down spins propagating in both directions, are fragile because the electronic states are susceptible to Anderson localization in the presence of weak disorder. By contrast, the quantum spin Hall edge states cannot be localized even for strong disorder. Here we will present one argument that this is the case [3]. Another proof will be given in Section 5.2.2.

Imagine an edge that is disordered in a finite region and perfectly clean outside that region. The exact eigenstates can be determined by solving the scattering problem relating incoming waves to those reflected from and transmitted through the disordered region. They will be characterized by a 2×2 unitary S matrix, which relates the incoming to outgoing states, $\Phi_{\text{out}} = S\Phi_{\text{in}}$, where Φ is a two component spinor consisting of the left and right moving edge states $\phi_{L\uparrow}$ and $\phi_{R\downarrow}$. Time reversal symmetry (to be discussed in more detail below) relates the left and right moving states by $\Phi_{\text{in,out}} \to \sigma^y \Phi_{\text{out,in}}^*$. Time reversal therefore imposes a constraint on the S matrix of the form $S = \sigma^y S^T \sigma^y$. It is straightforward to show that this requires the off diagonal component of S, which describes backscattering, to vanish. It follows that unless time reversal symmetry is broken, an incident electron is transmitted perfectly across the disordered region. Thus, eigenstates at any energy are extended, and at temperature $T = 0$ the edge state transport is ballistic. For $T > 0$ inelastic backscattering processes are allowed, which will, in general, lead to a finite conductivity.

The edge states can be explicitly seen in a lattice model that generalizes Haldane's model to include the intrinsic symmetry-allowed spin orbit interaction. We thus add to (26) a second neighbor spin dependent hopping amplitude proportional to $i\vec{s} \cdot (\mathbf{a}_1 \times \mathbf{a}_2)$, where $\mathbf{a}_{1,2}$ are the nearest neighbor bonds traversed. Such a term does not break any symmetries of the graphene

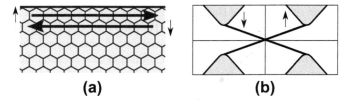

(a)   **(b)**

FIGURE 10   The edge states in graphene with the intrinsic spin orbit interaction have the spin filtered property that up and down spins propagate in opposite directions.

lattice, and can be represented as

$$\Delta H^{SO}(\mathbf{k}) = \lambda_{SO} s^z \vec{\sigma} \cdot \mathbf{d}^H(\mathbf{k}), \tag{34}$$

where $\mathbf{d}^H$ is given in (29). The resulting theory is simply two time reversed copies of Haldane's model, and on a semiinfinite strip has a spectrum shown in Fig. 10.

5.2 $\mathbb{Z}_2$ Topological Invariant

The fact that the edge states of the quantum spin Hall insulator are robust suggests that there must be a topological distinction between the quantum spin Hall insulator and an ordinary insulator. In this section we describe that topological invariant [4,42]. We will begin with a discussion of time reversal symmetry. We will then argue using the bulk-boundary correspondence that there are two and only two topological classes of time reversal invariant band structures. We will then discuss the physical meaning of the $\mathbb{Z}_2$ invariant as well as how to determine it.

5.2.1 Time Reversal Symmetry

The key to understanding this new topological class is to examine the role of $\mathcal{T}$ symmetry for spin 1/2 particles. $\mathcal{T}$ symmetry is represented by an antiunitary operator $\Theta = \exp(i\pi S_y/\hbar) K$, where S_y is the spin operator and K is complex conjugation. For spin 1/2 electrons, Θ has the property $\Theta^2 = -1$. This leads to an important constraint, known as Kramers' theorem, that all eigenstates of a $\mathcal{T}$ invariant Hamiltonian are at least twofold degenerate. This follows because if a nondegenerate state $|\chi\rangle$ existed then $\Theta|\chi\rangle = c|\chi\rangle$ for some constant c. This would mean $\Theta^2|\chi\rangle = |c|^2|\chi\rangle$, which is not allowed because $|c|^2 \neq -1$. In the absence of spin orbit interactions, Kramers' degeneracy is simply the degeneracy between up and down spins. In the presence of spin orbit interactions, however, it has nontrivial consequences.

A $\mathcal{T}$ invariant Bloch Hamiltonian must satisfy

$$\Theta \mathcal{H}(\mathbf{k}) \Theta^{-1} = \mathcal{H}(-\mathbf{k}). \tag{35}$$

One can classify the equivalence classes of Hamiltonians satisfying this constraint that can be smoothly deformed without closing the energy gap. The TKNN invariant is $n = 0$, but there is an additional invariant with two possible values $\nu = 0$ or 1 [4]. The fact that there are two and only two topological classes can be understood by appealing to the bulk-boundary correspondence.

5.2.2 Bulk Boundary Correspondence

In Fig. 11 we show plots analogous to Fig. 10(b) showing the electronic states associated with the edge of a $\mathcal{T}$ invariant 2D insulator as a function of the crystal momentum along the edge. Only half of the Brillouin zone $\Gamma_a = 0 < k_x < \Gamma_b = \pi/a$ is shown because $\mathcal{T}$ symmetry requires that the other half $-\pi/a < k < 0$ is a mirror image. As in Fig. 10(b), the shaded regions depict the bulk conduction and valence bands separated by an energy gap. Depending on the details of the Hamiltonian near the edge there may or may not be states bound to the edge inside the gap. If they are present, however, then Kramers' theorem requires they be twofold degenerate at the $\mathcal{T}$ invariant momenta $k_x = 0$ and $k_x = \pi/a$ (which is the same as $-\pi/a$). Away from these special points, labeled $\Gamma_{a,b}$ in Fig. 11, a spin orbit interaction will split the degeneracy. There are two ways the states at $k_x = 0$ and $k_x = \pi/a$ can connect. In Fig. 11(a) they connect pairwise. In this case the edge states can be eliminated by pushing all of the bound states out of the gap. Between $k_x = 0$ and $k_x = \pi/a$, the bands intersect E_F an even number of times. In contrast, in Fig. 11(b) the edge states cannot be eliminated. The bands intersect E_F an odd number of times.

Which of these alternatives occurs depends on the topological class of the bulk band structure. Since each band intersecting E_F at k_x has a Kramers partner at $-k_x$, the bulk-boundary correspondence relates the number N_K of Kramers pairs of edge modes intersecting E_F to the change in the $\mathbb{Z}_2$ invariants across the interface,

$$N_K = \Delta\nu \bmod 2. \qquad (36)$$

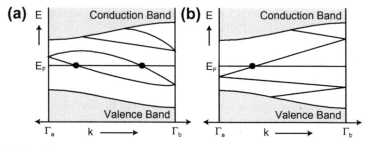

FIGURE 11 Electronic dispersion between two boundary Kramers degenerate points. In (a) the number of surface states crossing the Fermi energy E_F is even, whereas in (b) it is odd. An odd number of crossings leads to topologically protected metallic boundary states.

We conclude that a 2D topological insulator has topologically protected edge states. 3D topological insulators, discussed in Section 5.3, have protected surface states.

As discussed in Section 5.1, the edge states form a unique 1D conductor, which cannot be localized, even for strong disorder. Figure 11 provides a simple way of seeing why this must be the case [42]. Consider a cylinder with a large diameter, but treat the entire circumference as a single very large unit cell $a = 2\pi R$. Then the role of the momentum in Fig. 11 is played by the magnetic flux threading the cylinder, which gives a phase for the periodic boundary condition. Flux $\Phi = 0$ corresponds to $k = 0$, while $\Phi = \phi_0/2$ corresponds to $k = \pi/a$. In a clean system there will be many level crossings for $0 < k < \pi/a$ because the Brillouin zone has been folded back many times. However, in the presence of disorder all accidental degeneracies will be lifted. It is clear that when the unit cell is doubled and accidental degeneracies are removed, the folded Brillouin zone will keep its structure described in Fig. 11(a) and (b). Thus, for a topological insulator cylinder with $a = 2\pi R$, the spectrum "switches partners" as a function of magnetic flux. The spectrum is thus sensitive to the boundary conditions around the cylinder. This proves that every eigenstate must be extended, because localized states are insensitive to boundary conditions.

5.2.3 Physical Meaning of the Invariant
In Section 3.1, we interpreted the Chern number characterizing the integer quantum Hall effect in terms of Laughlin's argument and a 1D Thouless charge pump. The Chern number describes the change in the electric polarization when flux $\Phi = \phi_0$ is adiabatically threaded through the cylinder. An equivalent formulation is to deform the cylinder into a Corbino disk with a small hole threaded by flux. Laughlin's argument [33] then describes the binding of electric charge to the flux threading the hole. It is clear that this formulation is more general than the noninteracting electron framework that we have been using. The Laughlin argument can equally be applied to an interacting system, since the change in polarization is well defined in a many body setting. The $\mathbb{Z}_2$ invariant can be understood similarly [42].

Consider again a cylinder with a finite radius, so that the eigenstates associated with the ends are discrete. When this system is viewed as a 1D system, we wish to ask whether there is a Kramers degeneracy in the ground state associated with the ends. This would be the case if, for instance, there was an unpaired spin at the end. When there is time reversal symmetry, the existence of a Kramers degeneracy is a yes/no question. It is determined by whether the number of electrons is locally even or odd. We refer to this $\mathbb{Z}_2$ quantity as the "time reversal polarization," or the "local fermion parity." Like the polarization in Laughlin's argument, the time reversal polarization can be defined in an interacting system.

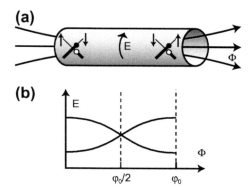

FIGURE 12 (a) When flux $\phi_0/2$ is threaded through a cylinder, the "time reversal polarization changes." (b) shows the evolution of the many body energy levels as a function of flux. When $\Phi = \phi_0/2$ the ground state is Kramers degenerate.

The $\mathbb{Z}_2$ topological invariant characterizes the *change* in the time reversal polarization when the flux Φ is changed from 0 to $\phi_0/2$. In the spin conserving case, the change in the Kramers degeneracy is easy to see because the $\phi_0/2$ flux insertion transfers "half" a spin up to the left and "half" a spin down to the right. Thus the eigenvalue of S_z associated with the end changes by $\hbar/2$, and due to Kramers' theorem, the degeneracy changes. Relaxing the S_z conservation (while preserving time reversal) prevents us from labeling the states with S_z, but the change in time reversal polarization remains well defined. The presence of Kramers degeneracy depends on whether the number of electrons is locally even or odd. Threading flux $\phi_0/2$ through the cylinder acts as a "pump" for fermion parity.

It is instructive to compare this interpretation with the edge state pictures in Fig. 11. If we view the cylinder as consisting of a single large unit cell in the azimuthal direction, then Fig. 11 describes the discrete end state spectrum as a function of flux, with $k = \Gamma_a = 0$ corresponding to $\Phi = 0$ and $k = \Gamma_b = \pi/a$ corresponding to $\Phi = \phi_0/2$. Suppose at $\Phi = 0$ there are no partially occupied Kramers pairs, so that the many body ground state is nondegenerate. Then at $\Phi = \phi_0$ there will be a single half-filled Kramers pair, which gives a Kramers degenerate many body state, when the $\mathbb{Z}_2$ invariant is nontrivial (see Fig. 12).

If we flatten the cylinder into a Corbino disk, then a flux $\phi_0/2$ piercing a topological insulator is associated with an odd fermion parity, but no net charge. This is an example of fractionalization, where the spin and charge of the electron decouple.

5.2.4 Formulas for the $\mathbb{Z}_2$ Invariant

There are several mathematical formulations of the $\mathbb{Z}_2$ invariant [4,7,10,42–47] ν. One approach [42] is to define a unitary matrix

$$w_{mn}(\mathbf{k}) = \langle u_m(\mathbf{k})|\Theta|u_n(-\mathbf{k})\rangle \tag{37}$$

built from the occupied Bloch functions $|u_m(\mathbf{k})\rangle$. Since Θ is anti-unitary and $\Theta^2 = -1, w^T(\mathbf{k}) = -w(-\mathbf{k})$. There are four special points Λ_a in the bulk 2D Brillouin zone where $\mathbf{k}$ and $-\mathbf{k}$ coincide, so $w(\Lambda_a)$ is antisymmetric. The determinant of an antisymmetric matrix is the square of its pfaffian, which allows us to define

$$\delta_a = \mathrm{Pf}[w(\Lambda_a)]/\sqrt{\mathrm{Det}[w(\Lambda_a)]} = \pm 1. \tag{38}$$

Provided $|u_m(\mathbf{k})\rangle$ is chosen continuously throughout the Brillouin zone (which is always possible), the branch of the square root can be specified globally, and the $\mathbb{Z}_2$ invariant is

$$(-1)^\nu = \prod_{a=1}^{4} \delta_a. \tag{39}$$

This formulation can be generalized to 3D topological insulators, and involves the eight special points in the 3D Brillouin zone.

The calculation of ν is considerably simpler if the crystal has extra symmetry. For instance, if the 2D system conserves the perpendicular spin S_z, then the up and down spins have independent Chern integers $n_\uparrow, n_\downarrow$. $\mathcal{T}$ symmetry requires $n_\uparrow + n_\downarrow = 0$, but the difference $n_\sigma = (n_\uparrow - n_\downarrow)/2$ defines a quantized spin Hall conductivity [52]. The $\mathbb{Z}_2$ invariant is then simply

$$\nu = n_\sigma \bmod 2. \tag{40}$$

While $n_\uparrow, n_\downarrow$ lose their meaning when S_z nonconserving terms (which are inevitably present) are added, ν retains its identity.

If the crystal has inversion symmetry there is another shortcut [10] to computing ν. At the special points Λ_a the Bloch states $u_m(\Lambda_a)$ are also parity eigenstates with eigenvalue $\xi_m(\Lambda_a) = \pm 1$. The $\mathbb{Z}_2$ invariant then simply follows from (39) with

$$\delta_a = \prod_m \xi_m(\Lambda_a), \tag{41}$$

where the product is over the Kramers pairs of occupied bands. This has proven useful for identifying topological insulators from band structure calculations [10,48–51].

For crystals without inversion symmetry determining the $\mathbb{Z}_2$ invariant is more difficult to implement numerically because (39) requires a continuous gauge, which is not provided by the computer. Efficient algorithms for numerically computing the $\mathbb{Z}_2$ invariant numerically have, however, been developed [53].

5.3 Topological Insulators in Three Dimensions

3D topological insulators will be discussed in detail in the following chapter. Here we will discuss them briefly in a manner that makes contact with our

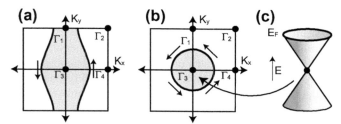

FIGURE 13 Fermi circles in the surface Brillouin zone for (a) a weak topological insulator and (b) a strong topological insulator. (c) In the simplest strong topological insulator the Fermi circle encloses a single Dirac point.

discussion of the quantum spin Hall insulator. A 3D topological insulator is characterized by four $\mathbb{Z}_2$ topological invariants $(\nu_0; \nu_1 \nu_2 \nu_3)$ [7–9]. They can be most easily understood by appealing to the bulk-boundary correspondence, discussed in Section 5.2.2. The surface states of a 3D crystal can be labeled with a 2D crystal momentum. There are four $\mathcal{T}$ invariant points $\Gamma_{1,2,3,4}$ in the surface Brillouin zone, where surface states, if present, must be Kramers degenerate (Fig. 13(a,b)). Away from these special points, the spin orbit interaction will lift the degeneracy. These Kramers degenerate points therefore form 2D *Dirac points* in the surface band structure (Fig. 13(c)). The interesting question is how the Dirac points at the different $\mathcal{T}$ invariant points connect to each other. Between any pair Γ_a and Γ_b, the surface state structure will resemble either Fig. 11(a) or (b). This determines whether the surface Fermi surface intersects a line joining Γ_a to Γ_b an even or an odd number of times. If it is odd, then the surface states are topologically protected. Which of these two alternatives occurs is determined by the four bulk $\mathbb{Z}_2$ invariants.

5.3.1 Weak Topological Insulator

The simplest nontrivial 3D topological insulators may be constructed by stacking layers of the 2D quantum spin Hall insulator. This is analogous to a similar construction for 3D integer quantum Hall states [54]. Consider a stack of weakly coupled layers of $\nu = n$ integer quantum Hall states. Each layer will have Hall conductivity e^2/h, leading to a 3D conductivity tensor,

$$\sigma_{\mu\nu} = \frac{e^2}{2\pi h} \epsilon_{\mu\nu\lambda} \mathbf{G}_\lambda, \tag{42}$$

where $\mathbf{G} = (2\pi/d)n\hat{\mathbf{n}}$ is a reciprocal lattice vector associated with the layers with separation d perpendicular to $\hat{\mathbf{n}}$. The edge states on each layer will form a chiral surface sheath. It is clear that when coupling between the layers is introduced the Hall conductivity and the surface states will remain, provided the bulk gap remains finite. The reciprocal lattice vector $\mathbf{G}$ is then specified by three integer Chern numbers, which characterize the three independent planes in momentum space.

Consider now a stack of weakly coupled 2D quantum spin Hall insulator layers. The helical edge states of the layers now become anisotropic surface states. A possible surface Fermi surface for weakly coupled layers stacked along the y direction is sketched in Fig. 13(a). In this figure a single surface band intersects the Fermi energy between Γ_1 and Γ_2 and between Γ_3 and Γ_4, leading to the nontrivial connectivity in Fig. 11(b). This layered state is referred to as a weak topological insulator, and has $\nu_0 = 0$. The indices $(\nu_1 \nu_2 \nu_3)$ can be interpreted as Miller indices describing the "mod 2" reciprocal lattice vector

$$\mathbf{G}^\nu = \nu_1 \mathbf{b}_1 + \nu_2 \mathbf{b}_2 + \nu_3 \mathbf{b}_3. \tag{43}$$

Here $\mathbf{b}_i$ are primitive reciprocal lattice vectors, and $\mathbf{G}^\nu$, which characterizes the layers, is defined modulo twice a reciprocal lattice vector.

Unlike the 2D helical edge states of a single layer, $\mathcal{T}$ symmetry does not protect these surface states. Though the surface states must be present for a clean surface, by breaking the translational symmetry, it is possible to eliminate the surface states without closing the bulk gap. The simplest way to see this is to imagine adding a dimerization (analogous to the SSH model, Fig. 4), which strongly couples pairs of layers, which results in a stack of trivial insulators.

Interestingly, however, there remain robust topological features associated with weak topological insulators. A line *dislocation* in a weak topological insulator is associated with protected 1D helical edge states [55]. This is easy to understand for weakly coupled layers in the case of an edge dislocation, where the dislocation involves a layer that ends on the dislocation line, and is associated with a helical edge state. Clearly, when the coupling between the layers is increased the edge state cannot disappear.

In addition, there has been interesting recent work characterizing the surface states of weak topological insulators in the presence of disorder [56–58]. The key point is that opening a gap by dimerization requires breaking the discrete lattice translation symmetry, which is not expected to occur in a macroscopic system. The surface of a weak topological insulator is thus expected to remain metallic, even for strong disorder.

5.3.2 Strong Topological Insulator

$\nu_0 = 1$ identifies a distinct phase, called a strong topological insulator, which cannot be interpreted as a descendant of the 2D quantum spin Hall insulator. ν_0 determines whether an even or an odd number of Kramers points is enclosed by the surface Fermi circle. In a strong topological insulator the surface Fermi circle encloses an *odd* number of Kramers degenerate Dirac points. The simplest case, with a single Dirac point (Fig. 13(b,c)), can be described by the Hamiltonian,

$$\mathcal{H}_{\text{surface}} = -i\hbar v_F \vec{\sigma} \cdot \vec{\nabla}, \tag{44}$$

where $\vec{\sigma}$ characterizes the spin. (For a surface with a mirror plane, symmetry requires $\vec{S} \propto \hat{z} \times \vec{\sigma}$.) The surface electronic structure of a topological insulator is similar to graphene, except rather than having four Dirac points (2 valley $\times$ 2 spin) there is just a single Dirac point.

The surface states of a strong topological insulator form a unique 2D topological metal [9, 10] that is essentially *half* an ordinary metal. Unlike an ordinary metal, which has up and down spins at every point on the Fermi surface, the surface states are not spin degenerate. Since $\mathcal{T}$ symmetry requires that states at momenta $\mathbf{k}$ and $-\mathbf{k}$ have opposite spin, the spin must rotate with $\mathbf{k}$ around the Fermi surface, as indicated in Fig. 13(b). This leads to a nontrivial Berry phase acquired by an electron going around the Fermi circle. $\mathcal{T}$ symmetry requires that this phase be 0 or π. When an electron circles a Dirac point, its spin rotates by 2π, which leads to a π Berry phase.

The Berry phase has important consequences for the behavior in a magnetic field and for the effects of disorder. In particular, in an ordinary 2D electron gas the electrical conductivity decreases with decreasing temperature, reflecting the tendency toward Anderson localization in the presence of disorder [59]. The π Berry phase changes the sign of the weak localization correction to the conductivity leading to weak *antilocalization* [60].

In fact, the electrons at the surface of a strong topological insulator cannot be localized even for strong disorder, as long as the bulk energy gap remains intact. The argument at the end of Section 5.2.2 can easily be generalized to 3D [61]. Consider a 3D topological insulator with periodic boundary conditions in two directions, but open boundary conditions in the third direction. This has the topology of a thickened torus with inside and outside surfaces, which was called a "Corbino doughnut" in Ref. [10], in analogy with the Corbino disk. There are now two independent fluxes associated with the two periodic directions, and the system has time reversal symmetry if these fluxes are either 0 or $\phi_0/2$. If we view the entire surface as a single unit cell, then the fluxes play the role of the two components of momentum in Fig. 13(b). This leads to sensitivity to boundary conditions described by Fig. 11. For the strong topological insulator, the odd number of crossings depicted in Fig. 13(b) persists when the unit cell is doubled. Therefore, the surface states of the topological insulator must be sensitive to the periodic boundary conditions, so they can not be localized.

6 RELATED TOPICS

We close by briefly mentioning some further applications of topological band theory. Some of these topics will be discussed in detail in later chapters.

6.1 Topological Crystalline Insulators

Spatial symmetries can modify and enhance the topological structure of band theory. The simplest example of this occurred in the SSH model, discussed in

Section 3.2, where the presence of inversion symmetry led to the topological distinction between the two dimerization patterns, which in the absence of inversion symmetry are topologically equivalent. A second example is the weak topological insulator, where translation symmetry plays an essential role. This leads to a broader class of topological band structures that are "less topological" than topological insulators (because disorder breaks all spatial symmetries), but nonetheless lead to interesting consequences.

A completely general classification of band structures with space group symmetries is a challenging math problem. Most progress to date has focused on specific examples. One class of topological invariants, which can exist when a system has mirror symmetry, is the "mirror Chern number" [48,62,63]. Consider the graphene model discussed in Section 5.1. This model has a mirror symmetry under $z \rightarrow -z$. It follows that eigenstates can be labeled with eigenvalues of that mirror operation. Importantly, since a mirror can be expressed as inversion times a 180° rotation, the eigenvalues of the mirror operator for spin 1/2 particles are $\pm i$, which means that the mirror operator is odd under time reversal. It is therefore possible to define Chern numbers $n_{\pm i}$ for the states with mirror eigenvalue $\pm i$. Time reversal symmetry dictates that $n_{+i} + n_{-i} = 0$, but the difference $n^M = (n_{+i} - n_{-i})/2$ defines a crystal symmetry protected integer topological invariant. For graphene, the $\mathbb{Z}_2$ invariant is simply $(-1)^{n^M}$.

The mirror Chern number can also be applied to 3D insulators [48,62,63]. In this case, mirror invariant planes in momentum space can be characterized by a mirror Chern number. This leads to consequences for the structure of the surface states. On surfaces that are perpendicular to a mirror plane (so that the surface retains the mirror symmetry) the surface must have gapless modes. A gap can be opened, however, if the mirror symmetry is broken. The situation is quite similar to a weak topological insulator, where the gapless surface states are protected by a discrete translation symmetry. In both cases, if disorder breaks the symmetry locally, but the symmetry is not macroscopically broken, then one expects the surface to remain conducting even in the presence of disorder. This is an interesting situation that warrants further exploration.

6.2 Topological Nodal Semimetals

Materials that are not insulators can also have topological aspects to their band structure [24]. One case of interest are nodal semimetals [64–66], in which the conduction band and valence band touch each other at points, leading to a Dirac-type low energy electronic structure. In 3D, the point touching of two nondegenerate bands, known as a Weyl point, is topologically protected. This is easy to understand by expanding the Hamiltonian around the degeneracy point. To linear order, for suitable coordinates k_i, the Hamiltonian can be written $H(\mathbf{k}) = v_{ij}\sigma_i k_j$. Importantly, in 3D, all three Pauli matrices are used, so that any perturbation, which may be written $u_0 I + \vec{u} \cdot \vec{\sigma}$, will only shift the location of the Weyl point and cannot open a gap. This topological protection can also be understood in terms of the Chern number $n = \text{sgn}(\det[v_{ij}])$ characterizing $H(\mathbf{k})$ on a sphere surrounding the Weyl point.

Time reversal symmetry requires Weyl points to come in pairs at $\pm\mathbf{K}$, with opposite Chern numbers, while inversion symmetry requires pairs at $\pm\mathbf{K}$ with the same Chern number. Thus, the presence of isolated Weyl points in a band structure requires breaking time reversal and/or inversion symmetry. The time reversal broken state exhibits an interesting anomalous Hall conductivity. In both cases, the surface exhibits interesting Fermi arcs that terminate on the projected Weyl points. Candidate materials and structures for Weyl semimetals have been proposed [64–66]. It will be interesting to observed these effects experimentally.

It is also possible to have nodal semimetals that are protected by point group symmetries. Dirac semimetals have fourfold degenerate band crossings at the Fermi level [67]. These are not topologically protected in the way Weyl points are, but can be protected by crystal symmetries which enforce the degeneracy. Such Dirac points occurred in a model system based on a diamond lattice, and candidate materials have been proposed.

6.3 Topological Superconductivity

Topological superconductivity is a beautiful subject that will be treated in detail elsewhere. Here we will just mention that considerations of topological band theory can also be used to classify superconductors. The Bardeen, Cooper, and Schrieffer (BCS) mean field theory of superconductivity is a noninteracting theory, which in addition to the usual terms includes anomalous terms of the form $\Delta \psi^\dagger \psi^\dagger$. These can be analyzed in terms of a one body Hamiltonian if the one body Hilbert space is artificially doubled to include both positive and negative energy states. This results in a Bloch-Bogoliubov de Gennes Hamiltonian $H(\mathbf{k})$ that is just like a Bloch Hamiltonian, except that it has an intrinsic particle-hole symmetry

$$H(\mathbf{k}) = -\Xi H(-\mathbf{k})\Xi^{-1}, \tag{45}$$

where Ξ is an antiunitary operator. This has a structure similar to time reversal symmetry in Eq. (35). Like time reversal symmetry it modifies the topological classifications of gapped Hamiltonians, leading to classes of topological superconductors. The bulk-boundary correspondence in topological superconductors leads to topologically protected boundary modes. The redundancy that was introduced by doubling the one body Hilbert space makes these boundary modes Majorana fermion modes.

Allowing for both time reversal and particle-hole symmetry leads to an elegant generalization of topological band theory [24,68–71]. There are 10 symmetry classes, which depend on the presence or absence of $\mathcal{T}$ symmetry (with $\Theta^2 = \pm 1$) and/or particle-hole symmetry (with $\Xi^2 = \pm 1$). The topological classifications, given by $\mathbb{Z}$, $\mathbb{Z}_2$, or 0, show a regular pattern as a function of symmetry class and dimensionality, dubbed the "10-fold way."

6.4 Topological Defects

Finally, we note that topological band theory can classify and characterize topological defects that carry protected gapless modes [72]. Let us illustrate this with a simple example. Consider a three-dimensional quantum Hall state, which can be viewed as a stack of layers of 2D quantum Hall states. For simplicity first consider the weakly coupled limit where the layers are independent. Consider now an edge dislocation, shown in Fig. 14, which occurs when one of the layers is terminated along a line. It is clear that line will be associated with a chiral edge state. Now imagine that coupling between the layers is turned on, but the bulk gap remains finite. The chiral edge state has nowhere to go and must remain. Its presence is guaranteed topologically in the same way the edge states in the 2D quantum Hall effect are guaranteed.

We thus have a topologically protected "boundary mode." For the 3D structure, what is the analog of the bulk? To analyze this it is useful to consider a large circle surrounding the dislocation line in real space. Far away from the dislocation the Hamiltonian varies slowly with position s along the circle. We thus have a one parameter *family* of bandstructures $H(\mathbf{k}, s)$, where s is defined on the circle. A circle that encloses a chiral edge state must be topologically distinct from a circle that does not enclose a chiral edge state. We are thus led to topologically classify *families* of gapped bandstructures $H(\mathbf{k}, s)$. Such a four parameter family of Hamiltonians is classified by an integer topological invariant called the second Chern number. In the layered quantum Hall state the second Chern number can be computed [72] and is given by $n = \mathbf{G} \cdot \mathbf{B}/2\pi$,

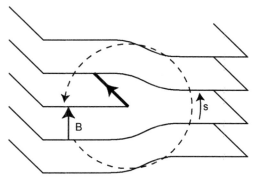

FIGURE 14 A dislocation line, with Burgers vector **B** in a layered 3D quantum Hall state, is associated with a gapless 1D chiral fermion mode.

where $\mathbf{B}$ is the Burgers vector characterizing the dislocation and $\mathbf{G}$, defined in (42), characterizes the quantum Hall state.

It is clear that this method of analysis is more general than this specific example. For example, a similar analysis explains the helical modes associated with a dislocation in a weak topological insulator. The analysis of D parameter families of Hamiltonians in d dimensions ($H(\mathbf{k},\mathbf{r})$ with $\mathbf{k} \in T^d$ and $\mathbf{r} \in S^d$) with or without time reversal symmetry leads to a generalization of the bulk-boundary correspondence that applies generally to the protected modes associated with $d - D - 1$ dimensional topological defects. When combined with the theory of topological superconductivity it leads to a generalization of the "ten-fold way" classification of insulators and superconductors.

ACKNOWLEDGMENTS

The author has benefited with discussions with numerous friends and colleagues. We particularly thank our principal collaborators, Liang Fu, Eugene Mele, and Jeffrey Teo. This work has been supported by the National Science Foundation Grant DMR 0906175.

REFERENCES

[1] Thouless DJ, Kohmoto M, Nightingale MP, den Nijs M. Phys Rev Lett 1982;49:405.
[2] Wen XG. Adv Phys 1995;44:405.
[3] Kane CL, Mele EJ. Phys Rev Lett 2005;95:226801.
[4] Kane CL, Mele EJ. Phys Rev Lett 2005;95:146802.
[5] Bernevig BA, Hughes TA, Zhang SC. Science 2006;314:1757.
[6] König M, Wiedmann S, Brne C, Roth A, Buhmann H, Molenkamp LW, et al. Science 2007;318:766.
[7] Moore JE, Balents L. Phys Rev B 2007;75:121306(R).
[8] Roy R. Phys Rev B 2009;79:195322. Available from: arXiv:cond-mat/0607531.
[9] Fu L, Kane CL, Mele EJ. Phys Rev Lett 2007;98:106803.
[10] Fu L, Kane CL. Phys Rev B 2007;76:045302.
[11] Hsieh D, Qian D, Wray L, Xia Y, Hor YS, Cava RJ, et al. Nature 2008;452:970.
[12] Xia Y, Qian D, Hsieh D, Wray L, Pal A, Lin H, et al. Nat Phys 2009;5:398.
[13] Zhang H, Liu CX, Qi XL, Dai X, Fang Z, Zhang SC. Nature Phys 2009;5:438.
[14] Halperin BI. Phys Rev B 1982;25:2185.
[15] Tsui DC, Stormer HL, Gossard AC. Phys Rev Lett 1982;48:1559.
[16] Laughlin RB. Phys Rev Lett 1983;50:1395.
[17] Bloch FZ. Physik 1929;52:555.
[18] Nakahara M. Geometry, topology and physics. Bristol: Adam Hilger; 1990.
[19] Jackiw R, Rebbi C. Phys Rev D 1976;13:3398.
[20] Su WP, Schrieffer JR, Heeger AJ. Phys Rev Lett 1979;42:1698.
[21] Volkov BA, Pankratov OA. Pisma Zh Eksp Teor Fiz 1985;42:145; JETP Lett 1985;42:178.
[22] Fradkin E, Dagotto E, Boyanovsky D. Phys Rev Lett 1986;57:2967.
[23] Kaplan DB. Phys Lett B 1992;288:342.
[24] Volovik GE. The universe in a helium droplet. Oxford: Clarendon Press; 2003.
[25] Berry MV. Proc R Soc Lond A 1984;392:45.
[26] Blount EI. Solid State Phys 1962;13:305.
[27] Zak J. Phys Rev Lett 1989;62:2747.
[28] King-Smith RD, Vanderbilt D. Phys Rev B 1993;47:1651.
[29] Resta R. Rev Mod Phys 1994;66:899.
[30] Thouless DJ. Phys Rev B 1983;27:6083.

[31] Niu Q, Thouless DJ. J Phys A 1984;17:2453.
[32] Von Klitzing K, Dorda G, Pepper M. Phys Rev Lett 1980;45:494.
[33] Laughlin RB. Phys Rev B 1981;23:5632.
[34] Haldane FDM. Phys Rev Lett 1988;61:2015.
[35] Novoselov KS, Geim AK, Morozov SV, Jiang D, Katsnelson MI, Grigorieva IV, et al. Nature 2005;438:197.
[36] Zhang Y, Tan YW, Stormer HL, Kim P. Nature 2005;438:201.
[37] Geim AV, Novoselov KS. Nature Mater 2007;6:183.
[38] Castro Neto AH, Guinea F, Peres NMR, Novoselov KS, Geim AK. Rev Mod Phys 2009;81:109.
[39] DiVincenzo DP, Mele EJ. Phys Rev B 1984;29:1685.
[40] Semenoff GW. Phys Rev Lett 1984;53:2449.
[41] Wu C, Bernevig BA, Zhang SC. Phys Rev Lett 2006;96:106401.
[42] Fu L, Kane CL. Phys Rev B 2006;74:195312.
[43] Fukui T, Hatsugai Y. J Phys Soc Jpn 2007;76:053702.
[44] Fukui T, Fujiwara T, Hatsugai Y. J Phys Soc Jpn 2008;77:123705.
[45] Qi XL, Hughes TL, Zhang SC. Phys Rev B 2008;78:195424.
[46] Roy R. Phys Rev B 2009;79:195321. Available from: <arXiv:cond-mat/0604211>.
[47] Wang Z, Qi XL, Zhang SC. New J Phys 2010;12:065007.
[48] Teo JCY, Fu L, Kane CL. Phys Rev B 2008;78:045426.
[49] Zhang H, Liu CX, Qi XL, Dai X, Fang Z, Zhang SC. Nature Phys 2009;5:438.
[50] Pesin DA, Balents L. Nature Phys 2010;6:376.
[51] Guo HM, Franz M. Phys Rev Lett 2009;103:206805.
[52] Sheng DN, Weng ZY, Sheng L, Haldane FDM. Phys Rev Lett 2006;97:036808.
[53] Soluyanov AA, Vanderbilt D. Phys Rev B 2011;83:235401.
[54] Kohmoto M, Halperin BI, Wu YS. Phys Rev B 1992;45:13488.
[55] Ran Y, Zhang Y, Vishwanath A. Nat Phys 2009;5:298.
[56] Ringel Z, Kraus YE, Stern A. Phys Rev B 2012;86:045102.
[57] Mong RSK, Bardarson JH, Moore JE. Phys Rev Lett 2012;108:076804.
[58] Liu CX, Qi XL, Zhang SC. Physica E 2012;44:906.
[59] Lee PA, Ramakrishnan TV. Rev Mod Phys 1985;57:2871985.
[60] Suzuura H, Ando T. Phys Rev Lett 2002;89:266603.
[61] Nomura K, Koshino M, Ryu S. Phys Rev Lett 2007;99:146806.
[62] Fu L. Phys Rev Lett 2011;106:106802.
[63] Hsieh TH, Lin H, Liu J, Duan W, Bansil A, Fu L. Nature Commun 2012;3:982.
[64] Wan X, Turner AM, Vishwanath A, Savrasov SY. Phys Rev B 2011;83:205101.
[65] Yang KY, Lu YM, Ran Y. Phys Rev B 2011;84:075129.
[66] Burkov AA, Balents L. Phys Rev Lett 2011;107:127205.
[67] Young SM, Zaheer S, Teo JCY, Kane CL, Mele EJ. Phys Rev Lett 2012;108:140405.
[68] Schnyder AP, Ryu S, Furusaki A, Ludwig AWW. Phys Rev B 2008;78:195125.
[69] Schnyder AP, Ryu S, Furusaki A, Ludwig AWW. AIP Conf Proc 2009;1134:10.
[70] Kitaev A. AIP Conf Proc 2009;1134:22.
[71] Qi XL, Hughes TL, Raghu S, Zhang SC. Phys Rev Lett 2009;102:187001.
[72] Teo JCY, Kane CL. Phys Rev B 2010;82:115120.

Chapter 2

Theory of Three-Dimensional Topological Insulators

Joel E. Moore[*,†]

[*]*Department of Physics, University of California, Berkeley, CA 94720, USA*
[†]*Materials Sciences Division, Lawrence Berkeley National Laboratory, Berkeley, CA 94720, USA*

Chapter Outline Head

1 INTRODUCTION

Topological phases of matter have been an active subject of study since the discovery of the integer quantum Hall effect (IQHE) in 1980. While making a two-dimensional electron gas (2DEG) and subjecting it to low temperatures and high magnetic fields is a nontrivial exercise, the reward for this effort is dramatic

Topological Insulators. http://dx.doi.org/10.1016/B978-0-444-63314-9.00002-0

indeed: quantum mechanics leads to an electronic state that is remarkably robust to impurities and thermal fluctuations. The lesson of classical chaos is that a little bit of uncertainty or disorder becomes magnified as a system becomes larger, but in topological phases the opposite is true. The conductance quantization in the Hall effect becomes more perfect, eventually better than one part per billion, as the system grows. After the IQHE discovery, increases in the quality of semiconductor 2DEGs and ever lower electron temperatures continued to yield ample reward as the Abelian and non-Abelian quantum Hall states were discovered within the next decade. More recently, the quantum spin Hall effect (QSHE) was observed in 2DEGs in (Hg,Cd) Te quantum wells in zero magnetic field. As explained earlier in this volume, the QSHE is driven by spin-orbit coupling and can be caricatured as two copies of the integer quantum Hall state related by time-reversal-invariance.

This chapter is about a new direction in research on topological phases that grew out of considering spin-orbit effects in bulk solids. The last 5 years have shown that there are several bulk three-dimensional (3D) materials ("topological insulators" [1]) with topological order in zero magnetic field and at relatively large energy scales. The remainder of this introduction gives a quick summary of the key properties of this phase and how they were experimentally confirmed. As there exist several extended reviews [2–4], the goal of this chapter is to present the essentials in an accessible fashion, without attempting to be comprehensive or follow the historical sequence of discoveries.

The observable features of a topological insulator in experiment result from its metallic surface state. It is easiest to explain why this surface metal is unusual by starting with a good crystalline surface, so that crystal momentum in the surface plane is conserved and we can think of a 2D band structure. First consider an ordinary 2D metal with time-reversal symmetry. If there is no spin-orbit coupling, then every band in the band structure is twofold degenerate at each momentum because all spin states have the same energy. Hence the Fermi level is crossed by an even number of bands, where now we count spin-up and spin-down bands separately. Turning up the spin-orbit coupling breaks this degeneracy at all the points in the Brillouin zone where $\mathbf{k} \neq -\mathbf{k}$. Figure 1(a) shows the case of Rashba spin-orbit coupling added to the standard kinetic term,

$$H = H_{\text{kin}} + H_{\text{SO}}, \tag{1}$$

$$H_{\text{kin}} = \frac{p^2}{2m}, \tag{2}$$

$$H_{\text{SO}} = \lambda(p_x \sigma_y - p_y \sigma_x), \tag{3}$$

where m is the mass, p the momentum, σ the spin Pauli matrices, and λ the strength of spin-orbit coupling. Note that at every energy the Fermi surface crosses an even number of Fermi sheets; this must always be the case in an isolated 2D electron system with time-reversal symmetry. (As an exercise, the reader is invited to show that a time-reversal-breaking Zeeman magnetic field

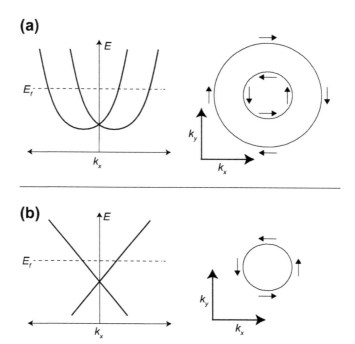

FIGURE 1 (a) Band structure cut and Fermi surface of a 2D quantum well with Rashba spin-orbit coupling. At each energy, there are an even number (either 0 or 2 in this example) of Fermi surface sheets. (b) Surface band structure cut and surface Fermi surface for a 3D topological insulator. When the Fermi energy lies in the gap of the bulk bands (not shown), there are an odd number of Fermi surface sheets at a single surface.

$H' = gB\sigma_z$ can open up a gap at $\mathbf{p} = 0$, leaving a single-sheet Fermi surface at some Fermi energies; this effect can be used to enable emergent Majorana fermions in conventional semiconductors.)

The surface of a 3D topological insulator gets around this rule because it is not an isolated 2D system but rather a boundary between inequivalent 3D systems: the 2D surface has time-reversal symmetry but, in the simplest case, a single-sheet Fermi surface as in Fig. 1(b), approximately described by the linear-in-momentum H_{SO}. There are surface electronic states at all energies in the bulk bandgap. This picture can be compared to the angle-resolved photoemission spectroscopy images of the electronic structure of Bi_2Se_3 and other materials in the chapter of this volume by M. Z. Hasan. A point where the Fermi surface shrinks from a circle to a single point, such as $\mathbf{k} = 0$ in the present example, is known as a Dirac point, and the neighboring band structure with zero effective mass (since energy is linear in momentum) is known as a Dirac cone.

The first important feature of this surface state is its stability. Because the surface state arises from bulk topology rather than surface physics, in the same way as the edge states of the quantum Hall effect are induced by the bulk, it cannot be eliminated by purely surface perturbations as long as those maintain time-reversal symmetry; time-reversal-breaking perturbations such as a magnetic field can induce a gap in the surface electronic structure (again, the reader can verify this for the case of Zeeman coupling). This symmetry condition is an interesting difference between topological insulators and the quantum Hall effect, in which case the edge state exists for any symmetry at the boundary. The nontrivial bulk topology of electronic wavefunctions induced by spin-orbit coupling that causes the surface state to exist is described in Section 2.

In real materials, different surfaces may have different odd numbers of Dirac cones, and other properties such as the velocities and degree of spin polarization can certainly vary as well. But all surfaces will be metallic, which explains a sense in which this state is "more three-dimensional" than the chiral metal obtained by stacking quantum Hall layers [5,6], as this stack will have insulating top and bottom surfaces. The remainder of this chapter is in large part about the origin and surprising consequences of the distinction between an even number and odd number of surface Fermi sheets. The following section discusses the theory of topological properties of bulk electronic wavefunctions in 3D materials, building on the theory of the quantum spin Hall effect in 2D. Consequences of time-reversal-breaking perturbations at the surface are then treated in Section 2, including a surprising connection to the magnetoelectric effect in all materials. Section 3 reviews the theory of surface transport in topological insulators and some recent experimental progress in this area. Section 4 discusses consequences of strong electronic interactions, including connections to superconductivity and magnetism, and includes descriptions of some recent theoretical developments and open problems.

2 TOPOLOGICAL PROPERTIES OF BAND STRUCTURES

The key difference between the surface states of topological insulators and "ordinary" surface states is that the former are forced by topological properties of the bulk electrons, rather than depending on microscopic details of the surface, provided that the surface does not break time-reversal symmetry. The first descriptions of topological insulators were in terms of these bulk topological properties: there are topological invariants of time-reversal-invariant band structures in 2D [7,8] and 3D [9–11] that are similar in spirit to the TKNN integers or Chern numbers [12] that underlie lattice models of the IQHE, including the influential honeycomb-lattice model of Haldane [13]. As the quantum spin Hall state or 2D topological insulator is discussed elsewhere in this volume, this section explains three approaches for the 3D topological invariants, concentrating on the physical implications rather than derivations. As in the 2D case, the topological invariants in a 3D topological insulator are not integers but $\mathbb{Z}_2$ invariants or parities, which take two possible values: even/odd, also known as "ordinary"/"topological."

The first two definitions given here [9, 10] of the four topological invariants of a 3D band structure can be stated simply in terms of earlier results on the 2D case. These two definitions are presented first, followed by a brief digression on actual materials and an introduction to the Berry connection or Berry phase. This introduction is motivated by the third definition [14], which concisely expresses the most important of the four invariants in terms of a geometrical quantity (the Chern-Simons form) built from the non-Abelian Berry connection. Definitions of the topological insulator state beyond single-particle physics are given later in Section 4; this was an important theoretical step that remains less physically transparent than the quantum Hall case, where a simple redefinition gives a topological invariant in terms of the many-electron wavefunction.

2.1 Building the 3D Topological Insulator from 2D

The essential facts from the 2D case can be summarized briefly. First, with time-reversal symmetry, bands naturally come in pairs because of the Kramers degeneracy at the time-reversal-invariant points in momentum space. Kramers degeneracy is the statement that a time-reversal-invariant Hamiltonian of spin-half particles has minimum degeneracy of at least 2 for each eigenvalue: every state is degenerate with and distinct from its time-reversal conjugate. One reason why the $\mathbb{Z}_2$ invariant took some time to be discovered is that only at certain special points in the Brillouin zone where $\mathbf{k} = -\mathbf{k}$ is the Bloch Hamiltonian time-reversal invariant; in general, the Hamiltonian at $\mathbf{k}$ is time-reversal conjugate to the Hamiltonian at $-\mathbf{k}$.

Each band pair in a time-reversal-symmetric 2D system is associated with a single $\mathbb{Z}_2$ invariant [7], and these are independent except for the constraint that in a finite-dimensional model the invariants sum to zero. (Addition in $\mathbb{Z}_2$ is given by the same rules as addition of odd and even numbers.) Without time-reversal symmetry, a non-degenerate band has an integer-valued invariant (the Chern number or TKNN integer [12]). When two bands become degenerate, only the total Chern number remains defined. All the Chern numbers vanish if the system has time-reversal symmetry. The $\mathbb{Z}_2$ invariant in 2D can thus be pictured as a subdivision of the zero-Chern-number class for the case of time-reversal symmetry.

The $\mathbb{Z}_2$ invariants in 2D can be understood in terms of a Pfaffian construction [7], as a topological obstruction [15], as a type of pumping [16–19], or in terms of homotopy of Hamiltonian spaces [9]. Probably the most powerful mathematical explanation of the $\mathbb{Z}_2$ invariants in 2D and 3D is by K-theory [20], which can be described roughly as an abstraction of homotopy theory of Hamiltonians beyond fixed dimension. (There is now a full classification [20, 21] of topological invariants in the presence of the most important symmetries of free-fermion systems, which lead to 10 symmetry classes; this classification is discussed in the chapter by Ludwig in this volume.) Here we confine ourselves to giving two descriptions of the four $\mathbb{Z}_2$ invariants of a 3D band structure that have been useful in practice, and a description of the associated surface state for a perfect boundary. In 2D as in 3D, the invariants are a property of a pair of bands, and the physical state of a band insulator is given by the $\mathbb{Z}_2$ sum over occupied bands.

In 3D, we will take the Brillouin zone to be the 3-torus with coordinates ranging from $-\pi/a$ to π/a in each direction. First, ignore time-reversal symmetry. There are three 2D Chern numbers for each band, which correspond to the xy, yz, and xz planes [22]. To understand this, note that any two xy planes are smoothly connected to each other, the xy Chern number of a non-degenerate band cannot change as k_z is changed, so all xy planes have the same Chern number [23]. The band similarly has one yz and one xz Chern number, and these three integers are independent of each other. These three topological invariants can all be thought of as layering 2D quantum Hall systems, leading to a "chiral metal" [5,6]; there is not a genuinely 3D topological invariant in the time-reversal-broken case corresponding to the quantum Hall effect.

Now consider the xy planes again in a time-reversal-invariant system (Fig. 2). The key difference from the previous case is that not all xy planes are equivalent: the $k_z = 0$ plane is brought back to itself under $\mathbf{k} \leftrightarrow -\mathbf{k}$, as is the $k_z = \pm\pi/a$ plane (because of periodic boundary conditions this is a single plane). But planes with other values of k_z are not brought to themselves and hence do not have the time-reversal symmetry of a 2D time-reversal-symmetric system. Hence there are two $\mathbb{Z}_2$ invariants for xy planes, one for the $k_z = 0$ plane and one for the $k_z = \pi/a$ plane. They could be equal but need not be; a smooth evolution of a 2D system that connects "ordinary" to "topological" without closing the gap by breaking time-reversal symmetry can be viewed as a recipe for making a 3D system with different invariants in the $k_z = 0$ and $k_z = \pi/a$. It turns out that the phase with different invariants in these two planes

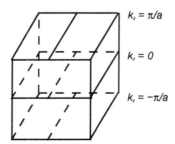

FIGURE 2 xy planes distinguished by time-reversal in a cubic 3D Brillouin zone of a material with time-reversal symmetry. The two inequivalent planes $k_z = 0$ and $k_z = \pm\pi/a$ are closed under time-reversal symmetry and can be thought of as Brillouin zones of a 2D time-reversal-invariant system. Other planes are not constrained by time-reversal symmetry because they are taken to inequivalent planes. Hence the 2D topological $\mathbb{Z}_2$ invariant is only defined for $k_z = 0$ and $k_z = \pm\pi/a$ and it is possible to interpolate smoothly between these as k_z changes without closing the gap.

is the "strong topological insulator," an essentially isotropic topological phase in 3D.

However, repeating this argument for the yz and xz suggests naively that there would be six $\mathbb{Z}_2$ invariants for the six special planes. A small amount of geometry [4,9] shows that there are two identities reducing the number of invariants from this construction to four: if in one of the three directions the $\mathbb{Z}_2$ invariant changes from "even" to "odd" between the two special planes, then the same must happen in the other two directions also. Note that this argument does not prove that there are only these four invariants. A useful terminology and notation for the four invariants is as follows [10]: a band is labeled by the four invariants $(\nu_0; \nu_1\nu_2\nu_3)$, where the first is the "strong" topological invariant (i.e., whether there is a change from 2D ordinary to topological between the two special planes in any direction) and the latter three are "weak" invariants corresponding to the three reference directions. In this notation, $\nu = 0$ is ordinary and $\nu = 1$ is topological. The term "topological insulator" was coined for these 3D phases as the connection to the quantum Hall effect is less clear than in the 2D case (the "quantum spin Hall state" as then known).

The weak invariants are like the quantum Hall case in that they can be understood as resulting from layering; the name "weak" indicates that they are less stable to disorder than in the quantum Hall case, because the rules of $\mathbb{Z}_2$ addition mean that combining an even number of topologically nontrivial layers returns the system to the topologically trivial case. The properties of dislocations [24] and transport [25,26] in weak topological insulators have recently been of interest, but as experimental searches for a purely weak topological insulator have not yet been successful, we will focus here on strong topological insulators and denote that phase simply as "topological insulator."

There are some other important differences between the 2D and 3D cases that are clear from the construction above. While it is possible to approach the 2D case by thinking of the "ordinary" and "topological" cases as deformable [15] to paired integer quantum Hall states $\nu = +n$ and $\nu = -n$, with even n in the ordinary case and odd n in the topological case, the 3D strong topological insulator cannot be realized if a spin component is conserved—it depends on the vectorial nature of spin. To see this, note that if one spin component were conserved (say S_z), then in each xy plane we could define Chern numbers for spin-up and spin-down, and these would have to be continuous. Looking ahead, this inability to construct the 3D topological insulator by pairing time-reversal-breaking states is one reason why fractional 3D states have been hard to write explicitly; in 2D, one can take two copies of the fractional quantum Hall effect, one for spin-up and one for spin-down.

An important step in understanding other physical consequences of the bulk band structure was taken by Fu et al. [10], who introduced an explicit model Hamiltonian on the diamond lattice and gave a characterization of the surface state (assuming a crystal with a smooth boundary) that results from these topological invariants. This work used a very direct and useful simplification of the four invariants if the system has an inversion symmetry [27]. In a 3D Brillouin zone, there are eight time-reversal-invariant points where $\mathbf{k} = -\mathbf{k}$. At each such point Γ_i, define $\delta_i = \pm 1$ as the following product over occupied band pairs:

$$\delta_i = \prod_{m=1}^{N} \xi_{2m}(\Gamma_i), \tag{4}$$

where $\xi_{2m}(\Gamma_i)$ is the parity eigenvalue ± 1 of the $2m$th occupied band (note that the two bands in the same Kramers pair have the same parity eigenvalue). Then the strong topological invariant ν_0 is determined by the product of the δ_i over the eight time-reversal-symmetric points: $(-1)^\nu = \prod_{i=1}^{8} \delta_i$.

The FKM diamond lattice model [10] has been used in some subsequent papers so we reproduce it here and add a time-reversal-breaking term, a staggered Zeeman field, for illustration purposes [28]:

$$H = \sum_{\langle ij \rangle} t_{ij} c_i^\dagger c_j + i \frac{4\lambda_{SO}}{a^2} \sum_{\langle\langle ij \rangle\rangle} c_i^\dagger \sigma \cdot \left(d_{ij}^1 \times d_{ij}^2 \right) c_j$$

$$+ \, \mathbf{h} \cdot \left(\sum_{i \in A} c_i^\dagger \sigma c_i - \sum_{i \in B} c_i^\dagger \sigma c_i \right). \tag{5}$$

In the first term, the nearest-neighbor hopping amplitude depends on the direction of the bond; let $t_{[111]} = 3t + m \cos\beta, t_{[\bar{1}11]} = t_{[1\bar{1}1]} = t_{[11\bar{1}]} = t$. The last term is the staggered Zeeman field with opposite signs on the two fcc sublattices A and B.

The second term describes spin-dependent hopping between pairs of second neighbors $\langle\langle ij \rangle\rangle$, where d_{ij}^1 and d_{ij}^2 are the bonds making up the pair and σ are the Pauli spin matrices. The linear size of the face-centered cubic (fcc) conventional cell is a. At half-filling, with $0 < m < 2t$ and λ_{SO} sufficiently large, the original model has a direct bandgap of $2m$ at $\beta = \pi$. Taking $|\mathbf{h}| = m \sin\beta$ and $\mathbf{h}$ in the [111] direction, the single parameter β keeps the gap constant and interpolates between the ordinary ($\beta = 0$) and the topological ($\beta = \pi$) insulator.

A crystalline surface (i.e., one where Bloch's theorem applies) of a strong topological insulator such as that described by the above Hamiltonian has an odd number of Dirac points in the surface band structure between the bulk valence and conduction bands. At these points the electron is effectively massless, as stated in the Introduction and shown in Fig. 1(b). This can be viewed as a generalization of the edge of the quantum spin Hall state, where the two directions along the 1D edge have opposite spin states; in the 2D surface state of a topological insulator, the spin and momentum are again locked, but the spin precesses smoothly through 2π as the electron momentum moves around a Fermi circle. Many of the interesting transport properties described below are consequences of this spin precession. One transport difference between the 1D edge and 2D surface state is fairly obvious: Kramers' theorem is sufficient to prohibit backscattering between the two points on the 1D edge Fermi surface by time-reversal-invariant perturbations, while in the 2D surface Fermi surface, only 180-degree backscattering is forbidden by this argument. This prevention of backscattering has been inferred in an STM experiment [29] using the quasiparticle interference technique.

2.2 Materials Considerations

Although the focus of this chapter is on basic theory, we list a few important materials in the experimental developments and refer the reader to the chapter of M. Z. Hasan for more information. The spin-orbit coupling must induce a phase transition in the electronic structure (from either a metal or ordinary insulator to a topological insulator), which suggests looking at heavy-element semiconductors or semimetals. The first topological insulator to be characterized in photoemission was the alloy $Bi_x Sb_{1-x}$, which is not ideal on several levels because of its intrinsic alloy disorder and the fact that the surface band structure is fairly complicated, with five sheets of the Fermi surface measured in the first experiment. Its $\mathbb{Z}_2$ invariants are (1; 111).

Most recent experiments are on the "next-generation" topological insulators $Bi_2 Se_3$ [30,31] (bandgap approximately 0.3 eV) or $Bi_2 Te_3$ [32] (bandgap approximately 0.15 eV), which are structurally similar layered compounds. The natural cleave plane shows a single Dirac cone in photoemission. While for many experiments these materials are sufficient to reveal predicted topological insulator behavior, they are plagued by a high degree of residual bulk conductivity at low temperature. The bulk contribution can complicate transport experiments, especially on large crystals. The intermediate compound $Bi_2 Te_2 Se$ has recently been shown by the Princeton and Osaka groups to have bulk conductivity that is several orders of magnitude smaller. A different approach to creating a 3D topological insulator is to strain 3D HgTe, which is a semimetal in crystal form but can be strained in thin films into a bona fide insulating state [33].

Many other topological insulator materials have been predicted based on electronic structure calculations, including some that might realize the interacting phases described later in this chapter, but the success rate of such predictions is currently middling. It has been pointed out that scalar relativistic effects or other spin-independent terms are the "cause" of band inversion and hence topological insulator behavior in a particular material, rather than the spin-dependent part (spin-orbit coupling). Indeed, including spin-orbit but not other relativistic terms for a material might lead to a metal or ordinary insulator rather than the topological insulator, and from the point of view of the energetics of band inversion, spin-orbit can be relatively unimportant [34]. However, spin-orbit remains essential because a spin-independent Hamiltonian can realize only the trivial "even" class of the $\mathbb{Z}_2$ invariants.

2.3 Berry Phases of Bloch Electrons and Chern-Simons Form of $\mathbb{Z}_2$ Invariant

We now return to theoretical discussion and explain how the geometry of Bloch wavefunctions leads naturally to a "vector potential in momentum space." The Abelian part of this vector potential underlies electrical polarization [35], the integer quantum Hall effect [12], and the intrinsic contribution to the anomalous Hall effect [36]. Topological insulators and a contribution to the magnetoelectric effect arise from the non-Abelian part when there are multiple occupied bands. Our starting point is a set of Bloch wavefunctions

$$\psi^a(\mathbf{r}) = e^{i\mathbf{k}\cdot\mathbf{r}} u_{\mathbf{k}}^a(\mathbf{r}) \tag{6}$$

at each point in momentum space. Here a is a band index running over occupied bands: $a = 1, \dots, 2N$, where we are looking forward to imposing time-reversal-invariance so that the number of occupied bands must be even.

Assuming that the wavefunctions are smoothly defined in some region, we define the Berry connection as

$$\mathcal{A}_\mu^{ab}(\mathbf{k}) = i \langle u_{a\mathbf{k}} | \partial_\mu | u_{b\mathbf{k}} \rangle. \tag{7}$$

Here a and b are band indices, so that the diagonal parts correspond to contributions within a single band, and off-diagonal terms are interband contributions. Note that the Kramers degeneracies at time-reversal-invariant points in the Brillouin zone mean that at least at these points there is a gauge symmetry $U(2)^N$, i.e., a freedom to redefine the wavefunctions. At other points, if there is no degeneracy, then the symmetry is just $U(1)^{2N}$. However, the Berry-phase expressions that are physically important typically have the full $U(2N)$ gauge symmetry, reflecting that a topological property is expected to be unchanged in going from the physical Hamiltonian to the "flattened Hamiltonian" $H'(\mathbf{k})$ where all occupied bands have energy -1 and empty bands have energy $+1$. Another frequently used object is the projection operator onto the occupied subspace, $P(\mathbf{k}) = (1 + H'(\mathbf{k}))/2$.

The Berry curvature is the "field strength" obtained from the Berry connection:

$$\mathcal{F}_{nn'}^{ab} = \partial^a \mathcal{A}_{nn'}^b - \partial^b \mathcal{A}_{nn'}^a - i[\mathcal{A}^a, \mathcal{A}^b]_{nn'}. \tag{8}$$

The electric polarization is the integral over the Brillouin zone of the trace over band indices of $\mathcal{A}$, although this formula is quite subtle [37]. The TKNN integer mentioned before is the Brillouin zone integral of the trace of $\mathcal{F}$, to which the last term (the commutator) does not contribute. In a metal, restricting the Tr$\mathcal{F}$ integral to the occupied states gives the intrinsic contribution to the anomalous Hall effect. A better way in practice to compute the Berry curvature is in terms of gauge-invariant projection operators onto the occupied subspace [38].

Since condensed matter physics normally begins with only an Abelian gauge field, i.e., electromagnetism, it is nice to see that the full non-Abelian Berry connection has physical significance, and not just in topological insulators. We assume in the following that there is no obstruction to defining wavefunctions smoothly throughout the Brillouin zone (for example, this would not be true if the Chern numbers were nonzero). Then an object well studied in the theory of non-Abelian gauge fields is the Chern-Simons form K whose differential is the second Chern form $\epsilon_{abcd}\mathcal{F}^{ab}\mathcal{F}^{cd}$. The integral of the Chern-Simons form over the Brillouin zone leads to another expression of the $\mathbb{Z}_2$ invariant. We follow convention in writing it in terms of an angular variable θ_{CS} with two possible values:

$$\theta_{CS} = -\frac{1}{4\pi}\,\epsilon_{abc} \int d^3k\,\mathrm{tr}\left[\mathcal{A}^a\partial^b\mathcal{A}^c - \frac{2i}{3}\mathcal{A}^a\mathcal{A}^b\mathcal{A}^c\right]. \tag{9}$$

The two values consistent with time-reversal-invariance are $\theta_{CS}=0$ (ordinary insulator) and $\theta_{CS}=\pi$. The mathematics of how this integral is quantized is interesting: non-Abelian gauge fields in 3D are like Abelian gauge fields in 1D in that there are "large" gauge transformations, i.e., those that cannot be smoothly deformed to the identity. Under a large gauge transformation, the integral for θ_{CS} changes by an integral multiple of 2π. Under time-reversal symmetry, the integral is odd, so there are only two inequivalent values: the even and odd multiples of 2π, which we label for simplicity as $\theta_{CS}=0$ and $\theta_{CS}=\pi$.

Briefly, the locally gauge-dependent Chern-Simons integrand in the above expression bears the same relationship to the gauge-invariant second Chern form as $\mathcal{A}$ does to $\mathcal{F}$. In the same way as the 2D integral of $\mathcal{F}$ can be related to pumping of polarization (the 1D integral of $\mathcal{A}$), where the two-dimensional space includes one momentum and one parameter, one can derive the existence of two values of θ_{CS} in a time-reversal-symmetric system from the integral quantization of the second Chern form in 4D, which underlies the four-dimensional Hall effect [39]. For further details the reader is referred to Ref. [14] or the chapter by Qi in this volume.

As the formula reproduces the other definitions of the $\mathbb{Z}_2$ invariant that are more easily computed, the reader may wonder whether it has an advantage other than its esthetic appeal. Indeed it does: the integral of the Chern-Simons form is more directly physical, e.g., gives part [14], but not all [40,41], of the orbital electron contribution to the linear magnetoelectric effect in all materials. We explain the physical consequences of θ_{CS} and its connection to the topological surface state in the following section.

2.4 Time-Reversal Breaking, Surface Hall Effect, and Magnetoelectric Response

The first definition we gave of a 3D topological insulator was as a time-reversal-symmetric insulator with an odd number of surface Dirac fermions. When a magnetic field is applied at the surface, a surface quantum Hall effect

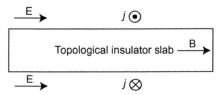

FIGURE 3   Illustration of the equivalence of surface Hall layers to bulk magnetoelectric term of type $\mathbf{E} \cdot \mathbf{B}$. Surface Hall layers lead to surface currents in an applied electrical field that generate a magnetic moment. The dual effect, that a magnetic field generates charge accumulation on Hall layers at the ends, is not shown. The magnitude of this effect depends on the surface Hall effect, which receives a fractional part from the bulk and a surface-dependent integer part in units of e^2/h. For the $\theta = \pi$ case realized in a topological insulator, the surface Hall effect is quantized to a half-integer, consistent with the behavior when an odd number of Dirac fermions are gapped by a magnetic field.

generically develops that is quantized as a *half-integer* times e^2/h [27]: $\sigma_{xy}^s = (e^2/h) \times (\ldots, -\frac{3}{2}, -\frac{1}{2}, \frac{1}{2}, \frac{3}{2}, \ldots)$. Note that as the other surface will also contribute a half-integer, a film shows a total Hall effect that is quantized to integers, as expected. The half-integer quantization of a single Diract fermion is consistent with the quantum Hall effect sequence in graphene (which has four Dirac fermions): $\sigma_{xy}^s = (e^2/h) \times (\ldots, -6, -2, 2, 6, \ldots)$.

A different way to interpret this surface Hall effect allows it to be connected to the angle θ_{CS} in the preceding section (Fig. 3). The electromagnetic description of a surface Hall effect is as a Chern-Simons term in the Abelian electromagnetic vector potential $A : \mathcal{L}_{CS} = \sigma_{xy} \epsilon^{abc} A_a \partial_b A_c$. As mentioned above in the non-Abelian case, the derivative of a Chern-Simons term can be expressed as the second Chern form, which in the electromagnetic case is just proportional to $\mathbf{E} \cdot \mathbf{B}$. In other words, a bulk term

$$\mathcal{L} = \theta \frac{e^2}{2\pi h} \mathbf{E} \cdot \mathbf{B}. \qquad (10)$$

This type of electromagnetic term is known as "axion electrodynamics" in the particle physics literature [42]. A variety of proposals to detect this half-quantization of the surface Hall effect ($\theta = \pi$) are reviewed in the chapter by Qi in this volume. As motivation for the following subsection on general magnetoelectric response, note that here we use θ to describe the electromagnetic response and θ_{CS} to describe the Chern-Simons integral result; these are equal for the case of topological insulators but not in general.

The physics of axion electrodynamics is quite subtle: to observe the corresponding magnetoelectric effect, the surface must be gapped by a time-reversal-breaking perturbation. The details of this perturbation pick out which of the values of θ (differing by $2\pi n, n \in \mathbb{Z}$) allowed by the bulk is actually observed. However, the perturbation must be weak enough that it does not perturb the bulk significantly, as then θ would no longer be quantized. The same situation occurs in the theory of polarization, where the bulk band structure determines the fractional part of polarization, while surface details can change this by an integer number of charges per transverse unit cell area.

Another physical consequence of a magnetoelectric term like $\mathbf{E} \cdot \mathbf{B}$ is that it leads to a magnetic response, which in this case is diagonal: an electrical polarization develops that is proportional to an applied magnetic field, and a magnetic moment that is proportional to an applied electric field. Such linear magnetoelectric effects have been studied for many years in materials that break both time-reversal and inversion symmetries. In general, there is a linear magnetoelectric response tensor

$$\frac{dP_i}{dB_j} = \frac{dM_j}{dE_i} = \alpha_{ij} = \tilde{\alpha}_{ij} + \alpha_\theta \delta_{ij}, \tag{11}$$

where $\tilde{\alpha}$ is traceless and

$$\alpha_\theta = \frac{\theta}{2\pi} \frac{e^2}{h} \tag{12}$$

is the scalar trace part (related to $\mathbf{E\dot{B}}$ discussed above) expressed in terms of the dimensionless parameter θ; α has units of conductance.

Now above we stressed the similarity between the Chern-Simons expression of the magnetoelectric effect and the simpler theory of electrical polarization. But a key difference is that the Berry phase is not the only contribution, even to the scalar diagonal contribution. The full orbital electronic contribution (i.e., freezing the lattice and considering only the orbital electron magnetic moment) has been obtained [40,41]. Here we quote the result and refer the reader to those works for further explanation, including the definition of the position operator on a periodic lattice:

$$\alpha_j^i = (\alpha_I)_j^i + \alpha_{CS}\delta_j^i, \tag{13a}$$

$$(\alpha_I)_j^i = \sum_{\substack{n \text{ occ} \\ m \text{ unocc}}} \int_{BZ} \frac{d^3k}{(2\pi)^3}$$

$$\text{Re}\left\{ \frac{\langle u_{n\mathbf{k}}|e\,\hat{r}_{\mathbf{k}}^i|u_{m\mathbf{k}}\rangle \langle u_{m\mathbf{k}}|e(\mathbf{v_k} \times \hat{r}_{\mathbf{k}})_j - e(\hat{r}_{\mathbf{k}} \times \mathbf{v_k})_j - 2i\partial H'_{\mathbf{k}}/\partial B^j|u_{n\mathbf{k}}\rangle}{E_{n\mathbf{k}} - E_{m\mathbf{k}}} \right\}, \tag{13b}$$

$$\alpha_{CS} = -\frac{e^2}{2\hbar}\epsilon_{abc}\int_{BZ}\frac{d^3k}{(2\pi)^3}\,\text{tr}\left[\mathcal{A}^a\partial^b\mathcal{A}^c - \frac{2i}{3}\mathcal{A}^a\mathcal{A}^b\mathcal{A}^c\right]. \tag{13c}$$

In a topological insulator (or a material with inversion symmetry), the ordinary part α vanishes by symmetry and the Chern-Simons part is quantized. In general magnetoelectric materials such as Cr_2O_3, both terms are present and the surface is naturally gapped without any further perturbation being required.

Work continues on evaluating these expressions when the magnetoelectric effect is not quantized in order to understand how large the orbital contribution is [43]. There has been recent progress in finding more efficient ways to evaluate topological invariants, especially in disordered non-interacting systems. An open problem where some mathematical insight would be helpful is to find an efficient evaluation (without needing to find a path in parameter space to a reference material) of the Chern-Simons integral (Eq. 9) when it is not quantized. The next section returns to the case of gapless Dirac surfaces of an unperturbed topological insulator and asks how the unusual properties of a single Dirac cone are manifested in transport.

3 TRANSPORT IN TOPOLOGICAL SURFACE STATES AND REAL-SPACE GEOMETRY

At first glance, transport in a topological insulator surface state with one Dirac cone appears similar to that in graphene, which has four Dirac cones as a result of spin and valley degeneracies. The two systems are generally similar in zero magnetic field if scattering between the Dirac cones in graphene can be neglected. There are several fundamental differences in the response to disorder and magnetic fields, and these differences become more pronounced in certain geometries such as nanowires or nanoribbons. A brief discussion of the superconducting case is postponed to the following section. Bardarson and the present author [44] have recently written a longer review, with a much more comprehensive citation list, devoted specifically to transport theory and experiment in topological insulator surfaces.

The nonzero bulk conductivity in current topological insulator materials places a premium on finding ways to "subtract out" bulk effects. One successful approach has been to look for quantum oscillations in a magnetic field and attempt to identify a distinct period associated with the surface Fermi level. Even though true quantum Hall plateaus with a mobility gap have not been observed, the motion of Landau levels through the chemical potential is observed, and can be fit to the expected result for a Dirac cone with velocity v,

$$|E_n| = \sqrt{(g\mu B/2)^2 + 2|nB|\hbar e^2 v}, \qquad (14)$$

where the first term is the Zeeman effect and the second gives the $E \sim \sqrt{B}$ dependence familiar to graphene. There remain some subtleties in the comparison between different experiments (see the chapter by Hasan for more discussion). Observing a gapped surface with clear half-integer quantization is one of the main goals of current research.

Aside from magnetism, another way that the Dirac cones in a topological insulator can disappear is if one surface is brought close to another, as it happens in a thin film. There is then a gap in the surface states: the gap can either result from tunneling between the two surfaces in an infinite slab, which is exponentially small once the surfaces are farther apart than the thickness of the surface state, or from quantization of the electronic motion around the circumference of a finite slab [45]. This thickness-dependent bandgap in a film has been observed experimentally [46] and could be useful for low-temperature thermoelectricity [47].

Another important feature of thin films is the following: the locking between spin and momentum in a topological insulator surface state means that a metallic current at the surface implies a spin density. This effect exists in ordinary spin-orbit-coupled quantum wells also, but there the existence of two sheets of the Fermi surface means that there is nearly a cancellation and the spin density produced for a typical current is quite small. In the topological insulator case, there is only one sheet of the Fermi surface at one surface (say, the top), but if there is another surface with the same velocity but the opposite spin direction (say, the bottom, if it is related by inversion symmetry and is at the same chemical potential), then there is again a cancellation. Applying an electrical bias or other asymmetry in a thin film can lead to a spin density for a given current that is an order of magnitude larger than in conventional quantum wells made from strong spin-orbit semiconductors [48].

One way to subtract out bulk effects is to consider the effects of a uniform magnetic field along the long axis of a topological insulator nanowire or nanoribbon, as in an experiment by Cui and collaborators [49]. Suppose that the wire is large enough that the magnetic field required to generate a flux quantum h/e through the cross-sectional area of the wire is weak in its effect on the surface metallic electrons: the surface electrons are affected only by the phase generated when they move around the magnetic flux. If the wire is perfectly clean, then the linear-response conductance along the nanowire at zero temperature and zero flux just counts the number of one-dimensional bands or "channels" that are partially filled, so that it increases as the chemical potential moves in either direction away from the Dirac point.

An interesting feature that comes out from simply calculating the bands of a Dirac electron is that whether flux increases or decreases the conductance (i.e., whether the magnetoconductance is positive or negative) oscillates as a function of the chemical potential. When disorder is added, two simple situations can be guessed based on the behavior of metallic cylinders. A short cylinder behaves like a single metallic ring and shows a periodicity h/e in its flux dependence as observed in Ref. [49]. A longer cylinder behaves like the ensemble average of metallic rings, and the periodicity is $h/(2e)$ as in the classic Sharvin-Sharvin experiment. These two types of magnetic field dependences are visible in the large conductance (because of either small disorder or large chemical potential) region of Fig. 4, which was obtained by numerical simulation of transport by a single disordered Dirac fermion on the surface of a cylinder [50].

An even more interesting feature occurs in the small-conductance region. There is a protected mode when the flux through the cylinder is precisely

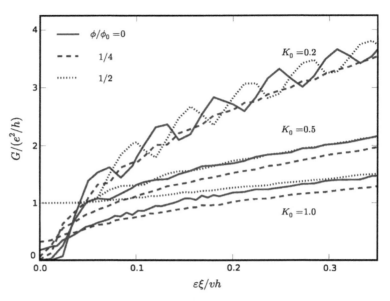

FIGURE 4 (From Ref. [50]) Conductance of a nanowire of topological insulator versus scaled chemical potential $e\xi/(vh)$ (relative to the Dirac point) for various disorder strengths K_0 and fluxes ϕ. The flux quantum $\phi_0 = h/e$. In the high-conductance regime, either h/e or $h/(2e)$ oscillations are seen, and the sign of the former is sensitive to the chemical potential. In the low-conductance regime (bottom left corner), there is a protected conductance e^2/h with one-half flux quantum through the nanowire.

one-half flux quantum (note that this is a special value in that the phase factor induced in going around the surface is -1, and time-reversal symmetry is satisfied). At this flux, the conductance is at least e^2/h and saturates to this value in a long wire where all other modes are localized. At other fluxes, there is no such protection and the conductance goes to 0. The intuitive reason for why this protected mode at one-half flux quantum exists is as follows: the factor -1 from the magnetic flux exactly compensates another -1 from the spin precession through 2π of an electron moving around the cylinder. Therefore the half-flux-quantum protection tests not only the Dirac spectrum but also the unique spin precession and anti-localization properties of the topological insulator surface state.

4 CONSEQUENCES OF STRONG ELECTRONIC INTERACTIONS

Until this point, the material presented in this chapter has largely concentrated on a non-interacting picture of topological insulators. Since real materials have interactions, it is worth asking, first, whether the non-interacting picture is a reasonable description, and second, whether adding strong interactions to topological insulators gives rise to new phases or other interesting physics. The answer to the first question was foreshadowed somewhat in the earlier discussion of the magnetoelectric effect: it is possible to define a bulk physical quantity in an interacting 3D insulator that determines whether the insulator is "ordinary" or "topological." However, this requires considerably more care than in the quantum Hall case, and it remains a challenging problem how best to predict whether a particular correlated insulator is ordinary or topological.

The answer to the second question, about new phases induced by strong interactions, is an active subject of current research, but there has already been considerable progress. Some of these examples are discussed elsewhere in this volume, as in the search for topological superconductors (or superfluids, which already exist in some cases via ^{3}He) that realize topologically nontrivial phases in the "tenfold way" table, or the possibility of unusual 3D semimetals enabled by spin-orbit coupling. In the later parts of this section, we briefly mention three areas of recent research that build on what has already been presented here: what happens to a 3D topological insulator when it is placed next to a superconductor; whether there are "fractional" topological insulators that bear the same relation to the topological insulator as the fractional quantum Hall effect does to the IQHE; and whether there are time-reversal-broken phases analogous to the topological insulator.

4.1 Defining the 3D Topological Insulator with Interactions

One question we have deferred until now is how to characterize the topological insulator beyond single-electron physics. This is considerably more challenging than in the integer quantum Hall effect case, where the mathematical integral that gives the non-interacting TKNN integer of a single-electron wavefunction [12] can be recast in terms of the many-body ground-state wavefunction's response to boundary fluxes [51]. This approach allows $\mathbb{Z}_2$ invariants to be defined for individual single-electron states in a disordered non-interacting system, but some new insight is required in the interacting case, for the following reason. The topological invariants described in Section 2 all use more than one single-electron state, which is consistent with the idea that the topological insulator results from the non-Abelian part of the Berry connection, unlike the IQHE, which comes from the Abelian part. The many-electron ground state can give rise to an Abelian Berry connection, but using only the unique ground state it is not possible to obtain a non-Abelian connection and imitate the various non-interacting definitions.

As the connection to the theory of magnetoelectricity in Section 3 suggests, it is possible to define the 3D topological insulator in terms of its bulk response to applied fields, and this approach will be pursued here. The key is to note that the electrical polarization is an Abelian Berry-phase quantity and hence generalizes immediately to interactions [52] in the same way as the Chern number [51]. So our basic strategy will be to compute the change in polarization induced by an applied magnetic field, using only bulk properties of the material. The key is to note, assuming that the unit cell is a parallelepiped for simplicity, that the quantum (ambiguity) of polarization in direction i, $\Delta P_i = e/\Omega_i$ involves the same geometrical quantity Ω_i, defined as the transverse unit cell area perpendicular to direction i, as the minimum field $B_i = h/(e\Omega_i)$ that can be applied while maintaining periodicity in the Hamiltonian [28]:

$$\Delta\left(\frac{dP_i}{dB_i}\right) = \frac{\Delta P_i}{B_i} = \frac{e^2}{h}. \tag{15}$$

It follows that there are two values of dP_i/dB_i that are compatible with time-reversal symmetry, 0 and $\frac{\pm e^2}{(2h)}$. One computes numerically which value obtains using the first-Chern expression of polarization change, which is valid in an interacting system [52]. In practice this is computed using a limiting process with a supercell of several original unit cells arranged transverse to the magnetic field direction, which allows the applied magnetic field to become increasingly small, and works in both the non-interacting case and simple Hubbard-type models.

However, integration of topological invariants with algorithms for strongly correlated materials is still at an early stage—at the moment, prediction of topological insulator behavior is still mostly taking place with methods that ultimately describe the excitations as free fermions. Toward this end, there have also been efforts to describe the topological invariant in terms of the single-electron Green's function [53,54]. Such a definition is conceivably simpler for computational purposes than computing a response, though some subtleties arise in the Green's function definition [55] that are absent in the quantum Hall case.

4.2 Superconducting Proximity Effect

Aside from their stability to non-magnetic disorder, the 2D and 3D topological surface states support unusual proximity effects when placed next to a conventional superconductor, here assumed to be a fully gapped s-wave material. In the three-dimensional case, consider the effect of adding the s-wave proximity term

$$H' = \sum_k \left[\Delta c_{k\uparrow} c_{-k\downarrow} + \text{h.c.}\right] \tag{16}$$

to the surface of a topological insulator. Normally, this term would act to pair up both spin states at each momentum on the Fermi surface. In a topological insulator, it efficiently pairs the one spin state available at each momentum—if we had instead coupled the s-wave superconductor to a perfectly spin-polarized "half-metal," the leading-order proximity effect would vanish. The result is a time-reversal-symmetric superconductor made from only one Fermi sheet [56,57]. Like the more familiar time-reversal-breaking superconductor with order parameter $p_x + ip_y$, this superconducting layer has nontrivial topological effects such as the emergence of Majorana fermions in magnetic vortex cores.

Majoranas are discussed in detail elsewhere in this volume, and appear in a variety of topological insulator and spin-orbit semiconductor systems, but it seems worthwhile mentioning some recent progress toward their realization. Experimentally, Josephson junctions have been created using superconducting leads deposited on top of thin Bi_2Se_3 crystals [58,59]. These demonstrate ambipolar transport and proximity effect [58], and the dependence of critical current on magnetic field shows unusual features that may indicate an unconventional Josephson current-phase relation [59], as expected when Majorana excitations are present. Another feasible geometry to generate Majoranas may be to put a magnetic flux along a topological insulator nanowire that lies on top of a superconductor and is made superconducting by the proximity effect [60].

A thin film of a three-dimensional topological insulator has a possible instability to "exciton condensation" driven by Coulomb interactions between the two surfaces [61]. We mention this here because the theory of exciton condensation is formally similar to that of superconductivity, but the order parameter is in the particle-hole channel rather than the particle-particle channel. The tendency of Dirac systems toward exciton condensation was pointed out previously [62] for the case of a bilayer of graphene, where the symmetry between electron and hole Fermi surfaces means that, at the proper electrical bias, exciton condensation requires only weak coupling to occur in principle, although the transition temperature can be small depending on system parameters. The exciton condensate made in a topological insulator thin film has some distinct features, including an effective half-integer charge in vortex cores, that might become detectable once it is possible to make thin films with strong interactions and no tunneling between the surfaces.

4.3 Fractional Topological Insulators

The experimental discovery of the IQHE led very rapidly to the observation of the fractional quantum Hall effect, and the electronic state on a fractional quantum Hall plateau is one of the most beautiful and profound objects in physics. To date, there are no observations of fractional analogs of time-reversal-invariant topological insulators, but at least in two dimensions it is clear that such states exist theoretically. Just as integer quantum Hall states can be paired to form a quantum spin Hall state, fractional quantum Hall states can be paired to form a fractional 2D topological insulator, and at least under some conditions this is predicted to be a stable state of matter [63]. However, there are several challenges in making this state an experimental reality: if one imagines the state in semiclassical terms, then spin-up and spin-down electrons are circling in opposite directions, and the most logical effect of Coulomb interactions is to form a Wigner crystal (an incompressible quantum solid rather than an incompressible quantum liquid). If the interactions between electrons of different spins could somehow be made weaker than those of the same spin, then a fractional state might result. Considerable theoretical effort is currently going into lattice models that might realize the fractional two-dimensional phase.

A fractional phase in three dimensions must necessarily be a more complex state. Recall that in the non-interacting case the 3D state, unlike the 2D state, cannot be realized using two subsystems related by time-reversal symmetry. Furthermore, in three dimensions pointlike particles have only bosonic or fermionic statistics according to a classic argument of Leinaas and Myrheim [64]: briefly, a physical state in 2D is sensitive to the history of how identical particles were moved around each other, while in 3D, all histories leading to the same final arrangement are equivalent and the state is sensitive only to the permutation of the particle labels that took place. In more mathematical terms, 2D statistics of point particles is described by the braid group, while 3D statistics of point particles is described by the permutation group. Fractional statistics can occur in 3D between pointlike and linelike objects, so a genuinely fractional 3D phase must have both types of excitations.

One approach to constructing a 3D fractional topological insulator, at least formally, uses "partons": the electron is broken up into three pieces, which each go into the "integer" topological insulator state, and then a gauge constraint enforces that the wavefunction actually be an allowed state of electrons [65,66]. The challenge is in understanding how new physical properties emerge from this gauging process. A candidate effective theory for integer and fractional topological insulators in either 2D or 3D, in the same sense as Chern-Simons theory is the effective theory for the quantum Hall effect [67], is a form of BF theory [68]. There are some subtleties in this description, especially in 3D; in 2D it is understood how different compactification conditions determine whether BF theory has a gapless edge, as in the paired Chern-Simons form relevant to topological insulators, or no gapless edge, as in the $\mathbb{Z}_2$ spin liquid phase [69]. In 3D the possible compactifications are less clear, but at the classical non-compact level 3D BF theory does allow a Dirac fermion surface state [68]. It remains unclear whether, for example, there is a realistic interaction potential that could be imposed on a fractionally filled $\mathbb{Z}_2$ 3D band in order to create a state described by the parton construction and/or BF theory. Considerable theoretical effort is currently being devoted to understanding the formal aspects and practical realization of both fractional quantum Hall and fractional topological insulator states.

4.4 Related States in Three Dimensions

The discovery of topological insulators has led to considerable effort to find other symmetries besides time-reversal that enable topological phases. The most stable phases to disorder arise in the "tenfold way" classification of free-fermion

Hamiltonians discussed in the chapter of Ludwig, including topological superconductors and superfluids. But there are many symmetries that are important in real materials, such as the 230 space groups or 1441 magnetic space groups, even if in principle these symmetries are no longer strictly defined once there are defects. There are recent works that combine topology with crystal group theory to search for either bulk Dirac points in three dimensions stabilized by crystalline symmetry [70] (rather than by fine-tuning to a transition between ordinary and topological insulators) and to develop a general theory of topological insulators with an arbitrary space group symmetry [71].

Instead of attempting any general approach, we will be content here just to illustrate how symmetries other than time-reversal might protect topological behavior. Note as a simple example that the $\mathbf{E} \cdot \mathbf{B}$ term can be equally well quantized by inversion symmetry rather than time-reversal symmetry. A slightly more subtle example is to consider the symmetry of an antiferromagnet, in which the combination of a translation and time-reversal is a symmetry even though neither transformation alone is a symmetry, depending on whether the surface preserves the symmetry or not, an antiferromagnetic topological insulator of this type has Dirac cones on some surfaces and gaps on others; the latter case is actually more interesting as step edges would support chiral gapless wires (essentially IQHE edge states) [72].

One proposed "topological crystalline insulator" phase [73] would have quadratic band touchings on some surfaces. Even a metal can have a topological surface state once crystal momentum is conserved, as now states at different momenta along the surface do not mix, and there could be an effective bulk bandgap at a particular value of k in which a surface state lives, even if there is no true bulk gap. Finally, the Floquet Hamiltonians that describe periodically driven systems have a topological classification of their own [74]. We hope that this brief introduction to the theory of 3D topological insulators has conveyed some of the current excitement of the field, a few of its past successes, and most of all the continuing interplay between theory and experiment.

ACKNOWLEDGMENTS

The author has had the good fortune to discuss the physics in this chapter with too many outstanding scientists to list by name. His work on various aspects of topological insulators has been supported by the National Science Foundation, Department of Energy, and Defense Advanced Research Projects Agency.

REFERENCES

[1] In this section, the phrase topological insulator without further qualifiers refers specifically to the 3D state with fermionic time-reversal symmetry.
[2] Hasan MZ, Kane CL. Rev Mod Phys 2010;82:3045.
[3] Qi X-L, Zhang S-C. Rev Mod Phys 2011;83:1057.
[4] Hasan MZ, Moore JE. Annu Rev Cond Mat Phys 2011;2:55.

[5]   Chalker JT, Dohmen A. Phys Rev Lett 1995;75:4496.
[6]   Balents L, Fisher MPA. Phys Rev Lett 1996;76:2782.
[7]   Kane CL, Mele EJ. Phys Rev Lett 2005;95:226801.
[8]   Bernevig BA, Zhang S-C. Phys Rev Lett 2006;96:106802.
[9]   Moore JE, Balents L. Phys Rev B 2007;75:121306(R).
[10] Fu L, Kane CL, Mele EJ. Phys Rev Lett 2007;98:106803.
[11] Roy R. Phys Rev B 2009;79:195322.
[12] Thouless DJ, Kohmoto M, Nightingale MP, den Nijs M. Phys Rev Lett 1982;49:405.
[13] Haldane FDM. Phys Rev Lett 1988;61:2015.
[14] Qi X-L, Hughes TL, Zhang S-C. Phys Rev B 2008;78:195424.
[15] Roy R. Phys. Rev. B 2009;79:195321.
[16] Fu L, Kane CL. Phys Rev B 2006;74:195312.
[17] Essin AM, Moore JE. Phys Rev B 2007;76:165307.
[18] Ran Y, Vishwanath A, Lee D-H. Phys Rev Lett 2008;101:086801.
[19] Qi X-L, Zhang S-C. Phys Rev Lett 2008;101:086802.
[20] Kitaev A. Periodic table for topological insulators and Superconductors 2009, arxiv:0901.0686
[21] Schnyder A, Ryu S, Furusaki A, Ludwig AWW. Phys Rev B 2008;78:195125.
[22] If the original Brillouin zone is not a 3-torus, then the appropriate planes can be defined as the image of the reference planes in the torus under the inverse of whatever smooth map is required to take the Brillouin zone to a 3-torus.
[23] For systems with a fixed finite number of bands, there can be additional accidental invariants, such as the Hopf invariant for a 3D system with 2 bands, which underlies the Hopf insulator. These invariants are generally not stable to disorder and do not necessarily give rise to gapless boundaries, and hence are ignored here.
[24] Ran Y, Zhang Y, Vishwanath A. Nature Phys 2009;5:298.
[25] Ringel Z, Kraus YE, Stern A. Phys Rev B 2012;86:045102.
[26] Mong RSK, Bardarson JH, Moore JE. Phys Rev Lett 2012;108:076804.
[27] Fu L, Kane CL. Phys Rev B 2007;76:045302.
[28] Essin AM, Moore JE, Vanderbilt D. Phys Rev Lett 2009;102:146805.
[29] Roushan P et al. Nature 2009;460:1106.
[30] Xia Y et al. Nature Phys 2009;5:398.
[31] Zhang H et al. Nature Phys 2009;5:438.
[32] Chen YL et al. Science 2009:178.
[33] Brüne C, Liu CX, Novik EG, Hankiewicz EM, Buhmann H, Chen YL, et al. Phys Rev Lett 2011;106:126803.
[34] Zhu Z, Cheng Y, Schwingenschlögl U. Phys Rev B 2012;85:235401.
[35] King-Smith RD, Vanderbilt D. Phys Rev B 1993;47:1651.
[36] Nagaosa N, Sinova J, Onoda S, MacDonald AH, Ong NP. Rev Mod Phys 2010;82:1539.
[37] Resta R. Ferroelectrics 1992;136:51.
[38] Avron JE, Seiler R, Simon B. Phys Rev Lett 1983;51:51.
[39] Zhang S-C, Hu J. Science 2001;294:823.
[40] Essin AM, Turner AM, Moore JE, Vanderbilt D. Phys Rev B 2010;81:205104.
[41] Malashevich A, Souza I, Coh S, Vanderbilt D. New J Phys 2010;12:053032.
[42] Wilczek F. Phys Rev Lett 1987;58:1799.
[43] Coh S, Vanderbilt D, Malashevich A, Souza I. Phys Rev B 2011;83:085108.
[44] Bardarson JH, Moore JE. Quantum interference and aharonov-bohm oscillations in topological insulators. Rep Prog Phys 2013;76:056501.
[45] Imura K-I, Takane Y, Tanaka A. Phys Rev B 2011;84:195406.
[46] Zhang T et al. Phys Rev Lett 2009;103:266803.
[47] Ghaemi P, Mong RSK, Moore JE. Phys Rev Lett 2010;105:166603.
[48] Yazyev OV, Moore JE, Louie SG. Phys Rev Lett 2010;105:266806.
[49] Peng H, Lai K, Kong D, Meister S, Chen Y, Qi X-L, et al. Nat Mat 2010;9:225.
[50] Bardarson JH, Brouwer PW, Moore JE. Phys Rev Lett 2010;105:156803.
[51] Niu Q, Thouless DJ, Wu Y-S. Phys Rev B 1985;31:3372.
[52] Ortiz G, Martin RM. Phys Rev B 1994;49:14202.
[53] Wang Z, Qi X-L, Zhang S-C. Phys Rev Lett 2010;105:256803.
[54] Gurarie V, Zee A. Int J Mod Phys B 2001;15:1225.

[55] Essin AM, Gurarie V. Phys Rev B 2011;84:125132.
[56] Sato M. Phys Lett B 2003;575:126.
[57] Fu L, Kane CL. Phys Rev Lett 2008;100:096407.
[58] Sacepe B, Oostinga JB, Li J, Ubaldini A, Couto NJG, Giannini E, et al. Nat Comm 2011;2:575.
[59] Williams JR, Bestwick AJ, Gallagher P, Hong SS, Cui Y, Bleich AS, et al. Phys Rev Lett 2012;109:056803.
[60] Cook A, Franz M. Phys Rev B 2011;84:201105.
[61] Seradjeh B, Moore JE, Franz M. Phys Rev Lett 2009;103:066402.
[62] Min H, Bistritzer R, Su J-J, MacDonald AH. Phys Rev B 2008;78:121401.
[63] Levin M, Stern A. Phys Rev Lett 2009;103:196803.
[64] Leinaas JM, Myrheim J. Nuovo Cim 1977;B37:1.
[65] Maciejko J, Qi X-L, Karch A, Zhang S-C. Phys Rev Lett 2010;105:246809.
[66] Swingle B, Barkeshli M, McGreevy J, Senthil T. Phys Rev B 2011;83:195139.
[67] Wen X-G. Int J Mod Phys B 1992;6:1711.
[68] Cho GY, Moore JE. Ann Phys 2011;326:1515.
[69] Hansson TH, Oganesyan V, Sondhi SL. Ann Phys 2004;313:497, arxiv:cond-mat/0404327.
[70] Young SM, Zaheer S, Teo JCY, Kane CL, Mele EJ, Rappe AM. Phys Rev Lett 2012;108:140405.
[71] Slager R-J, Mesaros A. Juricic V. The space group classification of topological band insulators: Zaanen J; 2012.
[72] Mong R, Essin AM, Moore JE. Phys Rev B 2010;81:245209.
[73] Fu L. Phys Rev Lett 2011;106:106802.
[74] Lindner NH, Refael G, Galitski V. Nat Phys 2011;7:490.

Chapter 3

Models and Materials for Topological Insulators

Chaoxing Liu[*] and Shoucheng Zhang[†]

[*]*Department of Physics, The Pennsylvania State University, University Park, Pennsylvania 16802-6300, USA*
[†]*Department of Physics, Stanford University, Stanford, CA 94305, USA*

Chapter Outline Head

---

1 INTRODUCTION

In this chapter we give a pedagogical introduction to topological insulators (TIs) through simple and yet realistic models. Although the word "topological" may indicate requirement of highly advanced mathematical tools, a physical understanding of TIs does not require too much mathematics. Indeed, there are two simple and intuitive prototype models which can be easily followed step by

Topological Insulators. http://dx.doi.org/10.1016/B978-0-444-63314-9.00003-2

step with conventional analytical methods. All the salient features of TIs, such as topological quantum phase transition, gapless surface, or edge states, etc., can be easily extracted from these two models.

The first model was introduced in 2006 by Bernevig, Hughes, and Zhang (BHZ) to describe quantum spin Hall (QSH) effect in HgTe quantum wells [8], soon after the establishment of the concept of 2D TIs [51,9]. The predicted QSH effect was soon observed in a series of beautiful transport experiments on HgTe quantum wells [61,91]. Excellent agreement between theory and experiment reveals the prediction power of the BHZ model, despite its simplicity. Later it turned out this model also serves as the low-energy effective model for other two-dimensional (2D) TIs, such as type-II AlSb/InAs/GaSb quantum wells [71]. The second model was introduced by Zhang et al. to describe the low-energy physics of 3D TI materials in the Bi_2Se_3 family [132,72]. Similar to the BHZ model of 2D TIs, this model is also a minimal model which captures all the basic features of 3D TI materials. The BHZ model and the Zhang et al. model establish the band inversion mechanism through spin-orbit coupling (SOC) for TIs. This mechanism provides guiding principles to search for other 2D/3D topological materials. Therefore we focus on the discussion of these two models in this chapter.

This chapter is organized as follows. First two prototype models for 2D and 3D TIs are introduced in Sections 1 and 2. Then the edge/surface states are solved explicitly in Section 3, followed by a discussion of their basic properties in Section 4. Finally the search principle for topological materials is outlined and different families of topological materials are discussed in Section 5.

2 HgTe QUANTUM WELLS AND THE BERNEVIG-HUGHES-ZHANG MODEL

HgTe and CdTe are typical II-VI group compound semiconductors, which possess the zinc-blende lattice structure with both the anion and cation atoms forming two interpenetrating face-centered-cubic lattices. Zincblende structure is usually formed due to the sp^3 hybridization. Therefore, similar to other common semiconductors such as GaAs, the bands near Fermi energy are a s-type band (Γ_6), and a p-type band split by SOC into a $J = 3/2$ band (Γ_8) and a $J = 1/2$ band (Γ_7), as shown in Fig. 1 (a). CdTe has an energy gap ($\sim$1.6 eV) and its band ordering is normal with s-type (Γ_6) bands above p-type (Γ_8,Γ_7) bands. In contrast, HgTe has an inverted band structure: Γ_6 bands are below Γ_8 bands with a negative energy gap of -0.3 eV. Due to the degeneracy between heavy-hole and light-hole bands at the Γ point ($k = 0$), the conduction (valence) band consists of light-hole (heavy-hole) Γ_8 bands and HgTe becomes a zero-gap semiconductor.

HgTe can form a quantum well (QW) structure with CdTe as a barrier, in which both the conduction and valence bands are split into sub-bands due to

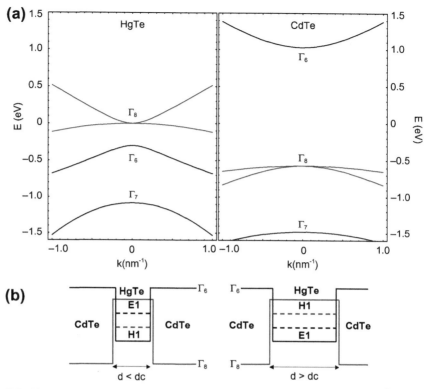

FIGURE 1 (a) Bulk band structure of HgTe and CdTe; (b) the energy levels of the lowest electron and heavy-hole sub-bands of the HgTe/CdTe quantum wells for the well thickness $d < d_c$ (left) and $d > d_c$ (right) [8].

the quantum confinement, as shown in Fig. 2. Let's consider a HgTe/CdTe QW along z direction, and the energy levels of QW sub-bands vary with the QW width d. As shown in Fig. 2, the heavy-hole Γ_8 sub-bands, denoted by H_n ($n = 1, 2, 3, \ldots$), decrease their energies with decreasing the QW width, while the electron Γ_6 sub-bands and light-hole Γ_8 sub-bands, denoted by E_n, increase their energies. Here we notice that electron Γ_6 bands are always hybridized with light-hole Γ_8 bands because both bands have the total angular momentum $\pm\frac{1}{2}$ along z direction. We focus on the behavior of the first electron sub-band E_1, which mainly consists of electron Γ_6 bands and the first heavy-hole sub-band H_1 which is derived from Γ_8 bands. For large QW width, the E_1 sub-band has lower energy than H_1 sub-band due to the band inversion between Γ_6 and Γ_8 bands in bulk HgTe. However when the well width is reduced, E_1 sub-band is shifted up while H_1 sub-band is shifted down, and for a thin QW, strong quantum confinement lifts E_1 sub-band above H_1 sub-band. Therefore, there is a critical

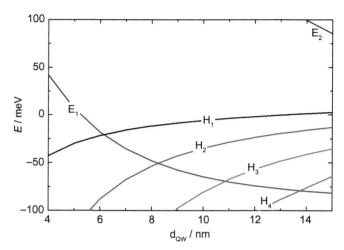

FIGURE 2 Energy levels of the HgTe/CdTe quantum wells as a function of quantum well width From [60].

width ($d_{QW} = d_c \sim 6.3$ nm in Fig. 2) where E_1 and H_1 sub-bands become degenerate. We expect the 2D band structure to be inverted when $d > d_c$ and normal when $d < d_c$. However in order to confirm this picture and understand the underlying physics, it is necessary to construct an effective model to describe the HgTe/CdTe QWs with the well width close to d_c.

The effective model can be constructed in the basis $\{|E_1+\rangle, |H_1+\rangle, |E_1-\rangle, |H_1-\rangle\}$, where $|E_1\pm\rangle$ and $|H_1\pm\rangle$ are degenerate since they are two sets of Kramers partners under time reversal (TR) symmetry. The form of the effective Hamiltonian can be determined by the symmetry properties of the basis $|E_1\pm\rangle$ and $|H_1\pm\rangle$. The opposite parities of $|E_1\pm\rangle$ and $|H_1\pm\rangle$ states require the matrix elements between them to be linear in $\mathbf{k}$ in the lowest order. Furthermore, rotation symmetry requires the matrix elements between $|E_1\pm\rangle$ and $|H_1\pm\rangle$ to take the form of $k_\pm = k_x \pm ik_y$ since the total angular momentum along the z direction is $\pm\frac{1}{2}$ for $|E_1\pm\rangle$ states and $\pm\frac{3}{2}$ for $|H_1\pm\rangle$ states. With these constraints, the effective Hamiltonian in the above basis is written as

$$\mathcal{H} = \begin{pmatrix} h(\mathbf{k}) & 0 \\ 0 & h^*(-\mathbf{k}) \end{pmatrix}, \tag{1}$$

$$h(\mathbf{k}) = \epsilon(\mathbf{k})\mathbb{I}_{2\times2} + d_a(\mathbf{k})\sigma^a, \tag{2}$$

where $\mathbb{I}_{2\times2}$ is the 2×2 identity matrix, and

$$\begin{aligned} \epsilon(\mathbf{k}) &= C - D\left(k_x^2 + k_y^2\right), \\ d_a(\mathbf{k}) &= \left(Ak_x, -Ak_y, \mathcal{M}(\mathbf{k})\right), \\ \mathcal{M}(\mathbf{k}) &= M - B\left(k_x^2 + k_y^2\right), \end{aligned} \tag{3}$$

where A, B, C, D, M are material parameters that depend on the QW geometry. The effective Hamiltonian (1), derived above from simple symmetry arguments, is exactly the BHZ model, which has also been confirmed by the more systematical perturbation method [60].

Since the BHZ model (1) is block diagonal with two blocks related to each other by TR symmetry, we can consider the physical property of one block $h(\mathbf{k})$ first. One block $h(\mathbf{k})$, as described by (2) and (3), is known as the quantum anomalous Hall (QAH) model, which possesses a non-zero and quantized Hall conductance [87,105]. The bulk energy spectrum of $h(\mathbf{k})$ is given by $E_{\pm} = \epsilon(k) \pm \sqrt{A^2(k_x^2 + k_y^2) + \mathcal{M}^2(\mathbf{k})}$, which can reproduce the energy dispersion when the well width is close to the critical value d_c. For $B = 0, h(\mathbf{k})$ is nothing but the massive Dirac Hamiltonian in $(2 + 1)$D with the mass M. Therefore, $\mathcal{M}(\mathbf{k})$ term in (2) is usually called the mass term, which describes the energy difference between the energy levels of E_1 and H_1 sub-bands at Γ point. For $d_{QW} > d_c$, the E_1 level falls below the H_1 level at the Γ point, and the mass M becomes negative. A pure massive Dirac model ($B = 0$) does not differentiate between a positive or negative mass M. Since we are dealing with a non-relativistic system, the B term is generally allowed. In order to make the distinction clear, we call M the Dirac mass, and B the Newtonian mass. The Newtonian mass B describes the usual non-relativistic mass term with quadratic dispersion relation. The relative sign between the Dirac mass M and the Newtonian mass B is crucial to determine whether the model describes a topological insulator state with protected edge states or not. To see this, we may consider the Hall conductance of the QAH model $h(\mathbf{k})$ (2), which can be determined by the Berry phase gauge field in the Brillouin zone and is given by the following formula:

$$\sigma_H = \frac{e^2}{h} \frac{1}{4\pi} \int dk_x \int dk_y \, \hat{\mathbf{d}} \cdot \left(\frac{\partial \hat{\mathbf{d}}}{\partial k_x} \times \frac{\partial \hat{\mathbf{d}}}{\partial k_y} \right), \tag{4}$$

which is e^2/h times the winding number of the unit vector $\hat{\mathbf{d}}(\mathbf{k}) = \mathbf{d}(\mathbf{k})/|\mathbf{d}(\mathbf{k})|$ around the unit sphere. The $\mathbf{d}(\mathbf{k})$ vector defined in Eq. (3) points north (south) at $k_x = k_y = 0$ for $M > 0(M < 0)$, and points south (north) at $\mathbf{k} \to \infty$ for $B > 0(B < 0)$. Consequently, the unit vector $\hat{\mathbf{d}}(\mathbf{k})$ has a skyrmion structure for $M/B > 0$ with winding number 1 and the system carries Hall conductance $\frac{e^2}{h}$, which leads to one chiral edge state propagating on the edge. When $M/B < 0$ the winding number is 0 and the system is a normal insulator. Therefore, a topological quantum phase transition occurs when the relative sign between the Dirac mass M and the Newtonian mass B is reversed. The BHZ Hamiltonian (1) is two copies of the QAH Hamiltonian (2). Although the total Hall conductance cancels due to TR symmetry, spin Hall conductance is non-zero and quantized in the inverted regime $M/B > 0$ when spin is conserved. Therefore, there is also a topological quantum phase transition for the BHZ Hamiltonian (1) when

M/B changes its sign. In the more general case without spin conservation, the topological phase transition still exists, which is protected by TR symmetry [51,52,9].

HgTe has a crystal structure of the zincblende type which lacks inversion symmetry, leading to a bulk-inversion-asymmetric (BIA) term in the Hamiltonian, given by [60]

$$
H_{BIA} = \begin{pmatrix} 0 & 0 & \Delta_e k_+ & -\Delta_z \\ 0 & 0 & \Delta_z & \Delta_h k_- \\ \Delta_e k_- & \Delta_z & 0 & 0 \\ -\Delta_z & \Delta_h k_+ & 0 & 0 \end{pmatrix} \tag{5}
$$

up to the linear order in **k**. This term plays an important role in determining the spin orientation of the helical edge state. The topological phase transition in the presence of the BIA term has been investigated recently [60,82]. In addition, structural inversion symmetry (SIA) can be broken by applying an electric field, leading to a SOC term of Rashba type in the effective Hamiltonian [92,101].

$$
H_{SIA} = \begin{pmatrix} 0 & 0 & i\xi_e k_- & 0 \\ 0 & 0 & 0 & 0 \\ -i\xi_e^* k_+ & 0 & 0 & 0 \\ 0 & 0 & 0 & 0 \end{pmatrix}. \tag{6}
$$

Here we recognize the SIA term as the k-linear Rashba term for the electron band; the k-cubic Rashba term for the heavy-hole is neglected. The parameters $\Delta_h, \Delta_e, \Delta_0, \xi_e$ depend on the quantum well geometry [92,60]. In Table 1, we give the parameters of the BHZ model for various values of d_{QW}.

TABLE 1 Material parameters for HgTe/CdTe quantum wells with different well thicknesses d. Here $\mathcal{E}_z$ denotes the electric field along z direction, which is required for a non-vanishing Rashba term.

$d(\text{Å})$	55	61	70
$A(\text{eV Å})$	3.87	3.78	3.65
$B(\text{eV Å}^2)$	−48.0	−55.3	−68.6
$D(\text{eV})$	−30.6	−37.8	−51.1
$M(\text{eV})$	0.009	−0.00015	−0.010
$\Delta_z(\text{eV})$	0.0018	0.0017	0.0016
Δ_e (eV Å)	−0.122	−0.125	−0.128
Δ_h (eV Å)	0.220	0.216	0.211
$\xi_e/(e\mathcal{E}_z)\ \text{Å}^2$	8.0×10^2	1.1×10^3	1.56×10^3

TABLE 2 Material parameters for InAs/GaSb/AlSb quantum wells with different well thicknesses d.

d_{GaSb} (Å)	100	100	100
d_{InAs} (Å)	84	90	100
A (eV Å)	0.72	0.62	0.37
B (eV Å^2)	-81.9	-78.3	-66.0
D (eV)	-21.6	-18.0	-5.8
M (eV)	0.0055	-0.00018	-0.0078
Δ_z (eV)	0.0003	0.00024	0.0002
Δ_e (eV Å)	0.0011	0.00096	0.00066
Δ_h (eV Å)	0.0006	0.0006	0.0006
ξ_e (eV Å)	-0.16	-0.13	-0.07

The BHZ model is not limited to the HgTe system. Actually a similar model can be obtained for type II InAs/GaSb/AlSb QWs [71]. In type II quantum wells, both InAs and GaSb layers consist of the well region and the AlSb layer serves as the barrier region. The key point is that the valence band edge of GaSb is 0.15 eV higher than the conduction band edge of the InAs layer. The E_1 electron sub-band is from InAs layer while the H_1 heavy-hole sub-band from GaSb layer, thus the relative position of E_1 and H_1 sub-bands can be tuned by controlling the thickness of InAs and GaSb layers, similar to the case of HgTe. The low-energy physics is again described by BHZ model (1), as well as BIA term (5) and SIA term (6), with the parameters shown in Table 2. The coefficient of BIA term is almost vanishing for InAs/GaSb QWs. In HgTe QWs, a finite Rashba term requires an external electric field $\mathcal{E}_z$ along z direction. In contrast, there is no structure inversion symmetry for InAs/GaSb QWs, and a finite Rashba term exists even for zero external electric field.

3 EFFECTIVE MODEL OF THE THREE-DIMENSIONAL TOPOLOGICAL INSULATOR

Similar to the case of 2D TI in HgTe QWs, a simple model Hamiltonian can be applied to a class of materials: Bi_2Se_3, Bi_2Te_3, and Sb_2Te_3 [132,72], which captures all the salient features of 3D TIs. Although many other families of 3D TIs have been proposed and some of them have also been confirmed experimentally, our discussion will be focused on the Bi_2Se_3 family in this section due to its simplicity.

Bi_2Se_3, Bi_2Te_3, and Sb_2Te_3 share the same rhombohedral crystal structure with space group D_{3d}^5 ($R\bar{3}m$) and have five atoms in one unit cell. As shown in Fig. 3(a), the crystal structure of Bi_2Se_3 consists of a layered structure where one individual layer forms a triangular lattice. Five-atom layers, known as quintuple

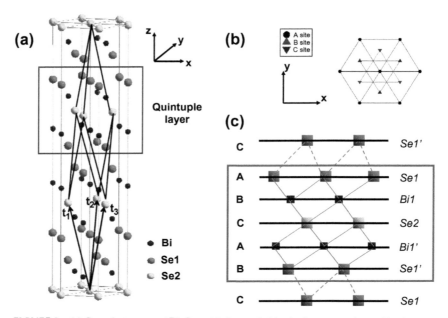

FIGURE 3 (a) Crystal structure of Bi_2Se_3 with three primitive lattice vectors denoted by $t_{1,2,3}$. A quintuple layer with Se1-Bi1-Se2-Bi1'-Se1' is indicated by the red box. (b) Top view along the z direction. Triangular lattice in one quintuple layer has three inequivalent positions, denoted by A, B, and C. (c) Side view of the quintuple layer structure. Along the z direction, Se and Bi atomic layers are stacked in the sequence $\cdots$-C(Se1')-A(Se1)-B(Bi1)-C(Se2)-A(Bi1')-B(Se1')-C(Se1)-$\cdots$. The Se1 (Bi1) layer is related to the Se1' (Bi1') layer by inversion, where Se2 atoms play the role of inversion center. (For interpretation of the references to color in this figure legend, the reader is referred to the web version of this book.) Adapted from [132].

layers, are stacked along the z direction. There are five atoms in one unit cell with two equivalent Se atoms denoted by Se1 and Se1', two equivalent Bi atoms denoted by Bi1 and Bi1', and a third Se atom denoted by Se2, as shown in Fig. 3(b). The coupling between the atomic layers within a quintuple layer is strong, while that between quintuple layers is much weaker, and predominantly of the van der Waals type. The primitive lattice vectors $t_{1,2,3}$ and rhombohedral primitive unit cells are shown in Fig. 3(a). The crystal has the threefold rotation symmetry with the trigonal axis defined as the z axis, the twofold rotation symmetry with the binary axis defined as the x axis, and the reflection symmetry with a bisectrix axis in the reflection plane, defined as the y axis. Furthermore, there is inversion symmetry for the crystal with the inversion center of Se2 site. Under inversion, Bi1 is mapped to Bi1' and Se1 is mapped to Se1'.

To get a better understanding of the band structure and orbitals involved, we start from the atomic energy levels and then consider the effects of crystal field splitting and SOC on the energy eigenvalues at the Γ point in momentum space. This is summarized schematically in three stages (I), (II), and (III) [Fig. 4(a)].

The states near the Fermi energy are primarily from p orbitals, so we neglect the s orbitals and focus on the p orbitals of Bi (electronic configuration $6s^2 6p^3$) and Se ($4s^2 4p^4$). In stage (I), we consider chemical bonding between Bi and Se atoms within a quintuple layer, which corresponds to the largest energy scale in this problem. First, we can recombine the orbitals in a single unit cell according to their parity. This results in three states (two odd, one even) from each Se p orbital and two states (one odd, one even) from each Bi p orbital. The formation of chemical bonds hybridizes the states on the Bi and Se atoms, and pushes down all the Se states and lifts all the Bi states. In Fig. 4(a), these five hybridized states are labeled as $\left| P1_{x,y,z}^{\pm} \right\rangle$, $\left| P2_{x,y,z}^{\pm} \right\rangle$, and $\left| P0_{x,y,z}^{-} \right\rangle$, where the superscripts $\pm$ denote the parity of the corresponding states. In stage (II), we consider the effect of crystal field splitting between different p orbitals. According to the point group symmetry, the p_z orbital is split from the p_x and p_y orbitals while the latter two remain degenerate. After this splitting, the energy levels closest to the Fermi energy turn out to be the p_z levels $\left| P1_z^{+} \right\rangle$ and $\left| P2_z^{-} \right\rangle$. In the last stage (III), we take into account the effect of SOC. The atomic SOC Hamiltonian is given by $H_{SO} = \lambda \mathbf{L} \cdot \mathbf{S}$, with $\mathbf{L}, \mathbf{S}$ the orbital and spin angular momentum, respectively, and λ the strength of SOC. The SOC Hamiltonian mixes spin and orbital angular momenta while preserving the total angular momentum. This leads to a level repulsion between $\left| P1_z^{+}, \uparrow \right\rangle$ and $\left| P1_{x+iy}^{+}, \downarrow \right\rangle$, and between similar combinations. Consequently, the energy of the $\left| P1_z^{+}, \uparrow (\downarrow) \right\rangle$ state is pushed down by the effect of SOC, and the energy of the $\left| P2_z^{-}, \uparrow (\downarrow) \right\rangle$ state is pushed up, as shown in stage (III) of Fig 4(a). It should be noted that due to SOC, both spin and orbital angular momentum are not conserved and only the total angular momentum is a good quantum number. The state $\left| P1_z^{+}, \uparrow \right\rangle (\left| P1_z^{+}, \downarrow \right\rangle)$ is mixed with the state $\left| P1_{x+iy}^{+}, \downarrow \right\rangle (\left| P1_{x-iy}^{+}, \uparrow \right\rangle)$. Therefore, in the following, we use the notation $\left| P1_z^{+}, \frac{1}{2} \left(-\frac{1}{2}\right) \right\rangle$ instead of $\left| P1_z^{+}, \uparrow (\downarrow) \right\rangle$, where $\pm \frac{1}{2}$ denote the total angular momentum. If SOC is larger than a critical value $\lambda > \lambda_c$, these two energy bands are inverted, similar to the case of HgTe. To illustrate this inversion process explicitly, the energy levels $\left| P1_z^{+} \right\rangle$ and $\left| P2_z^{-} \right\rangle$ have been calculated [132] for a model Hamiltonian of $Bi_2 Se_3$ with artificially rescaled atomic SOC parameters $\lambda(Bi) = x\lambda_0(Bi), \lambda(Se) = x\lambda_0(Se)$, as shown in Fig. 4(b). Here $\lambda_0(Bi) = 1.25$ eV and $\lambda_0(Se) = 0.22$ eV are the actual values of the SOC strength for Bi and Se atoms, respectively [114]. From Fig. 4(b), one can clearly see that a level crossing occurs between $\left| P1_z^{+} \right\rangle$ and $\left| P2_z^{-} \right\rangle$ when the SOC strength is about 60% of its actual value. Since these two levels have opposite parity, the inversion between them drives the system into a TI phase.

An effective model can be constructed to describe the topological phase transition due to the band inversion between $\left| P1_z^{+} \right\rangle$ and $\left| P2_z^{-} \right\rangle$, similar to the BHZ model for HgTe QWs discussed in the last section. Starting from the four low-lying states $\left| P1_z^{+}, \pm \frac{1}{2} \right\rangle$ and $\left| P2_z^{-}, \pm \frac{1}{2} \right\rangle$ at the Γ point, such a Hamiltonian can be constructed by the theory of invariants [113] at a finite wavevector $\mathbf{k}$. The

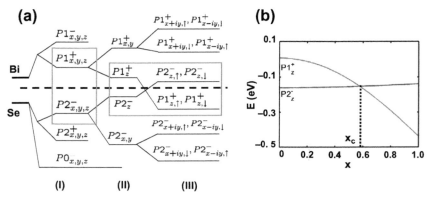

FIGURE 4 (a) Schematic picture of the evolution from the atomic $p_{x,y,z}$ orbitals of Bi and Se into the conduction and valence bands of Bi_2Se_3 at the Γ point. The three different stages (I–III) represent the effect of turning on chemical bonding, crystal field splitting, and SOC, respectively (see text). The blue dashed line represents the Fermi energy. (b) The energy levels $|P1_z^+\rangle$ and $|P2_z^-\rangle$ of Bi_2Se_3 at the Γ point versus an artificially rescaled atomic SOC $\lambda(Bi) = x\lambda_0(Bi) = 1.25x$ [eV], $\lambda(Se) = x\lambda_0(Se) = 0.22x$ [eV] (see text). A level crossing occurs between these two states at $x = x_c \simeq 0.6$. (For interpretation of the references to color in this figure legend, the reader is referred to the web version of this book.) Adapted from [132].

important symmetries of the system are TR symmetry T, inversion symmetry I, threefold rotation symmetry C_3 around the z axis and twofold rotation symmetry along y axis C_2. In the basis $\{|P1_z^+, \frac{1}{2}\rangle, |P2_z^-, \frac{1}{2}\rangle, |P1_z^+, -\frac{1}{2}\rangle, |P2_z^-, -\frac{1}{2}\rangle\}$, the representation of these symmetry operations is given by $T = i\sigma^y\mathcal{K} \otimes \mathbb{I}_{2\times2}, I = \mathbb{I}_{2\times2} \otimes \tau_3, C_3 = \exp\left(i\frac{\pi}{3}\sigma^z \otimes \mathbb{I}_{2\times2}\right)$, and $C_2 = i\sigma_x \otimes \tau_z$, where $\mathbb{I}_{n\times n}$ is the $n \times n$ identity matrix, $\mathcal{K}$ is the complex conjugation operator, and $\sigma^{x,y,z}$ and $\tau^{x,y,z}$ denote the Pauli matrices in the spin and orbital space, respectively. By requiring these symmetries and keeping only terms up to quadratic order in k, we obtain the following generic form of the effective Hamiltonian:

$$H_0' = \epsilon_k\mathbb{I}_{4\times4} + \mathcal{M}(\mathbf{k})\mathbb{I}_{2\times2} \otimes \tau_z + A_1 k_z\mathbb{I}_{2\times2} \otimes \tau_y$$
$$+A_2(k_y\sigma_x \otimes \tau_x - k_x\sigma_y \otimes \tau_x) \tag{7}$$
$$= \epsilon_0(\mathbf{k})\mathbb{I}_{4\times4} +$$

$$\begin{pmatrix} \mathcal{M}(\mathbf{k}) & -iA_1k_z & 0 & iA_2k_- \\ iA_1k_z & -\mathcal{M}(\mathbf{k}) & iA_2k_- & 0 \\ 0 & -iA_2k_+ & \mathcal{M}(\mathbf{k}) & -A_1k_z \\ -iA_2k_+ & 0 & -A_1k_z & -\mathcal{M}(\mathbf{k}) \end{pmatrix}, \tag{8}$$

with $k_{\pm} = k_x \pm ik_y, \epsilon_0(\mathbf{k}) = C + D_1k_z^2 + D_2k_\perp^2$, and $\mathcal{M}(\mathbf{k}) = M - B_1k_z^2 - B_2k_\perp^2$ $(k_\perp^2 = k_x^2 + k_y^2)$. This Hamiltonian is the same as that presented by Zhang et al.

TABLE 3 Material parameters in the effective Hamiltonian (10) for Bi_2Te_3, Bi_2Se_3, and Sb_2Te_3.

	Bi_2Se_3	Bi_2Te_3	Sb_2Te_3
A_1 (eV Å)	2.26	0.30	0.84
A_2 (eV Å^2)	3.33	2.87	3.40
C (eV)	−0.0083	−0.18	0.001
D_1 (eV Å^2)	5.74	6.55	−12.39
D_2 (eV Å^2)	30.4	49.68	−10.78
M (eV)	−0.28	−0.30	−0.22
B_1 (eV Å^2)	−6.86	−2.79	−19.64
B_2 (eV Å^2)	−44.5	−57.38	−48.51

in the Ref. [132], which can be shown by performing the unitary transformation

$$U_1 = \begin{pmatrix} 1 & 0 & 0 & 0 \\ 0 & -i & 0 & 0 \\ 0 & 0 & 1 & 0 \\ 0 & 0 & 0 & i \end{pmatrix}, \tag{9}$$

and the Hamiltonian is transformed into

$$H_0 = U_1 H_0' U_1^\dagger = \epsilon_{\mathbf{k}} \mathbb{I}_{4\times4} +$$

$$\begin{pmatrix} \mathcal{M}(\mathbf{k}) & A_1 k_z & 0 & A_2 k_- \\ A_1 k_z & -\mathcal{M}(\mathbf{k}) & A_2 k_- & 0 \\ 0 & A_2 k_+ & \mathcal{M}(\mathbf{k}) & -A_1 k_z \\ A_2 k_+ & 0 & -A_1 k_z & -\mathcal{M}(\mathbf{k}) \end{pmatrix}. \tag{10}$$

The parameters in the effective model can be extracted from the systematic $k \cdot p$ theory [72], as is shown in Table 3. The effective Hamiltonian shows a gap for Bi_2Se_3 and Sb_2Te_3, while the gap collapses at large k_z for Bi_2Te_3 due to $|D_1| > |B_1|$. In order to keep the band gap, the simplest way is to introduce the high order terms as an energy cut-off for large k_z. For Bi_2Te_3, we can keep the fourth order terms of k_z with the form

$$H^{(4)} = \epsilon_{\mathbf{k}}^{(4)} + \mathcal{M}^{(4)}(\mathbf{k}) \mathbb{I}_{2\times2} \otimes \tau_z \tag{11}$$

with $\epsilon_{\mathbf{k}}^{(4)} = H_1 k_z^4$ and $\mathcal{M}(\mathbf{k})^{(4)} = -L_1 k_z^4$. The parameters H_1 and L_1 can be extracted from the perturbation theory, given by $H_1 = -25.7$ eV Å^4 and $L_1 = -125$ eV Å^4, and we recover a finite gap for Bi_2Te_3, which is required for a well-defined bulk topology.

We can obtain another useful form for the effective Hamiltonian by choosing the local Wannier basis instead of the basis of bonding and anti-bonding states, which corresponds to applying a unitary transformation τ_x to the Hamiltonian (8), as done in Ref. [47]. Although these three forms of the Hamiltonian are equivalent to each other, the physical observables may take different forms in different basis. For example, the surface states have a non-trivial total angular momentum texture, which will be discussed in detail in the following section. In the basis of Hamiltonian (8), the total angular momentum operators are given by $\vec{J} = \frac{\hbar}{2}\mathbb{I}_{2\times2} \otimes \vec{\sigma}$, while in the basis of Hamiltonian (10), the total angular momentum operators are changed to $J_z = \frac{\hbar}{2}\mathbb{I}_{2\times2} \otimes \sigma_z$ and $J_{x,y} = \frac{\hbar}{2}\tau_{x,y} \otimes \sigma_y$.

Except for the identity term $\epsilon_0(\mathbf{k})$, the Hamiltonian (10) is similar to a 3D Dirac model with uniaxial anisotropy along the z direction, but with the crucial difference that the mass term is $\mathbf{k}$ dependent. From the fact that $M, B_1, B_2 < 0$ we can see that the order of the bands $|P1_z^+, \frac{1}{2}(-\frac{1}{2})\rangle$ and $|P2_z^-, \frac{1}{2}(-\frac{1}{2})\rangle$ is inverted around $\mathbf{k} = 0$ compared with large $\mathbf{k}$, which correctly characterizes the topologically non-trivial nature of the system. In addition, the Dirac mass M, i.e., the bulk insulating gap, is ~ 0.3 eV, which allows the possibility of having a room-temperature TI. Such an effective Dirac model can be used for further theoretical study of the Bi_2Se_3 system, as long as low-energy properties are concerned.

The above Hamiltonian (8) preserves a continuous in-plane rotation symmetry, however the underlying crystal structure has only a discreted threefold rotation symmetry. Therefore, some new terms in cubic (k^3) order are allowed, which lead to a hexagonal wrapping of the Fermi surface for surface states [31,72]. The hexagonal Fermi surface has important consequences for experiments on TIs such as surface state quasiparticle interference [137,2,3,64]. More quantitative description of Bi_2Se_3 family of materials requires an eight-band model [72].

4 HELICAL EDGE/SURFACE STATE OF 2D/3D TOPOLOGICAL INSULATORS

The main feature of 2D/3D TIs is the existence of the edge/surface state. The main advantage of the 2D/3D TI models introduced in the last two sections lies in the fact that the edge/surface states can be solved for a finite slab geometry explicitly. The edge/surface states show a distinct helical nature. In 2D, two states with opposite spin polarization counterpropagate at a given 1D edge. The edge states come in Kramers doublets, and TR symmetry ensures that the crossing of their energy levels at time reversal invariant momenta cannot be gapped. In 3D, the surface state consists of a single massless Dirac fermion. The 2D massless Dirac fermion is also helical, in the sense that the electron spin points perpendicularly to the momentum, forming a lefthanded helical texture in momentum space. TR invariant single-particle perturbations cannot introduce a gap for the surface state. In the following, we will study the edge/surface state and its effective Hamiltonian based on the 2D/3D TI model Hamiltonian with open boundary conditions.

Let's first consider the BHZ model Hamiltonian (1) of 2D TI defined on the half-space $x > 0$ in the xy plane or the four band model (10) on $z > 0$. In this configuration, translation symmetry is still preserved along y direction for 2D case and in the xy plane for 3D case, so we can divide the model Hamiltonian into two parts,

$$\hat{H} = \tilde{H}_0 + \tilde{H}_1 \tag{12}$$

in which $\tilde{H}_0$ is the Hamiltonian at $k_y = 0(k_x = k_y = 0)$ and $\tilde{H}_1$ determines the dispersion along y direction (xy plane) for 2D (3D) semi-infinite configuration. For 2D TI,

$$\tilde{H}_0 = \tilde{\epsilon}(k_x) + \begin{pmatrix} \tilde{M}(k_x) & Ak_x & 0 & 0 \\ Ak_x & -\tilde{M}(k_x) & 0 & 0 \\ 0 & 0 & \tilde{M}(k_x) & -Ak_x \\ 0 & 0 & -Ak_x & -\tilde{M}(k_x) \end{pmatrix}, \tag{13}$$

$$\tilde{H}_1 = -Dk_y^2 + \begin{pmatrix} -Bk_y^2 & iAk_y & 0 & 0 \\ -iAk_y & Bk_y^2 & 0 & 0 \\ 0 & 0 & -Bk_y^2 & iAk_y \\ 0 & 0 & -iAk_y & Bk_y^2 \end{pmatrix}, \tag{14}$$

with $\tilde{\epsilon}(k_x) = C - Dk_x^2$ and $\tilde{M}(k_x) = M - Bk_x^2$, while for 3D TI,

$$\tilde{H}_0 = \tilde{\epsilon}(k_z) + \begin{pmatrix} \tilde{M}(k_z) & A_1k_z & 0 & 0 \\ A_1k_z & -\tilde{M}(k_z) & 0 & 0 \\ 0 & 0 & \tilde{M}(k_z) & -A_1k_z \\ 0 & 0 & -A_1k_z & -\tilde{M}(k_z) \end{pmatrix}, \tag{15}$$

$$\tilde{H}_1 = D_2k_\perp^2 + \begin{pmatrix} -B_2k_\perp^2 & 0 & 0 & A_2k_- \\ 0 & B_2k_\perp^2 & A_2k_- & 0 \\ 0 & A_2k_+ & -B_2k_\perp^2 & 0 \\ A_2k_+ & 0 & 0 & B_2k_\perp^2 \end{pmatrix}, \tag{16}$$

with $\tilde{\epsilon}(k_z) = C + D_1k_z^2$ and $\tilde{M}(k_z) = M - B_1k_z^2$. Our strategy is to solve the $\tilde{H}_0$ exactly to obtain the eigenwavefunctions of the edge (surface) states at $k_y = 0(k_x = k_y = 0)$ as basis in 2D (3D) TI and then project the Hamiltonian $\tilde{H}_1$ into the subspace spanned by this basis. $\tilde{H}_0$ of Eqs (13) and (15) are identical, with the parameters A, B, C, D, M in Eq. (13) replaced by A_1, B_1, C, D_1, M in Eq. (15). Therefore, we will take 2D case as an example and similar procedure can also be applied to 3D case.

For such a semi-infinite system, k_x needs to be replaced by the operator $-i\partial_x$ and the Schrödinger equation of $\tilde{H}_0$ is given by

$$\tilde{H}_0(k_x \to -i\partial_x)\Psi(x) = E\Psi(x). \tag{17}$$

Since $\tilde{H}_0$ is block-diagonal, the eigenstates have the form

$$\Psi_+(x) = \begin{pmatrix} \psi_0 \\ 0 \end{pmatrix}, \quad \Psi_-(x) = \begin{pmatrix} 0 \\ \psi_0 \end{pmatrix}, \tag{18}$$

where $\mathbf{0}$ is a two-component zero vector, and $\Psi_+(x)$ is related to $\Psi_-(x)$ by TR. For the edge states, the wave function $\psi_0(x)$ is localized at the edge and satisfies the Schrödinger equation

$$\left(\tilde{\epsilon}(-i\partial_x) + \begin{pmatrix} \tilde{M}(-i\partial_x) & -iA_1\partial_x \\ -iA_1\partial_x & -\tilde{M}(-i\partial_x) \end{pmatrix} \right) \psi_0 = E\psi_0, \tag{19}$$

which has been solved analytically for open boundary conditions using different methods [60,140,68,74]. In order to find the conditions of the existence of edge states, we briefly review the derivation of the explicit form of the edge states by neglecting $\tilde{\epsilon}$ for simplicity [60].

Neglecting $\tilde{\epsilon}$, the Schrödinger equation (19) has particle-hole symmetry. Therefore, we expect that a special edge state with $E = 0$ can exist. With the wave function ansatz $\psi_0 = \phi e^{\lambda x}$, the above equation can be simplified to

$$\left(M + B\lambda^2 \right) \tau_y \phi = A\lambda\phi. \tag{20}$$

Therefore the two-component wave function ϕ should be an eigenstate of the Pauli matrix τ_y. Defining a two-component spinor $\phi_\pm$ by $\tau_y\phi_\pm = \pm\phi_\pm$, the Eq. (20) is simplified to a quadratic equation for λ. If λ is a solution for ϕ_+, then $-\lambda$ is a solution for ϕ_-. Consequently, the general solution is given by

$$\psi_0(x) = (ae^{\lambda_1 x} + be^{\lambda_2 x})\phi_+ + (ce^{-\lambda_1 x} + de^{-\lambda_2 x})\phi_-, \tag{21}$$

where $\lambda_{1,2}$ satisfy

$$\lambda_{1,2} = \frac{1}{2B}\left(A \pm \sqrt{A^2 - 4MB} \right). \tag{22}$$

The coefficients a,b,c,d can be determined by imposing the open boundary condition $\psi(0) = 0$. Together with the normalizability of the wave function in the region $x > 0$, the open boundary condition leads to an existence condition for edge states: $\Re\lambda_{1,2} < 0 (c = d = 0)$ or $\Re\lambda_{1,2} > 0 (a = b = 0)$, where $\Re$ stands for the real part. As we see from the Eq. (22), these conditions can only be satisfied in the inverted regime when $M/B > 0$. Furthermore, one can show that when $A/B < 0$, we have $\Re\lambda_{1,2} < 0$, while when $A/B > 0$, we have $\Re\lambda_{1,2} > 0$. Therefore, the wave function for the edge states at the Γ point is given by

$$\psi_0(x) = \begin{cases} a\left(e^{\lambda_1 x} - e^{\lambda_2 x}\right)\phi_+, & A/B < 0; \\ c\left(e^{-\lambda_1 x} - e^{-\lambda_2 x}\right)\phi_-, & A/B > 0. \end{cases} \tag{23}$$

The sign change of A/B affects the spin polarization of edge states, which is key to determine the helicity of the topological edge states. Another important quantity characterizing the edge states is their decay length, which is defined as $l_c = \max\left\{|\Re\lambda_{1,2}|^{-1}\right\}$.

The effective edge model can be obtained by projecting the bulk Hamiltonian onto the edge states Ψ_+ and Ψ_- defined in (18). This procedure leads to a 2×2 effective Hamiltonian defined by $H_{\text{edge}}^{\alpha\beta} = \langle\Psi_\alpha|\left(\tilde{H}_0 + \tilde{H}_1\right)|\Psi_\beta\rangle$. In 2D TI case, the effective Hamiltonian of helical edge states is written as

$$H_{\text{edge}} = vk_y\sigma^z \tag{24}$$

to leading order in k_y. For HgTe QWs, we have $A \simeq 3.6$ eV Å [60], and the Dirac velocity of the edge states is given by $v = A/\hbar \simeq 5.5 \times 10^5$ m/s. In 3D TI case, the effective Hamiltonian of helical surface states yields

$$H_{\text{surf}}(k_x, k_y) = C + A_2\left(\sigma^x k_y - \sigma^y k_x\right) \tag{25}$$

to the first order in $\mathbf{k}$. For $A_2 = 4.1$ eV Å, the velocity of the surface states is given by $v = A_2/\hbar \simeq 6.2 \times 10^5$ m/s, which agrees reasonably with *ab initio* results [132] with $v \simeq 5.0 \times 10^5$ m/s.

5 PHYSICAL PROPERTIES OF TOPOLOGICAL EDGE/SURFACE STATES

In this section, we will discuss the basic properties of topological edge/surface states based on the explicit expressions for the wave function (18) and (23), as well as the effective Hamiltonian (24) and (25). The salient feature of edge/surface states is the spin-momentum locking [116,90], which means that spin, orbital angular momentum or total angular momentum forms a texture in the whole momentum space. It should be clarified that although we use the terminology "spin" here for "spin-momentum locking," only total angular momentum is a good quantum number and different spin components are mixed due to strong SOC. Therefore, for a fixed momentum $\mathbf{k}$, different spin components, as well as different orbital components, can mix with each other. However, one spin component usually dominates over the opposite one, so the surface state still have spin polarization and forms a real spin texture in the whole momentum space. We will distinguish the different textures of spin, orbitals, and total angular momentum and show them explicitly in this section. Due to spin-momentum locking, it is easy to see that the backscattering from $\mathbf{k}$ to $-\mathbf{k}$ is forbidden, which indicates the topological stability of edge/surface states. This stability is deeply related to a fundamental symmetry in physics, TR symmetry, which can also be shown explicitly in the present model. As a physical consequence, non-trivial topological response occurs when TR is broken.

5.1 Total Angular Momentum, Spin, and Orbital Texture of Helical Edge/Surface States

For 2D HgTe QWs, the BHZ model (1) is block diagonal with one block for spin up and the other block for spin down. Consequently, two branches of a helical edge state have opposite spin. For 3D TI Bi_2Se_3, the 2D surface states have in-plane rotation symmetry, so we need to consider orbital angular momentum besides spin. First we distinguish the difference between total angular momentum, real spin, and orbital momentum by carefully looking at the basis of surface states. For the basis $\left| P1_z^+, \frac{1}{2}\left(-\frac{1}{2}\right) \right\rangle$ and $\left| P2_z^-, \frac{1}{2}\left(-\frac{1}{2}\right) \right\rangle$ of the four band model (8), $\left| \pm \frac{1}{2} \right\rangle$ denote total angular momentum, including both orbital and spin angular momentum, along the direction normal to the surface (z direction). The Pauli matrices in the surface effective Hamiltonian (25) are for total angular momentum, which form a lefthanded helical texture.

It is often asked what is the texture for real spin. In order to answer this question, we need to decompose the surface wave function into the atomic p orbitals and real spin. It turns out that spin and p orbital are coupled to each other, forming a complicated spin-orbital coupled texture. The spin textures for p_x, p_y, and p_z orbitals are shown in Fig 5(b), (c) and (d), respectively. The p_z orbital (Fig 5(d)) has the same lefthanded helical texture as the texture of total angular momentum. For the in-plane orbitals, the spin texture takes the form of $(S_x(\mathbf{k}), S_y(\mathbf{k})) \propto -(\sin\theta, \cos\theta)$ for p_x orbital and $(S_x(\mathbf{k}), S_y(\mathbf{k})) \propto (\sin\theta, \cos\theta)$ for p_y orbital, where $\mathbf{k} = (k, \theta)$ in the polar coordinates. It is clear that the spin textures for p_x and p_y orbitals are opposite. Since the basis wave function of the surface state has the same amplitude of p_x and p_y orbitals, the real spin texture is dominated by p_z orbitals, leading to a lefthanded helical texture, as shown in Fig 5 (a). The spin texture of helical surface states has been observed in the spin-resolved angle-resolved photoemission spectroscopy (ARPES) measurement [46] and the orbital texture is measured in a circularly or linearly polarized ARPES experiment [110,98].

One physical consequence of the helical spin texture is the general relation between charge current density $\mathbf{j}(\mathbf{x})$ and spin density $\mathbf{S}(\mathbf{x})$ on the surface of TIs [90], $\mathbf{j}(\mathbf{x}) = v\mathbf{S}(\mathbf{x}) \times \hat{\mathbf{z}}$. Consequently with long range Coulomb interactions, the surface plasmon generally carries spin, giving rise to a hybridized "spin-plasmon" collective mode [90,12].

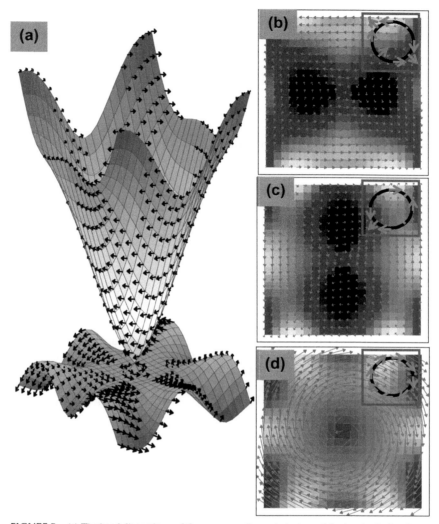

FIGURE 5 (a) The band dispersion and the corresponding spin texture of the topological surface states of Bi_2Se_3. The spin textures for p_x, p_y, and p_z orbitals are shown in (b), (c), and (d), respectively. The insets of (b), (c), and (d) show the spin textures schematically. The background colors of (b), (c), and (d) show the amplitudes of p_x, p_y, and p_z orbitals in the surface states. Here red color indicates a large amplitude and blue color means a small amplitude. (For interpretation of the references to color in this figure legend, the reader is referred to the web version of this book.)

5.2 Topological Stability of Edge/Surface State

For opposite momenta, the spin directions are opposite for helical edge/surface states of 2D/3D TIs. Consequently for ordinary non-magnetic impurities, the backscattering from momentum $\mathbf{k}$ to $-\mathbf{k}$ disappears due to the orthogonality of two spin states. This argument can be generalized to other spin-flip scattering cases of TR invariant impurities. Since the Pauli matrices in the edge/surface effective Hamiltonian (24) and (25) denote angular momentum $\frac{1}{2}$, the TR operator takes the form of

$$T = i\sigma_y K \tag{26}$$

in the basis $\Psi_\pm$ given by (18), where K represents complex conjugation. This TR operator is anti-unitary and satisfies $T^2 = -1$, which is a general condition for the TR operator of a half-odd-integer spin system. In the present system, the degenerate points at $k = 0$ of the Dirac cone for the edge/surface Hamiltonian (24) and (25) are due to the Kramers' theorem, which states that any single-particle eigenstate of a spin $\frac{1}{2}$ Hamiltonian with TR symmetry must have a degenerate partner. The general momentum independent perturbation terms

$$H_{\text{mass}} = \sum_\alpha m_\alpha \sigma_\alpha, \tag{27}$$

with $\alpha = x, y, z$ can be added to the Hamiltonian (24). For the edge Hamiltonian (24) of 2D TI, m_x and m_y terms in (27) can induce a gap and can be regarded as mass terms, while m_z term only shifts the position of the degenerate point. For the surface Hamiltonian (25) of 3D TI, only m_z term can induce a gap while m_x and m_y terms shift the position of the Dirac cone. With the TR operator (26), we have $T H_{\text{mass}} T^{-1} = -H_{\text{mass}}$, so H_{mass} is a TR symmetry breaking perturbation, which is forbidden in a TR symmetric case. Therefore, we can see explicitly how the degenerate point of edge/surface Dirac cone is protected by TR symmetry. In the presence of interactions, it can be shown that any $2n + 1$-particle backscatterings are forbidden and only $2n$-particle backscatterings are allowed [116, 121].

For the edge state of a 2D TI, the absence of backscatterings leads to dissipationless transport. Since the contact resistance still exists, this leads to $\frac{2e^2}{h}$ conductance in a two-terminal measurement, as confirmed in the transport experiment in HgTe/CdTe QWs [61, 60]. This edge state effective theory is non-chiral, and is qualitatively different from the usual spinless or spinful Luttinger liquid theories. It can be considered as a new class of 1D critical theories, dubbed a "helical liquid" [116]. For the surface state of 3D TI, the absence of backscatterings leads to the anti-localization of 2D surface transport [32, 75], which has been observed experimentally [83, 16, 15, 39]. However, the scatterings of other angles are still allowed, which can induce resistance during the transport of surface electrons.

5.3 Topological Response of TR Breaking Mass Term

As shown in the above, the Hamiltonian (27) can gap the edge/surface states due to the breaking of TR symmetry. The gapped edge/surface states lead to non-trivial topological responses. In the following, we will discuss two examples of the topologically non-trivial response: fractional charges of 1D helical edge states [85] and half quantum Hall effect of 2D surface states of 3D TIs [32,86].

5.3.1 Fractional Charge of 1D Helical Edge States

We begin with the effective Hamiltonian (24) of 1D helical edge states with TR breaking terms (27). Physically the mass term (27) can be interpreted as the exchange coupling between the total angular momentum of surface electrons and magnetic impurities [33,54]. The mass m_α is regarded as a field with a spatial and temporal dependence. According to the work of Goldstone and Wilczek [34], at zero temperature the ground-state charge density $j_0 \equiv \rho$ and current $j_1 \equiv j$ in a background field $m_a(x,t)$ is given by

$$ j_\mu = \frac{1}{2\pi} \frac{1}{\sqrt{m_\alpha m^\alpha}} \epsilon^{\mu\nu} \epsilon^{\alpha\beta} m_\alpha \partial_\nu m_\beta, \alpha, \beta = x, y, $$

with $\mu, \nu = 0, 1$ corresponding to the time and space components, respectively, and m_z does not enter the long-wavelength charge-response equation. If we parameterize the mass terms in terms of an angular variable θ, i.e., $m_x = m\cos\theta, m_y = m\sin\theta$, the response equation is simplified to

$$ \rho = \frac{1}{2\pi} \partial_x \theta(x,t), \quad j = -\frac{1}{2\pi} \partial_t \theta(x,t). \tag{28} $$

It is remarkable that the above charge and current response equations only rely on the angle θ of the direction of magnetization and have no dependence on the magnitude of magnetization. In a ferromagnetic domain wall, the angle θ of magnetization is changed by π. Consequently, a half-charge $\pm\frac{e}{2}$ is localized at the boundary. When θ is rotated from 0 to 2π adiabatically, a quantized charge e is pumped through the 1D helical edge state [85].

5.3.2 Half Quantum Hall Effect on 2D Surface of 3D TIs

Next we will discuss 2D surface states, described by the effective Hamiltonian (25), with the TR breaking terms (27). Our discussion follows Ref. [86,88]. Since m_x and m_y terms only shift the Dirac cone and do not open a mass gap, we will focus on m_z term by setting $m_x = m_y = 0$ in the following. As discussed in Section 2, the Hall conductance of a generic two-band Hamiltonian (1) is determined by Eq. (4), which is the winding number of the unit vector $\hat{\mathbf{d}}(\mathbf{k}) = \mathbf{d}(\mathbf{k})/|\mathbf{d}(\mathbf{k})|$ on the Brillouin zone. The perturbed surface state Hamiltonian (25) with (27) corresponds to a vector

$$ \mathbf{d}(\mathbf{k}) = \left(A_2 k_y, - A_2 k_x, m_z \right). \tag{29} $$

We notice that this $\hat{\mathbf{d}}(\mathbf{k})$ vector is quite similar to that of the quantum anomalous Hall Hamiltonian, which is given by (3). The only difference is that the mass term m_z is a constant in (27) while it has quadratic dependence of momentum k in (3). Therefore, the unit vector $\hat{\mathbf{d}}(\mathbf{k})$ in (4) forms a skyrmion configuration for (2) when $M/B > 0$, which leads to $\frac{e^2}{h}$ Hall conductance, but it has a "meron" configuration for (29), which yields a winding number $\pm 1/2$ and a half Hall conductance

$$\sigma_H = \frac{m_z}{|m_z|} \frac{e^2}{2h}. \tag{30}$$

Although the direct calculation shows a half Hall conductance for the gaped topological surface states, in contrast to the integer Hall conductance in quantum anomalous Hall model, one needs to be careful about this result. Both the surface Hamiltonian (29) and the QAH model (2) are low-energy effective models, which are only accurate near $k = 0$ and the derivation from Dirac-type model at large momenta is not included in the calculation. The latter may contribute to the Hall conductance. Actually it can be proved that in any 2D lattice model, the Hall conductance must be quantized in integer units of $\frac{e^2}{h}$ [105, 30]. One may wonder if the Hall conductance will be changed and go back to integer units of $\frac{e^2}{h}$ once we expand the effective Hamiltonian of the topological surface states to the higher orders.

However, the answer is surprising. The surface of a 3D topological insulator is different from any 2D lattice model, in the sense that such contributions from large momenta vanish due to the requirement of TR symmetry and the half Hall conductance is precise up to any order in the expansion. This conclusion can be obtained from the following generic argument based on TR symmetry. Consider the jump in Hall conductance at $m_z = 0$, which is given by $\Delta \sigma_H = \sigma_H(m_z \to 0^+) - \sigma_H(m_z \to 0^-) = \frac{m_z}{|m_z|} \frac{e^2}{h}$. This jump is independent of the large-momentum corrections, which are the same for $\sigma_H(m_z \to 0^+)$ and $\sigma_H(m_z \to 0^-)$. On the other hand, since the surface theory with $m_z = 0$ is TR invariant, TR transforms the system with mass m_z to that with mass $-m_z$. Consequently, from TR symmetry we have

$$\sigma_H(m_z \to 0^+) = -\sigma_H(m_z \to 0^-).$$

Together with the condition $\Delta \sigma_H = \frac{m_z}{|m_z|} \frac{e^2}{h}$, we see that the half Hall conductance given by Eq. (30) is robust, and the contribution from large-momentum corrections must vanish. By comparison, in a 2D QAH Hamiltonian (2) with (3), the Hamiltonian with mass M is not the TR conjugate of that with mass $-M$, and the above argument does not apply. Therefore, the half Hall conductance is a unique property of the surface states of 3D topological insulators which is determined by the bulk topology.

6 TOPOLOGICAL INSULATOR MATERIALS

The understanding of the above two models for 2D and 3D TIs provides us with the basic principle to search for topological materials. First, both 2D and 3D TI models are of Dirac type with linear $\mathbf{k}$ coupling between the conduction and valence bands. In a system with inversion symmetry, it requires that conduction and valence bands have opposite parities. There are two possibilities for opposite parities. One possibility is the different orbital characters for conduction and valence bands. For example, in HgTe, Γ_6 and Γ_8 bands have s and p orbitals respectively. The other one is to consider a non-Bravais lattice with more than one atoms in the unit cell. The bonding and anti-bonding states are formed with opposite parities. For example, in graphene, although only p_z orbital is taken into account, there are two atoms in one unit cell and the bonding and anti-bonding states lead to the formation of Dirac cone. In more general case, both orbitals and lattice structures play important roles to form Dirac type of Hamiltonian together.

Another condition for TIs is "band inversion," namely $MB > 0$ for Hamiltonian (1) and $MB_1 > 0$ for Hamiltonian (8), which is induced by SOC, as clearly seen in Fig 4(b). Therefore, it is more likely to find topological insulators in materials which consist of covalent compounds with narrow band gaps and heavy atoms with strong SOC. SOC is not the only factor because band gap generally relies on various material parameters. Therefore TIs can even be realized in some materials with relatively small SOC. This provides us various methods, such as strain, to engineer topological phases in realistic materials.

The conditions of linear coupling and "band inversion" can serve as our basic mechanism and guiding principles. We shall discuss different families of materials separately below, putting emphasis on different lattice structures and orbitals and their relation to the non-trivial topological phases. HgTe and Bi_2Se_3 provide two different examples of topological materials. Many other verified or proposed materials can be related to these two examples. Therefore, we will first discuss these two classes and the related materials. Then we will discuss other types of materials beyond these two classes. More discussion about topological materials can be found in [128, 80].

6.1 Bi_2Se_3 Family and Related Materials

Both conduction and valence bands of Bi_2Se_3 family consist of p orbitals. Therefore, the opposite parities between conduction and valence bands originate from bonding and anti-bonding states. The band inversion occurs only at Γ point due to strong SOC, which can be described by a simple effective Hamiltonian as shown in Section 3. Actually a large class of materials can be described by this simple model. The discovery of the Bi_2Se_3 family of TIs [132, 118, 18] inspired a search for new topological materials guided by requiring strong SOC. Among normal main group elements in the periodic table, the lower right corner elements Bi, Pb, and Tl are cations that have the strongest atomic SOC, and I, Br, Te, Se are anions that have strongest SOC [114]. Naturally Pb and Tl-based chalcogenides, which are similar to the Bi_2Se_3 family, attracted the earliest attention. On the other hand, new variations are synthesized by mixing element compositions from the Bi_2Se_3 family, generating excellent TI materials with much better bulk insulating properties than their parent materials.

6.1.1 Bi_2Se_3 Family

Bi_2Se_3, Bi_2Te_3, and Sb_2Te_3 are the most extensively studied 3D topological materials. The topological surface states with a single Dirac cone have been observed through ARPES [118,18] and scanning tunneling microscopy (STM) [2,137,20,37,3], confirming the theoretical prediction [132,118]. Furthermore, spin-resolved ARPES measurements indeed observed the left-handed helical spin texture of the massless Dirac fermion [46]. The absence of backscattering for TR invariant impurities has been confirmed through the interference pattern in the momentum space by STM [2,137,64]. However, unlike the surface-sensitive measurement, such as ARPES and STM, it is much harder to extract transport properties of topological surface states in Bi_2Se_3 family of materials due to the large residual bulk carriers and sample quality [88,38]. For instance, Bi_2Se_3 is always n-type because of Se vacancies. Bi_2Te_3 is usually rather metallic (n- or p-type) due to Bi-Te anti-site defects which are promoted by the weakly polarized bond between Bi and Te with similar electronegativity. Sb_2Te_3 is found to be p-type due to the Sb-Te anti-sites. These crystal defects not only introduce large bulk carriers, but also influence the sample quality and lower the carrier mobility. There are extensive efforts, involving material doping [18,45,42,15,4,89,112], nanostructuring [83,94,120], and electric gating [16,59], to reduce the bulk carriers. Furthermore, new ternary or even quaternary compounds by mixing Bi_2Se_3, Bi_2Te_3, and Sb_2Te_3 have been fabricated with the aim to improve the bulk insulating property and the sample quality [103,119,102,135,58,4].

Superconductivity has also been found in this class of materials. In Bi_2Se_3, Cu intercalation in the van der Waals gaps between QLs is found to induce superconductivity at $T_c = 3.8$ K in $Cu_xBi_2Se_3$ for $0.12 \leq x \leq 0.15$ [43,115]. Regardless of doping, Bi_2Te_3 is recently reported to become superconducting with $T_c \sim 3$K under pressure $3 \sim 6$ GPa [136].

6.1.2 $TlBiSe_2$ Family

Thallium(Tl)-based chalcogenides $TlBiQ_2$ and $TlSbQ_2$ (Q = Te, Se, S) have rhombohedral crystal structure ($D_{3d}^5 - R\bar{3}m$), similar to Bi_2Se_3. Take $TlBiSe_2$ as an example. Atoms are placed in layers with the sequence of -Tl-Te-Bi-Te- and the primitive unit cell contains four atoms. The coupling between every two atomic layers is always strong, so the crystal structure is essentially 3D. Based on *ab initio* calculations [125], $TlBiSe_2, TlBiTe_2$, and $TlSbSe_2$ are found to be strong TIs with an energy gap ~ 0.2 eV, TlBiTe$_2$ is an indirect gap semimetal, and TlBiS$_2$, TlSbTe$_2$, and TlSbS$_2$ are small gap insulators near the topological transition boundary. Soon after the theoretical prediction [125,65,26], TlBiSe$_2$ [97,62,19] and TlBiTe$_2$ [19] have been confirmed by ARPES experiments to be 3D topological insulators with non-trivial surface states. Spin-resolved ARPES measurements [122] have confirmed the spin-momentum locking feature of the helical surface states. A topological phase transition was observed [122] in TlBi(S$_{1-x}$Se$_x$), which can viewed as the mixture of a strong TI TlBiSe$_2$ and a trivial insulator TlBiS$_2$ and lies on the transition boundary. By increasing the Se composition x, a transition from a trivial insulator to a TI with gapless surface states is found in ARPES experiments. Furthermore superconductivity has also been found in p-doped TlBiTe$_2$ [40], and its origin is still unclear [49].

6.1.3 LaBiTe$_3$ Family

By simply substituting one cation atom with a rare earth element in the Bi$_2$Se$_3$ type of materials, a new class of stoichiometric compound LnBT$_3$ (Ln = rare earth elements, B = Bi,Sb, and T = Te,Se) can be obtained, which shares the same layered crystal structure as Bi$_2$Se$_3$. LnBiTe$_3$ and LnSbTe$_3$ (Ln = rare earth elements La and Y that do not have f electrons) are already synthesized in experiments [76], and *ab initio* calculations [127] predict LaBiTe$_3$ to be a TI with an energy gap about 0.07 $\sim$ 0.12 eV), whereas the other three materials to be trivial insulators or semimetals. GdBiTe$_3$ is a stoichiometric magnetic material with the magnetic moments originating from Gd atoms and a single quintuple layer of GdBiTe$_3$ is recently predicted to be a QAH insulator [134].

6.1.4 PbBi$_2$Se$_4$ Family

By inserting a double-layer -Pb-Se- into the quintuple layer of Bi$_2$Se$_3$, a new compound PbBi$_2$Se$_4$ with stacking sequence Se-Bi-Se-Pb-Se-Bi-Se is obtained. Extension to this double-layer filling idea, various layered chalcogenides can be derived as A$_n$B$_{2m}$X$_{3m+n}$ (A = Pb, Sn, and Ge, B = Bi and Sb, X = Te and Se). Most of these compounds have already been synthesized in experiments. For example, Ge$_2$Sb$_2$Te$_5$ (called as GST) has been extensively studied as phase-change materials and used for data storage in the industry [117]. Recently both ab initio calculations [124,50,55] and ARPES measurements [55] have revealed that many compounds among A$_n$B$_{2m}$X$_{3m+n}$ chalcogenides are TIs. Pb-based compounds Pb$_n$Bi$_2$Se$_{3+n}$ and Pb$_n$Sb$_2$Te$_{3+n}$ are found to be TIs for $n = 1,2$, but trivial insulators for n $\geq$ 3 [124,50]. In addition, PbBi$_4$Te$_7$ [124,25], GeBi$_2$Te$_4$ [124,55,77], SnBi$_2$Te$_4$, SnSb$_2$Te$_4$, and YbBi$_2$Te$_4$ [77] are also found to be TIs. In experiments, only GeBi$_2$Te$_4$ has been confirmed by ARPES [124].

6.2 HgTe and Related Materials

Unlike the Bi_2Se_3 family with only p orbitals, in HgTe both the s-type Γ_6 band and p-type Γ_8 bands are important for the low energy physics. Therefore, HgTe represents an example of another class of topological materials, in which different orbitals play an essential role. Since $s-p$ bonding is common in many conventional semiconductors and other materials, various topological materials belong to this category, which include Heusler compounds, HgSe, $\beta - HgS$, Chalcopyrite semiconductors, $\alpha(\beta) - Ag_2$ Te, skutterudites and filled skutterudites, etc.

6.2.1 HgTe Quantum Wells and Strained Bulk HgTe, HgSe, β-HgS

As we see in Section 2, bulk HgTe has an inverted band structure between p-type Γ_8 and s-type Γ_6 bands. According to the parity criteria it is topologically non-trivial. However, the bulk gap is zero, because the heavy-hole and light-hole are degenerate at Γ point $(k = 0)$ due to the cubic symmetry. There are two ways open the energy gap. One is to fabricate HgTe QWs, where an energy gap opens and band inversion persists above a critical QW thickness. This corresponds to the 2D TI, which is described by the BHZ model in Section 2. The QSH effect has been confirmed by a series of transport experiment in HgTe QWs [61,10,91,60]. HgTe QWs are the first experimentally verified topological insulators [8,61].

The other is to break the cubic symmetry and to lift the heavy-hole and light-hole degeneracy, e.g., by applying strain, turning it into a 3D TI [21,32]. With the atomic tight-binding model calculation, Dai et al. [21] have shown that HgTe indeed becomes a strong TI after applying uniaxial strain. Recent transport experiment of quantum Hall effect in 70 nm thick HgTe film [11] has shown odd plateaus, which provides the first evidence of the topological surface state in the transport measurement. Further ARPES experiment supports this result and indicates the HgTe as a strong TI.

Similar situation happens for HgSe [29,79], which possesses a negative band gap -0.20 eV from the experiment. Naturally we expect that HgSe is a strong topological insulator under strain. For another family member β-phase (zincblende) HgS, local-density-functional (LDA) [23,13] and LDA-based GW [29] calculations found that HgS has Γ_6 and Γ_8 as valence bands and Γ_7 as the conduction band with a finite energy gap due to the negative SOC, which indicates β-HgS as a 3D TI with highly anisotropic surface states [107]. However, recent GW calculations based on a corrected-LDA method showed that Γ_6 lies in the conduction band without inversion [79].

6.2.2 AlSb/InAs/GaSb Quantum Wells

The "broken gap" type-II AlSb/InAs/GaSb quantum wells display an "inverted" phase similar to HgTe/CdTe QWs and are predicted to exhibit the QSH effect

[71]. The band inversion also happens between E_1 electron sub-band and H_1 heavy-hole sub-band, similar to HgTe QWs. The only difference is that E_1 and H_1 sub-bands are from two different layers: E_1 state is localized in the InAs layer while H_1 state stays in the GaSb layer. Therefore, InAs/GaSb QWs provide an example of the possibility to engineer the inverted band structure through artificial heterostructures. Recent transport experiments have confirmed the QSH effect in InAs/GaSb QWs [56,57].

6.2.3 Heusler Compounds

Heusler compounds [41,35], a class of ternary materials with XYZ (half-Heusler) or X_2YZ stoichiometrical composition, share similar structural and electronic properties as the zincblende semiconductors, such as HgTe. For example, the YXYZ half-Heuslers can be understood as stuffing the $(YZ)^{n-}$ zincblende sublattice with an X^{n+} ion. Due to the diversity of Heusler materials, the energy gap can be tuned by the difference in electronegativity of the constituents and the lattice constant in a wide range from about 4 eV to (LiMgN) down to zero (LaPtBi). Using *ab initio* calculations two independent groups [14,67,66] predicted that tens of Heusler compounds with a zero gap show similar band inversion to that of HgTe, which are confirmed by the more sophisticated *ab initio* calculations, such as optimized exchange-correlation potential [1,28] and GW approach [106]. Strain can be applied to drive these materials into TI phases. Another advantage of the diversity of Heusler materials is that the topological phases may coexist with other phases, such as magnetism or superconductivity. For example, LnPtBi may produce TIs with coexisting magnetism (Ln is rare earth element, Nd, Sm, Gd, Tb, and Dy) and particularly LaPtBi could be a topological superconductor [14,67]. Recently an ARPES experiment has shown the evidence of surface band structures of three Heuslers LnPtBi (Ln = Lu, Dy, and Gd) [69].

6.2.4 Chalcopyrite Semiconductors

The ternary ABC_2 chalcopyrite compounds of composition I-III-IV2 or II-IV-V2 are isoelectronic analogs of II–V or III–V binary semiconductors, respectively. The crystal structure (space group $D_{2d}^{12}-I\bar{4}2d$) is described as a doubled zincblende structure with small structural distortions. Due to the structural similarity to their binary analogs, some chalcopyrites may become TIs like HgTe with the advantage that a finite energy gap is opened due to the breaking of cubic symmetry from the structural distortions. Ternary compounds, such as $AuTlS_2$, and quaternary compounds, such as $Cu_2HgPbSe_4$, are predicted to be TIs with the energy gaps around 10 meV $\sim$ 50 meV [27,17,111].

6.2.5 α- and β-Ag_2Te

The α-Ag_2Te has the antifluorite structure and can be viewed as a half-Heusler XYZ structure with X = Y = Ag and Z = Te. *ab initio* calculation shows that

α-Ag$_2$Te is a HgTe-type TI [138]. At low temperature (below 417 K), a structure phase transition occurs and drives the system into the β-phase with structural distortion. A gap ($\sim$80 meV) is opened and the system becomes a true TI with a highly anisotropic surface Dirac cone.

6.2.6 Skutterudites and Filled Skutterudites

Skutterudite [95] has a *bcc* lattice with two voids in a unit cell. Similar to packing the zincblende lattice in Heusler compounds, filled skutterudites (FSs) can be obtained by filling the voids with a variety of atoms including lanthanides (La, Ce, Pr, Nd, Sm, Eu, Gd, Tb). Recent theoretical calculation [100] predicts that Skutterudite materials, RhSb3 and IrSb3, are zero-gap TIs, similar to HgTe, while CoSb3 is near the transition boundary between trivial insulator and topological insulator. Different from HgTe, the inversion occurs between a *p* orbital and a *d* orbital in this class of materials. For filled skutterudites [126], two Ce-based FSs CeOs$_4$As$_{12}$ and CeOs$_4$Sb$_{12}$ are TIs with the band inversion occurring between a *d* orbital and a *f* orbital. *d* and *f* orbitals are expected to bring new phenomena, such as Kondo effect, magnetism, and superconductivity, into these materials, which coexists with the non-trivial topological property.

6.3 Other Materials

6.3.1 Bi$_x$Sb$_{1-x}$ and Related Materials

Bi$_x$Sb$_{1-x}$ alloy is the first experimentally verified 3D TI. The crystal structures of Bi and Sb are both of rhombohedral A7 type with space group $R\bar{3}m$ (similar to the Bi$_2$Se$_3$ family). The A7 structure can be viewed as a fcc NaCl structure distorted along the (1 1 1) direction. Triangular lattices in the (1 1 1) plane get paired and form bilayers. Pure Bi and Sb are group-V semimetals having a direct energy gap throughout the Brillouin zone but a negative indirect gap due to band overlap. When Bi is doped with Sb composition x, there are two main modifications to the band structure of Bi. First, the valence band maximum is pushed below the conduction band minimum, resulting in a real insulator in certain x range 0.07–0.22. Second, Sb dopants reduce the SOC in Bi and induce a band inversion around $x \sim 0.04$. Subsequently, Bi$_{1-x}$Sb$_x$ becomes a true TI. ARPES measurement has shown the complicated surface dispersion [44], which is also well reproduced by theoretical calculations [104,133]. Spin-resolved ARPES [46] and STM [93] have also been utilized to confirm the non-trivial topological nature of the surface states. Bilayer and multilayer Bi thin films are also predicted to be 2D TIs with 1D helical edge state [81,73] and recent ARPES and STM experiments have also shown the evidence of the helical edge states [130].

6.3.2 Graphene, Silicene, and Related Materials

In 2005, Kane and Mele [51,52] showed that when SOC is introduced into a graphene model, a material with a honeycomb lattice structure, the QSH effect can be realized. However, later calculation shows that the intrinsic SOC in graphene is insignificantly small [78,131], which will be easily overcome by other effects. Therefore, Kane-Mele model is mainly of academic interests and is hard to be realized in graphene system. Recently, the idea of Kane and Mele is applied to silicene, a buckled honeycomb lattice of Si, possibly realized on the silver surface in recent experiments [5,22,63], and the QSH phase with a spin-orbit energy gap of 1.55 meV is found from *ab initio* calculations [70].

6.3.3 PbTe, SnTe, and Related Materials

PbTe and SnTe are narrow gap semiconductors with a rocksalt structure. SnTe and PbTe have inverted band structures at four equivalent L points of the fcc Brillouin zone. Due to the even number of times of inversion, both SnTe and PbTe are not strong TIs. However, recently it has been shown that SnTe is a crystalline topological insulator with the surface states protected by mirror symmetry [48], which has been observed in ARPES measurement [123]. With the strain along (1 1 1) direction, one L point (called T) can be distinguished from the other three L points. Therefore (1 1 1) uniaxial strain can drive $Pb_{1-x}Sn_x$ Te into a strong TI phase [32]. Furthermore, superconductivity has also been found in $Sn_{1-x}In_x$ Te [96], which might be a realization of a topological superconductor.

6.3.4 Correlated Materials With d or f Orbitals

One of the future directions in this field to discover the topologically non-trivial phase in the correlated materials with d or f orbitals. Several different classes of materials have been proposed along this direction.

The first example is the Ir-based materials due to the strong SOC of Ir atoms. The QSH effect has been proposed in Na_2IrO_3 [99], and topological Mott insulator phases have been proposed in Ir-based pyrochlore oxides $Ln_2Ir_2O_7$ with Ln = Nd, Pr [84,108,36,129], as well as another two pyrochlores $Cd_2Os_2O_7$[36,53] and $Cd_2Re_2O_7$ [53].

Furthermore, a topological structure has also been considered in Kondo insulators, with a possible realization in SmB_6 and CeNiSn [24], as well as in $CeOs_4As_{12}$ and $CeOs_4Sb_{12}$, which contains $4f$ electrons [6,7]. Another correlated material α-Fe_2O_3 with corundum structure is predicted to be a possible topological magnetic insulator [109]. More recently, several compounds of Am mono-pnictides AmX (X = N, P, As, Sb, and Bi) and Pu mono-chalcogenides PuY (Y = Se and Te) are predicted to be topological Mott insulators [139], which is due to the band inversion between $6d$ and $5f$ orbitals driven by the interaction.

ACKNOWLEDGMENTS

This work is supported by the Department of Energy, Office of Basic Energy Sciences, Division of Materials Sciences and Engineering, under contract DE-AC02-76SF00515, by the ARO under Grant No. W911NF-09-1-0508 and by the Defense Advanced Research Projects Agency Microsystems Technology Office, MesoDynamic Architecture Program (MESO) through the Contract No. N66001-11-1-4105.

REFERENCES

[1] Al-Sawai W, Lin H, Markiewicz RS, Wray LA, Xia Y, Xu S-Y, et al. Phys Rev B 2010;82:125208.
[2] Alpichshev Z, Analytis JG, Chu JH, Fisher IR, Chen YL, Shen ZX, et al. Phys Rev Lett 2010;104:016401.
[3] Alpichshev Z, Analytis JG, Chu JH, Fisher IR, Kapitulnik A. 2010. arXiv:1003.2233.
[4] Analytis JG, McDonald RD, Riggs SC, Chu JH, Boebinger GS, Fisher IR. Nat Phys 2010;6:960.
[5] Aufray B, Kara A, Vizzini S, Oughaddou H, Léandri C, Ealet B, et al. Appl Phys Lett 2010;96:183102.
[6] Bauer E, Slebarski A, Freeman E, Sirvent C, Maple MJ. Phys Condens Matter 2001;13:4495.
[7] Baumbach R, Ho P, Sayles T, Maple M, Wawryk R, Cichorek T, et al. Proc Natl Acad Sci 2008;105(45):17307.
[8] Bernevig BA, Hughes TL, Zhang SC. Science 2006;314:1757.
[9] Bernevig BA, Zhang SC. Phys Rev Lett 2006;96:106802.
[10] Brüne Christoph, Roth Andreas, Buhmann Hartmut, Hankiewicz Ewelina M, Molenkamp Laurens W, Maciejko Joseph, et al. Spin polarization of the quantum spin hall edge states. Nat Phys 2012;8(6):485–490.
[11] Brüne C, Liu CX, Novik EG, Hankiewicz EM, Buhmann H, Chen YL, et al. Phys Rev Lett 2011;106:126803.
[12] Burkov AA, Hawthorn DG. 2010. arXiv:1005.1654.
[13] Cardona M, Kremer RK, Lauck R, Siegle G, Muñoz A, Romero AH. Phys Rev B 2009;80(19):195204.
[14] Chadov S, Qi XL, Kübler J, Fecher GH, Felser C, Zhang SC. Nat Mater 2010;9:541.
[15] Checkelsky JG, Hor YS, Liu M-H, Qu D-X, Cava RJ, Ong NP. Phys Rev Lett 2009;103:246601.
[16] Chen J, Qin HJ, Yang F, Liu J, Guan T, Qu FM, et al. Phys Rev Lett 2010;105(17):176602.
[17] Chen S, Gong XG, Duan C-G, Zhu Z-Q, Chu J-H, Walsh A, et al. Phys Rev B 2011;83:245202.
[18] Chen YL, Analytis JG, Chu JH, Liu ZK, Mo S-K, Qi XL, et al. Science 2009;325:178.
[19] Chen YL, Liu ZK, Analytis JG, Chu J-H, Zhang HJ, Yan BH, et al. Phys Rev Lett 2010;105:266401.
[20] Cheng P, Song C, Zhang T, Zhang Y, Wang Y, Jia J-F, et al. Phys Rev Lett 2010;105(7):076801.
[21] Dai X, Hughes TL, Qi X-L, Fang Z, Zhang S-C. Phys Rev B 2008;77:125319.
[22] De Padova P, Quaresima C, Ottaviani C, Sheverdyaeva P, Moras P, Carbone C, et al. Appl Phys Lett 2010;96(26):261905.
[23] Delin A. Phys Rev B 2002;65(15):153205.
[24] Dzero M, Sun K, Galitski V, Coleman P. Phys Rev Lett 2010;104:106408.
[25] Eremeev SV, Koroteev YM, Chulkov EV. JETP Lett 2010;92:161–5.
[26] Eremeev SV, Koroteev YM, Chulkov EV. JETP Lett 2010;91:594.
[27] Feng W, Xiao D, Ding J, Yao Y. Phys Rev Lett 2011;106(1):016402.
[28] Feng W, Xiao D, Zhang Y, Yao Y. Phys Rev B 2010;82:235121.
[29] Fleszar A, Hanke W. Phys Rev B 2005;71(4):045207.
[30] Fradkin E, Dagotto E, Boyanovsky D. Phys Rev Lett 1986;57:2967.
[31] Fu L. Phys Rev Lett 2010;104(5):056402.
[32] Fu L, Kane CL. Phys Rev B 2007;76:045302.

[33] Gao J, Chen W, Xie XC, Zhang FC. Phys Rev B 2009;80:241302.
[34] Goldstone J, Wilczek F. Phys Rev Lett 1981;47:986.
[35] Graf T. Felser C. Prog Solid State Chem: Parkin S; 2011.
[36] Guo HM, Franz M. Phys Rev Lett 2009;103:206805.
[37] Hanaguri T, Igarashi K, Kawamura M, Takagi H, Sasagawa T. Phys Rev B 2010;82(8):081305.
[38] Hasan MZ, Kane CL. Rev Mod Phys 2010;82(4):3045.
[39] He H-T, Wang G, Zhang T, Sou I-K, Wong GKL, Wang J-N, et al. Phys Rev Lett 2011;106:166805.
[40] Hein RA, Swiggard EM. Phys Rev Lett 1970;24:53.
[41] Heusler F, Starck W, Haupt E. Verh DPG 1903;5:220.
[42] Hor YS, Richardella A, Roushan P, Xia Y, Checkelsky JG, Yazdani A, et al. Phys Rev B 2009;79:195208.
[43] Hor YS, Williams AJ, Checkelsky JG, Roushan P, Seo J, Xu Q, et al. Phys Rev Lett 2010;104(5):057001.
[44] Hsieh D, Qian D, Wray L, Xia Y, Hor YS, Cava RJ, et al. Nature 2008;452:970.
[45] Hsieh D, Xia Y, Qian D, Wray L, Dil JH, Meier F, et al. Nature 2009;460:1101.
[46] Hsieh D, Xia Y, Wray L, Qian D, Pal A, Dil JH, et al. Science 2009;323:919.
[47] Hsieh TH, Fu L. Phys Rev Lett 2012;108:107005. http://dx.doi.org/10.1103/PhysRevLett.108.107005.
[48] Hsieh TH, Lin H, Liu J, Duan W, Bansil A, Fu L. Nat Commun 2012;3:982.
[49] Jensen JD, Burke JR, Ernst DW, Allgaier RS. Phys Rev B 1972;6:319.
[50] Jin H, Song J-H, Freeman AJ, Kanatzidis MG. 2010. arXiv:1007.5480.
[51] Kane CL, Mele EJ. Phys Rev Lett 2005;95:226801.
[52] Kane CL, Mele EJ. Phys Rev Lett 2005;95:146802.
[53] Kargarian M, Wen J, Fiete GA. Phys Rev B 2011;83(16):165112.
[54] Kharitonov M. 2010. e-print arXiv:1004.0194.
[55] Kim J, Kim J, Jhi S-H. Phys Rev B 2010;82:201312.
[56] Knez I, Du RR, Sullivan G. Phys Rev B 2010;81:201301(R).
[57] Knez I, Du R-R, Sullivan G. Phys Rev Lett 2011;107:136603.
[58] Kong D, Chen Y, Cha JJ, Zhang Q, Analytis JG, Lai K, et al. Nat Nanotechnol 2011;6:705.
[59] Kong D, Dang W, Cha JJ, Li H, Meister S, Peng H, et al. Nano Letters 2010;10(6):2245, [pMID: 20486680]. http://dx.doi:10.1021/nl101260j.
[60] König M, Buhmann H, Molenkamp LW, Hughes TL, Liu C-X, Qi XL, et al. J Phys Soc Jpn 2008;77:031007.
[61] König M, Wiedmann S, Brüne C, Roth A, Buhmann H, Molenkamp L, et al. Science 2007;318:766.
[62] Kuroda K, Ye M, Kimura A, Eremeev SV, Krasovskii EE, Chulkov EV, et al. Phys Rev Lett 2010;105:146801.
[63] Lalmi B, Oughaddou H, Enriquez H, Kara A, Vizzini S, Ealet B, et al. Appl Phys Lett 2010;97:223109.
[64] Lee WC, Wu C, Arovas DP, Zhang SC. Phys Rev B 2009;80:245439.
[65] Lin H, Markiewicz RS, Wray LA, Fu L, Hasan MZ, Bansil A. Phys Rev Lett 2010;105:036404.
[66] Lin H, Wray L, Xia Y, Xu S-Y, Jia S, Cava R, et al. Single-dirac-cone z2 topological insulator phases in distorted li2agsb-class and related quantum critical li-based spin-orbit compounds. 2010. e-print arXiv:1004.0999.
[67] Lin H, Wray LA, Xia Y, Xu S, Jia S, Cava RJ, et al. Nat Mater 2010;9:546.
[68] Linder J, Yokoyama T, Sudb A. Phys Rev B 2009;80:205401.
[69] Liu C, Lee Y, Kondo T, Mun ED, Caudle M, Harmon BN, et al. Phys Rev B 2011;83:205133.
[70] Liu C-C, Feng W, Yao Y. Phys Rev Lett 2011;107:076802.
[71] Liu C-X, Hughes TL, Qi X-L, Wang K, Zhang S-C. Phys Rev Lett 2008;100:236601.
[72] Liu C-X, Qi X-L, Zhang H, Dai X, Fang Z, Zhang S-C. Phys Rev B 2010;82:045122.
[73] Liu Z, Liu C-X, Wu Y-S, Duan W-H, Liu F. J. Phys Rev Lett 2011;107:136805. http://dx.doi.org/physRevLett.107.136805.
[74] Lu HZ, Shan WY, Yao W, Niu Q, Shen SQ. Phys Rev B 2010;81:115407.
[75] Lu H-Z, Shi J, Shen S-Q. Phys Rev Lett 2011;107:076801.

[76] Madelung O, Rossler U, Schulz M. Ternary compounds, organic semiconductors in Landolt-Boenstein, condensed matter, vol. III/41E. Berlin: Springer; 2000.

[77] Menshchikova T, Eremeev S, Koroteev Y, Kuznetsov V, Chulkov E. JETP Lett 2011;93:15–0.

[78] Min H, Hill J, Sinitsyn N, Sahu B, Kleinman L, MacDonald A. Phys Rev B 2006;74:165310.

[79] Moon C-Y, Wei S-H. Phys Rev B 2006;74(4):045205.

[80] Muechler L, Zhang H, Chadov S, Yan B, Casper F, Kuebler J, et al. Angew Chem Int Ed 2012;51(29):7221.

[81] Murakami S. Phys Rev Lett 2006;97:236805.

[82] Murakami S, Iso S, Avishai Y, Onoda M, Nagaosa N. Phys Rev B 2007;76:205304.

[83] Peng H, Lai K, Kong D, Meister S, Chen Y, Qi XL, et al. Nat Mater 2010;9:225.

[84] Pesin DA, Balents L. Nat Phys 2010;6:376.

[85] Qi X-L, Hughes T, Zhang S-C. Nat Phys 2008;4:273.

[86] Qi X-L, Hughes T, Zhang S-C. Phys Rev B 2008;78:195424.

[87] Qi XL, Wu YS, Zhang SC. Phys Rev B 2006;74:085308.

[88] Qi X-L, Zhang S-C. Rev Mod Phys 2011;83:1057.

[89] Qu D-X, Hor YS, Xiong J, Cava RJ. Ong NP. 2009;329:821.

[90] Raghu S, Chung SB, Qi XL, Zhang SC. Phys Rev Lett 2010;104:116401.

[91] Roth A, Brüne C, Buhmann H, Molenkamp LW, Maciejko J, Qi X-L, et al. Science 2009;325:294.

[92] Rothe DG, Reinthaler RW, Liu C-X, Molenkamp LW, Zhang SC, Hankiewicz EM. New J Phys 2010;12:065012.

[93] Roushan P, Seo J, Parker CV, Hor YS, Hsieh D, Qian D, et al. Nature 2009;460:1106.

[94] Sacepe B, Oostinga J, Li J, Ubaldini A, Couto N, Giannini E, et al. Nat Commun 2011;2:575.

[95] Sales Brian C. Filled skutterudites. Handbook on the Physics and Chemistry of Rare Earths, vol.33. Amsterdam: Elsevier; 2003. p. 1–34.

[96] Sasaki S, Ren Z, Taskin AA, Segawa K, Fu L, Ando Y. Odd-parity pairing and topological superconductivity in a strongly spin-orbit coupled semiconductor. 2012. e-print arXiv:1208.0059.

[97] Sato K, Loss D, Tserkovnyak Y. 2010. e-print arXiv:1003.4316.

[98] Scholz MR, Sanchez-Barriga J, Marchenko D, Varykhalov A, Volykhov A, Yashina LV, et al. High spin polarization and circular dichroism of topological surface states on bi2te3. 2011. arXiv:1108.1053.

[99] Shitade A, Katsura H, Kuneš J, Qi X-L, Zhang S-C, Nagaosa N. Phys Rev Lett 2009;102:256403.

[100] Smith JC, Banerjee S, Pardo V, Pickett WE. Phys Rev Lett 2011;106(5):056401.

[101] Ström A, Johannesson H, Japaridze GI. Phys Rev Lett 2010;104:256804.

[102] Taskin AA, Ren Z, Sasaki S, Segawa K, Ando Y. Phys Rev Lett 2011;107:016801. http://dx.doi.org/10.1103/PhysRevLett.107.016801.

[103] Taskin AA, Segawa K, Ando Y. Phys Rev B 2010;82:121302.

[104] Teo JCY, Fu L, Kane CL. Phys Rev B 2008;78:045426.

[105] Thouless DJ, Kohmoto M, Nightingale MP, den Nijs M. Phys Rev Lett 1982;49:405.

[106] Vidal J, Zhang X, Yu L, Luo J-W, Zunger A. Phys Rev B 2011;84:041109.

[107] Virot F, Hayn R, Richter M, van den Brink J. Metacinnabar (β-HgS): a strong 3D topological insulator with highly anisotropic surface states. Phys Rev Lett 2011;106(23):236806.

[108] Wan X, Turner AM, Vishwanath A, Savrasov SY. Phys Rev B 2011;83:205101. http://dx.doi.org/10.1103/PhysRevB.83.205101.

[109] Wang J, Li R, Zhang S-C, Qi X-L. Phys Rev Lett 2011;106(12):126403.

[110] Wang YH, Hsieh D, Pilon D, Fu L, Gardner DR, Lee YS, et al. Phys Rev Lett 2011;107:207602.

[111] Wang YJ, Lin H, Das T, Hasan MZ, Bansil A. New J Phys 2011;13:085017.

[112] Wang Z, Lin T, Wei P, Liu X, Dumas R, Liu K, et al. Appl Phys Lett 2010;97:042112.

[113] Winkler R. Spin-orbit coupling effects in two-dimensional electron and hole systems. Springer; 2003.

[114] Wittel K, Manne R. Theoret Chim Acta (Berl) 1974;33:347.

[115] Wray LA, Xu S, Xia Y, Hsieh D, Fedorov AV, Hor YS, et al. Nat Phys 2011;7:32.

[116] Wu C, Bernevig BA, Zhang SC. Phys Rev Lett 2006;96:106401.

[117] Wuttig M, Yamada N. Nat Mater 2007;6:824.

[118] Xia Y, Wray L, Qian D, Hsieh D, Pal A, Lin H, et al. Nat Phys 2009;5:398.
[119] Xiong J, Petersen AC, Qu D, Cava RJ, Ong NP. Physica E 2012;44:917.
[120] Xiu F, He L, Wang Y, Cheng L, Chang L-T, Lang M, et al. Nat Nanotechnol 2011;6:216–21.
[121] Xu C, Moore J. Phys Rev B 2006;73:045322.
[122] Xu S, Xia Y, Wray LA, Jia S, Meier F, Dil JH, et al. Science 2011;332:560.
[123] Xu S-Y, Liu C, Alidoust N, Qian D, Neupane M, Denlinger JD, et al. Observation of topological crystalline insulator phase in the lead tin chalcogenide pb1-xsnxte material class. 2012. arXiv:1206.2088.
[124] Xu S-Y, Wray LA, Xia Y, Shankar R, Petersen A, Fedorov A, et al. Discovery of several large families of topological insulator classes with backscattering-suppressed spin-polarized single-dirac-cone on the surface. 2010. arXiv:1007.5111.
[125] Yan B, Liu C-X, Zhang H, Yam CY, Qi XL, Frauenheim T. Europhys Lett 2010;90:37002.
[126] Yan B, Müchler L, Qi X-L, Zhang S-C, Felser C. Phys Rev B 2012;85:165125.
[127] Yan B, Zhang H-J, Liu C-X, Qi X-L, Frauenheim T, Zhang S-C. Phys Rev B 2010;82:161108.
[128] Yan B, Zhang S-C. Rep Prog Phys 2012;75(9):096501.
[129] Yang B-J, Kim YB. Topological insulators and metal-insulator transition in the pyrochlore iridates 2010. arXiv:1004.4630.
[130] Yang F, Miao L, Wang ZF, Yao M-Y, Zhu F, Song YR, et al. Phys Rev Lett 2012;109:016801.
[131] Yao Y, Ye F, Qi X-L, Zhang S-C, Fang Z. Phys Rev B 2007;75:041401.
[132] Zhang H, Liu C-X, Qi X-L, Dai X, Fang Z, Zhang S-C. Nat Phys 2009;5:438.
[133] Zhang H, Liu C-X, Qi X-L, Deng X-Y, Dai X, Zhang S-C. Phys Rev B 2009;80:085307.
[134] Zhang H-J, Zhang X, Zhang S-C. Quantum anomalous hall effect in magnetic topological insulator gdbite3. 2011. arXiv:1108.4857.
[135] Zhang J, Chang C-Z, Zhang Z, Wen J, Feng X, Li K, et al. Nat Commun 2011;2:574.
[136] Zhang J, Zhang S, Weng HM, Zhang W, Yang L, Liu Q, et al. Proc Nat Acad Sci 2011;108:24.
[137] Zhang T, Cheng P, Chen X, Jia J-F, Ma X, He K, et al. Phys Rev Lett 2009;103:266803.
[138] Zhang W, Yu R, Feng W, Yao Y, Weng H, Dai X, et al. Phys Rev Lett 2011;106(15):156808.
[139] Zhang X, Zhang H, Felser C, Zhang S-C. 2011. arXiv:1111.1267.
[140] Zhou B, Lu H-Z, Chu R-L, Shen S-Q, Niu Q. Phys Rev Lett 2008;101:246807.

Field-Theory Foundations of Topological Insulators

Xiao-Liang Qi

Department of Physics, Stanford University, Stanford, CA 94305, USA

Chapter Outline Head

1 INTRODUCTION TO TOPOLOGICAL FIELD THEORY

Although topological insulators can be realized in non-interacting electron models, and the topological invariant can also be defined in purely non-interacting band theory, it is still important to figure out whether topological insulators are still well-defined in interacting systems. For this purpose, the topological field theory approach to topological insulators (and other topological states) has been developed [36], which characterizes topological insulators by their generic topological properties rather than single electron properties and band structure. The topological field theory approach not only provides a generic definition of topological insulators in the presence of interactions, but also directly predicts physically observable topological properties that can qualitatively distinguish topological insulators from trivial states.

Topological Insulators. http://dx.doi.org/10.1016/B978-0-444-63314-9.00004-4

Generally speaking, the topological field theory approach applies not only to time-reversal invariant topological insulators but also to other topological states of matter. To gain better understanding of the precise meaning of topological field theory approach, we take quantum Hall effect [51] as an example. Consider a generic two-dimensional insulator with the Hamiltonian $H\left[c_{i\alpha}, c_{i\alpha}^{\dagger}\right]$ with i labeling lattice sites and α labeling the spin and/or other band indices. The Hamiltonian may contain electron hopping terms and interaction terms. The details of the Hamiltonian are not important for the discussion. What is important for defining the quantum Hall effect is the coupling to an external electromagnetic field A_μ, with $\mu = x, y, t$ the space-time indices. The coupling is determined by the gauge principle. For example, a term in the kinetic energy $-tc_i^{\dagger}c_{i+\hat{x}}$ will be replaced by $-tc_i^{\dagger}c_{i+\hat{x}}e^{iA_{ix}}$. Denote the Hamiltonian of the coupled system as $H\left[c_{i\alpha}, c_{i\alpha}^{\dagger}, A_\mu\right]$. We can write down the action correspondingly as (here and below we will use the imaginary time (Euclidean) formulation)

$$S\left[c_{i\alpha}(\tau), \bar{c}_{i\alpha}(\tau), A_\mu(\tau)\right] = \int_0^\beta \left[\sum_{i,\alpha} \bar{c}_{i\alpha}\partial_\tau c_{i\alpha} + H\right] d\tau. \tag{1}$$

From this action we can define an effective action by integrating out the fermions:

$$e^{-S_{\text{eff}}[A_\mu]} = \int Dc_{i\alpha}(\tau)D\bar{c}_{i\alpha}(\tau)e^{-S[c_{i\alpha}(\tau), \bar{c}_{i\alpha}(\tau), A_\mu(\tau)]}. \tag{2}$$

The effective action $S_{\text{eff}}[A_\mu]$ determines the response of this electron system to external electromagnetic field to all orders. More explicitly, the average value of the charge current (and density) J_μ is given by the response equation:

$$\langle J_\mu[A_\mu]\rangle = \left\langle \frac{\delta H}{\delta A_\mu}\right\rangle = \frac{\delta S_{\text{eff}}}{\delta A_\mu} \tag{3}$$

with $\langle \rangle$ denoting the average value in the thermal equilibrium of the fermion system. The effective action can be expanded in the powers of A_μ:

$$S_{\text{eff}} = \frac{1}{2}\int d^3x d^3x' \Pi_{\mu\nu}(x,x')A_\mu(x)A_\nu(x') + o(A^2). \tag{4}$$

Here we have taken the long wavelength limit in which case A_μ is considered as a field defined in the continuum, and x denotes the space-time coordinate. The coefficient $\Pi_{\mu\nu}(x,x')$ is the linear response function of the fermion system, which has to be consistent with gauge invariance $\exp\{-S_{\text{eff}}[A_\mu]\} = \exp\{-S_{\text{eff}}[A_\mu + \partial_\mu\varphi]\}$ for any gauge transformation function $\varphi(x)$. Now we can discuss the general property of the effective action S_{eff} instead of calculating it for a specific system. For an insulator, electron states are all gapped for a given configuration of electromagnetic field (as long as the electromagnetic field is weak enough so that the electron excitation gap is not closed), and all connected correlation functions are short ranged in space-time. Therefore the process in Eq. (2) of integrating out electrons to obtain the effective action is well-defined and has no singularity involved. The resulting effective action should be local in $A_\mu(x)$, which means $\Pi_{\mu\nu}(x,x')$ is short-ranged in the distance $|x - x'|$. The range of the response function $\Pi_\mu(x,x')$ is determined by the electron gap and some characteristic velocity scale. For example if the electrons are described by a massive Dirac Hamiltonian with the speed of light v and a mass gap E_g, the range of the correlation is $l = \hbar v/E_g$ [2]. When we consider the effective action at the scale much larger than l, the kernel $\Pi_\mu(x,x')$ can be approximated by a gradient expansion:

$$
\begin{aligned}
\Pi_{\mu\nu}(x,x') &= \int \frac{d^3 p}{(2\pi)^3} e^{ip(x-x')} \Pi_{\mu\nu}(p) \\
&= \sum_{n=0}^{\infty} \Pi^{(n)}_{\mu\nu; \tau_1 \tau_2 \ldots \tau_n} \int \frac{d^3 p}{(2\pi)^3} e^{ip(x-x')} p^{\tau_1} p^{\tau_2} \ldots p^{\tau_n} \\
&= \sum_{n=0}^{\infty} \Pi^{(n)}_{\mu\nu; \tau_1 \tau_2 \ldots \tau_n} (-i)^n \partial^{\tau_1} \partial^{\tau_2} \ldots \partial^{\tau_n} \delta(x - x').
\end{aligned}
\tag{5}
$$

Here we have assumed translation symmetry of the system so that $\Pi_{\mu\nu}(x,x') = \Pi_{\mu\nu}(x - x')$. Each term here has to be consistent with the gauge symmetry. For example, the gauge symmetry requires the first term $\Pi^{(0)}_{\mu\nu} = 0$. Usually the leading term which preserves the gauge symmetry, translation symmetry, and rotation symmetry is the Maxwell term

$$
\Pi^{(2)}_{\mu\nu; \tau_1 \tau_2} = \frac{1}{2}(g_{\mu\nu} g_{\tau_1 \tau_2} - g_{\mu\tau_1} g_{\nu\tau_2})
$$

$$
\int d^3 x d^3 x' \Pi^{(2)}_{\mu\nu; \tau_1 \tau_2} \partial^{\tau_1} \partial^{\tau_2} \delta(x - x') A^\mu(x) A^\nu(x') = -\frac{1}{4} \int d^3 x F_{\mu\nu} F^{\mu\nu}.
\tag{6}
$$

In $(2+1)$ dimensions, there is a special gauge invariant term that is of lower order than the Maxwell term, which is known as the Chern-Simons term:

$$
\Pi^{(1)}_{\mu\nu; \tau} = -\frac{\sigma_H}{2} \epsilon_{\mu\nu\tau}
$$

$$
S_{CS} = -i \int d^3 x d^3 x' \Pi^{(2)}_{\mu\nu; \tau} \partial^\tau \delta(x - x') A^\mu(x) A^\nu(x')
$$

$$
= i \frac{\sigma_H}{2} \int d^3 x \epsilon_{\mu\nu\tau} A^\mu \partial^\nu A^\tau.
\tag{7}
$$

Here we have denoted the prefactor by $-\sigma_H/2$ since the equation of motion (3) given by this term is $J^\mu = \sigma_H \epsilon^{\mu\nu\tau} \partial_\nu A_\tau$. In particular, $J_x = \sigma_H E_y$ which tells us that σ_H has the physical meaning of Hall conductivity. Naively one may worry that the Lagrangian density $\epsilon_{\mu\nu\tau}A^{\mu}\partial^{\nu}A^{\tau}$ of the Chern-Simons term is not gauge invariant. However, such gauge non-invariance turns out to be the essential property of this term which makes it a topological term. Upon a variation $A^{\mu} \rightarrow A^{\mu} + \delta A^{\mu}$, the variation of the Chern-Simons term is $\delta\left(\epsilon_{\mu\nu\tau}A^{\mu}\partial^{\nu}A^{\tau}\right) = 2\left(\delta A^{\mu}\right)\epsilon_{\mu\nu\tau}\partial^{\nu}A^{\tau} - \partial_{\mu}\left(\epsilon_{\mu\nu\tau}A^{\nu}\delta A^{\tau}\right)$. The second term is a total derivative. In the gauge transformation $\delta A^{\mu} = \partial^{\mu}\varphi$, the change of Chern-Simons term is $\delta S_{CS} = i\sigma_H \int d^3x \partial^{\mu}\varphi \epsilon_{\mu\nu\tau}\partial^{\nu}A^{\tau}$. As long as φ is single-valued, the integrand is a total derivative and $\delta S_{CS} = 0$, so that the gauge invariance is preserved. However, the gauge invariance is more subtle when we consider the compactness of the electromagnetic $U(1)$ gauge field. In a gauge transformation $A_{\mu} \rightarrow A_{\mu} + \partial_{\mu}\varphi$, the electron operators transform as $c_{\alpha}(x) \rightarrow e^{i\varphi(x)}c_{\alpha}(x)$. Therefore the gauge transformation is well-defined as long as the $U(1)$ phase variable $e^{i\varphi(x)}$ is single-valued. If the space-time manifold has a nontrivial fundamental group, one can consider a gauge transformation in which $\varphi(x)$ is not single-valued, but $e^{i\varphi(x)}$ is. For example, consider the theory on $S^2 \times S^1$ where the spatial manifold is a sphere S^2 and the time direction is a circle S^1, as is illustrated in Fig. 1. The phase φ can have a nontrivial winding around the time-direction, and the single-valueness of $e^{i\varphi}$ is preserved as long as $\int dt\partial_t\varphi = 2\pi n$ is quantized in unit of 2π. In such a case the gauge transformation of the Chern-Simons action is nontrivial. If we denote the time to be $t \in [0,T]$ with periodic boundary condition $\varphi(x,y,t) = \varphi(x,y,t+T)$, we can choose a branch-cut where φ jumps by $2\pi n$. For example, take the branch-cut at $t = 0$. The gauge transformation of the action is

$$\delta S_{CS} = i\sigma_H \int d^2x \int_0^T dt\partial_{\mu}\left(\epsilon^{\mu\nu\tau}\varphi\partial_{\nu}A_{\tau}\right)$$

$$= i\sigma_H \int_{S^2} d^2x F_{xy}(x,y,0)\left(\varphi(x,y,T) - \varphi(x,y,0)\right)$$

$$= i\sigma_H 4\pi^2 nm. \qquad (8)$$

In the last step we have used the fact that the total flux of F_{xy} on the spatial manifold is always quantized $\int_{S^2} d^2x F_{xy}(x,y,0) = 2\pi m$, $m \in \mathbf{Z}$. m is known as the Chern number of the gauge field.

Therefore we see that the Chern-Simons term is not really gauge invariant under "large" gauge transformations generated by non-single-valued $\varphi(x)$. Naively, we would conclude that the term cannot exist. However, the gauge invariance is restored if δS_{CS} is integer times 2π, since the only requirement is that $e^{-S_{CS}}$ is gauge invariant. This leads to the condition of $\sigma_H = k/2\pi$ with k an integer. (When the physical units are restored, $\sigma_H = k\frac{e^2}{\hbar}$.) Therefore we have demonstrated that the Chern-Simons term must have a quantized coefficient in any gauge invariant system. This property distinguishes the Chern-Simons term from the ordinary terms in the action such as the Maxwell term, since the coefficient of the latter can always be continuously tuned. The quantization of the coefficient σ_H provides a very generic reason of the Hall conductance

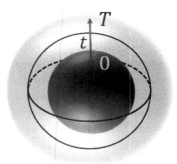

FIGURE 1 Illustration of the space-time manifold $S^2 \times S^1$. At each time $t \in [0, T]$ the space is a sphere S^2, and the spheres with $t = 0$ and $t = T$ should be identified.

quantization observed in integer quantum Hall states, which can be considered as the Laughlin's gauge argument [24] reformulated in field theory language. Since the coefficient is always quantized, it must be insensitive to perturbations and it can only be changed abruptly if the description above fails, which only occurs if the system becomes gapless so that the locality of the effective action is lost. Therefore from the very general discussion here we learn that the Hall conductance quantum k is a *topological invariant* of $(2+1)$-d insulators, which can only change when the system experiences a quantum phase transition.

The discussion above illustrates the way the topological field theory approach works for each topological state of matter. Since the topological term such as Chern-Simons term discussed here is written for an external field (which is the electromagnetic field A_μ in the above example), more precisely the theory should be called a *topological response theory*. The procedure of developing a topological response theory can be summarized as follows:

1. **Introduce the external fields which couple to the topological state**. In the QH example, the field we used is the electromagnetic field. For some other topological states, other external fields such as gravitational field have been considered.
2. **Write down possible topological terms of the external field that are consistent with symmetries of the system.** The topological terms are defined as those which cannot be tuned continuously and only contain parameters that take discrete values, as long as the symmetries of the system are respected. For the QH system the only symmetry is charge conservation, which leads to the gauge invariance of the effective action $S_{\text{eff}}[A_\mu]$ for the external field. The only topological term is the Chern-Simons term, which has an integer coefficient as we have shown above.

After the two steps above are finished, we obtain the topological classification of the underlying states of matter in terms of the discrete coefficient(s) of the topological term(s). The discrete coefficients in the topological terms can be called topological order parameters. Two states with different values of the topological order parameter cannot be adiabatically connected to each other. On the other hand, two states with the same value of topological order parameter may or may not belong to the same phase, since there may be other topological terms associated with other external fields that have not been considered. In other words, for two states of matter, having different topological order parameters is the sufficient but not necessary condition that they belong to topologically distinct phases.

It should be noticed that the discussion above only applies to insulating states with a unique ground state. If there are more than one ground states, after integrating over the electron states with energy above the gap there are still nontrivial degrees of freedom left, and one cannot repeat the discussion above without referring to these residual degrees of freedom. In such cases one has to introduce a theory which describes both the external fields and the residual degrees of freedom (sometimes called the "topological Hilbert space"). A simple example is the $1/m$ fractional quantum Hall state [59], which can be described by the following effective theory [58, 67]:

$$\int Dc^{\dagger} Dc\, e^{-S[c^{\dagger}, c, A]} = \int Da_{\mu}\, e^{-S_{\text{eff}}[a_{\mu}, A_{\mu}]}$$

$$S_{\text{eff}}[a_{\mu}, A_{\mu}] = \frac{m}{2\pi} \epsilon^{\mu\nu\tau} a_{\mu} \partial_{\nu} a_{\tau} + \frac{1}{4\pi} \epsilon^{\mu\nu\tau} a_{\mu} \partial_{\nu} A_{\tau}. \qquad (9)$$

Here a_{μ} is an effective gauge field which describes the low energy degrees of freedom in the fractional quantum Hall system. a_{μ} is not an external field but is dynamical. If one naively integrates over a_{μ}, one obtains the Chern-Simons term of A_{μ} with the coefficient of $\frac{1}{m}\frac{1}{2\pi}$, which explains the fractionally quantized Hall conductance. However as we discussed above, a Chern-Simons term with fractional coefficient is not consistent with gauge invariance, which is why the system has to be described by the effective theory (9) rather than by the topological response theory containing only A_{μ}. The effective theory (9) with A_{μ} and a_{μ} is also a topological field theory since it only contains discrete parameters. To distinguish such topological field theories from the topological response theories discussed above, we can refer to them as *dynamical topological field theories*. They describe nontrivial topological degrees of freedom of the system and their coupling to the external fields.

In the next section, we will discuss the application of the topological field theory description to the new topological insulators with time-reversal symmetry. Since the time-reversal invariant topological insulators have no ground state degeneracy, we will mainly discuss topological response theories, but we will also discuss dynamical topological field theories which are important in interacting topological insulators, and topological superconductors.

2 TFT DESCRIPTION OF 3D TIME-REVERSAL INVARIANT TI

In this section we will apply the topological response theory approach to $(3+1)$-d topological insulators. We will first follow the general procedure discussed in the previous section and obtain the topological term from general symmetry and topology consideration, and then present a microscopic derivation of the topological term in non-interacting TIs. In the last subsection the physical consequences of the TFT will be discussed.

2.1 Obtaining the Topological Term from General Considerations

Following the general procedure discussed in the last section, the first step of developing a TFT description is to identify the external fields coupled to the topological insulator. The natural choice is still the electromagnetic field A_μ. In $(3 + 1)$-d there is no Chern-Simons term, and the lowest order terms of the electromagnetic field contain two derivatives. Since we are interested in topological terms that are independent of any continuous parameters, they should be insensitive to translation and rotation symmetry breaking. Therefore we focus on the translation and rotation invariant terms. There are two types of such terms. The first one is the Maxwell term

$$S_1 = \int d^4x \frac{1}{2e^2}(\mathbf{E}^2 + \mathbf{B}^2) = \frac{1}{4e^2}\int d^4x F_{\mu\nu}F_{\sigma\tau}g^{\mu\sigma}g^{\nu\tau}. \tag{10}$$

Here $F_{\mu\nu} = \partial_\mu A_\nu - \partial_\nu A_\mu$ is the field strength, and the metric $g^{\mu\nu}$ is the Lorentz metric in which the speed of light is taken to be that in the material. The second one is the magneto-electric term

$$S_2 = \frac{i\theta}{4\pi^2}\int d^4x \mathbf{E} \cdot \mathbf{B} = \frac{i\theta}{32\pi^2}\int d^4x \epsilon^{\mu\nu\sigma\tau}F_{\mu\nu}F_{\sigma\tau}. \tag{11}$$

If there is no further symmetry requirement, none of the two terms are topological terms, since the coefficients e and θ can both vary continuously. Notice that we have defined the coefficient of S_2 to be $i\theta$ for later convenience. For now θ can be considered as a complex number although it will be restricted to real values later. Now we consider the condition of time-reversal symmetry. Under time-reversal, the gauge potentials transform as $A_i \to -A_i, i = x, y, z$, and $A_t \to A_t$. Therefore $\mathbf{E} \to \mathbf{E}$ and $\mathbf{B} \to -\mathbf{B}$, so that the Maxwell term S_1 is even and S_2 is odd under time-reversal. Therefore, it seems that time-reversal symmetry requires $\theta = 0$ and e remains arbitrary, which means there is still no topological term with discrete coefficient. However, similar to the Chern-Simons term discussion in the last section, we should consider the partition function rather than the action. The contribution of S_2 to the partition function is

$$e^{-S_2} = \exp\left[-i\frac{\theta}{32\pi^2}\int d^4x \epsilon^{\mu\nu\sigma\tau}F_{\mu\nu}F_{\sigma\tau}\right] \equiv e^{-i\theta C_2} \tag{12}$$

in which

$$C_2 = \frac{1}{32\pi^2} \int d^4x \epsilon^{\mu\nu\sigma\tau} F_{\mu\nu} F_{\sigma\tau} \tag{13}$$

is the integral of the second Chern class of the electromagnetic gauge field, which is quantized to integer values for any closed space-time manifold.[1] Time-reversal changes the sign of C_2 so that time-reversal invariance requires $e^{i\theta C_2} = e^{-i\theta C_2}$ which leads to the condition $\theta = \pi n, n \in \mathbb{Z}$. Therefore we see that the time-reversal symmetry allows nonzero values of θ, and restricts θ to discrete values. This demonstrates that the term S_2 becomes a topological term when time-reversal symmetry is preserved. Furthermore, due to the quantization of C_2, θ and $\theta + 2\pi$ always correspond to the same partition function, which means they describe the same physical system. Consequently there are only two inequivalent values of θ which are 0 and π modulo 2π. This leads to the conclusion that the 3D TI with time-reversal symmetry is classified by Z_2. The Z_2 nontrivial and trivial insulators are described by $\theta = \pi$ and $\theta = 0$ respectively [36]. It should be noticed that the Z_2 classification not only characterizes the number of topological classes, but also describes the algebra of the topological invariant. Consider an insulator which contains two decoupled sub-systems (such as two separate set of bands with no coupling between them). If each sub-system contributes $\theta = \pi$ to the topological term S_2, the total θ will be $2\pi \simeq 0$ so that the system is trivial. This is consistent with the Z_2 multiplication rule.

In summary, only using the general conditions of charge conservation and time-reversal symmetry, we conclude that the 3D TRI insulators are classified by Z_2 and the Z_2 nontrivial TI is described by the topological response theory

$$S_{\text{topo}} = \frac{i}{32\pi} \int d^4x \epsilon^{\mu\nu\sigma\tau} F_{\mu\nu} F_{\sigma\tau}. \tag{14}$$

Similar to the quantum Hall case, this theory also contains topological information about the nontrivial boundary states. For a manifold with a boundary, the second Chern class (13) is not quantized any more, so that time-reversal symmetry is broken even if $\theta = \pi$ modulo 2π. For $\theta = \pi$ the change of the action under time-reversal is

$$\delta S_{\text{topo}} = 2S_{\text{topo}} = \frac{1}{16\pi} \int_{\mathcal{M}} d^4x \epsilon^{\mu\nu\sigma\tau} F_{\mu\nu} F_{\sigma\tau} = \frac{1}{4\pi} \int_{\partial\mathcal{M}} d^3x \epsilon^{ijk} A_i \partial_j A_k. \tag{15}$$

---

[1] It should be clarified that the second Chern class and the second Chern number are defined differently in mathematics. The second Chern number vanishes for Abelian gauge field. The second Chern class for Abelian gauge field is non-vanishing, but its integral is determined by the first Chern numbers in two-dimensional cocycles (i.e., non-contractible two-dimensional surfaces) of the manifold. For example, $C_2 = 0$ if the space-time manifold is a sphere S^4, and it can be nonzero if the manifold is $S^2 \times S^2$ or T^4.

Such a change of the action breaks time-reversal symmetry. Consequently, if the open-boundary system is time-reversal invariant, there must be nontrivial surface states which have an "anomaly" so that a term exactly opposite to the term above is generated under time-reversal transformation. The surface theory that provided such an anomaly is known, and is simply a Dirac fermion coupled to the gauge field

$$S_{\text{surf}} = \int_{\partial \mathcal{M}} d^3x \, \psi^\dagger [(\partial_\tau - iA_\tau) - v\mathbf{n} \cdot \sigma \times (\mathbf{p} - \mathbf{A})]\psi. \tag{16}$$

The Dirac fermion theory is time-reversal invariant, but any ultraviolet cut-off of this theory breaks the time-reversal symmetry and leads to a Chern-Simons term with the half coefficient $\sigma_H = \frac{1}{4\pi}$. Consequently the change of this Chern-Simons term under time-reversal can exactly cancel the bulk contribution and restore the time-reversal symmetry of the whole system. This type of anomaly is known as the parity anomaly in high energy physics [32,18,1], although for our purpose it is better to call it time-reversal anomaly.

2.2 Microscopic Derivation of the Topological Order Parameter

The discussion in the last subsection provides the general reason for the existence of the Z_2 topological term, which applies to generic insulating systems with or without interaction (as long as the ground state is still unique). The general discussion, however, does not determine the value of θ in a specific system. To relate the general TFT description to the microscopic understanding of TI discussed in previous chapters, it is important to calculate θ in a given microscopic model of TI. In this subsection we first review the microscopic calculation of θ in non-interacting TI's and then later discuss the interacting case. The microscopic formula for θ was obtained in Ref. [36] and an alternative derivation was given in Ref. [9]. The approach of Ref. [36] is based on the idea of dimensional reduction: First, the topological response theory is obtained for a 4D generalization of integer quantum Hall states, proposed in the earlier work of Zhang and Hu [68]. The topological response theory has the form of a second Chern-Simons term $\mathcal{L} = \frac{C_2}{24\pi^2} \epsilon^{\mu\nu\sigma\tau\pi} A_\mu \partial_\nu A_\sigma \partial_\tau A_\tau$. Different from the 2D case, the 4D quantum Hall state is time-reversal invariant [68,36]. Second, the 3D topological insulator is obtained by a *dimensional reduction* of the 4D quantum Hall state, which consists of compactifying one of the spatial dimensions in the 4D state. When the compactified dimension is small enough, all fluctuations that are not uniform along that direction are gapped, and the 4D state is reduced to 3D. This is a simple example of the dimensional reduction procedure studied extensively in high energy physics, initiated by the Kaluza-Klein theory [20,23]. The time-reversal invariance of the 4D quantum Hall state allows the dimensional reduction to reach the 3D time-reversal invariant topological insulator state, and the dimensional reduction of the second Chern-Simons term correctly reproduces the θ-term in Eq. (11), with θ the flux of the gauge field threaded in the loop of the extra dimension. The dimensional reduction procedure not only provides a natural derivation of the microscopic formula for θ for band insulators, but also uncovers a generic relationship between topological states in different spatial dimensions, which we will discuss in Section 3.

Despite the conceptual importance and convenience of the dimensional reduction approach, the topological response theory of each topological state can also be understood without extending the spatial dimension. Considering that the dimensional reduction approach has been presented clearly in Refs. [36,40], here we will present a slightly different derivation of the topological term in 3D topological insulators which is equivalent to the dimensional reduction approach but does not directly involve higher dimensions. The purpose of this approach is to provide a derivation of the topological response theory in a more conventional condensed-matter language, and to understand the general structure of topological response theories from a different angle.

Since the topological term (14) is quadratic in electromagnetic field, naively one would expect the topological order parameter θ to be a linear response coefficient similar to the quantum Hall conductivity calculated in the last section. However, θ cannot be calculated in this way, because the value of C_2 is quantized, and thus have to stay invariant under a small variation of A_μ. Indeed, the variation of the action is

$$
\begin{aligned}
\delta S[A_\mu] &= \frac{i\theta}{16\pi^2} \int d^4x \epsilon^{\mu\nu\sigma\tau} F_{\mu\nu} (\partial_\sigma \delta A_\tau - \partial_\tau \delta A_\sigma) \\
&= \frac{i\theta}{8\pi^2} \int d^4x \partial_\sigma (\epsilon^{\mu\nu\sigma\tau} F_{\mu\nu} \delta A_\tau) = 0,
\end{aligned}
\tag{17}
$$

where we have used the Bianchi identity $\partial_\sigma F_{\mu\nu} \epsilon^{\mu\nu\sigma\tau} = 0$ and the condition that δA_τ is single-valued. Since the action has no continuous variation, the equation of motion (3) vanishes and one cannot obtain nontrivial linear response from the topological term.

To find a solution to this problem, one can consider the open-boundary problem, as the action is not quantized and has a nontrivial variation at the boundary. However, the presence of a sharp boundary is not convenient for the perturbation theory. One will have to study directly the open-boundary problem with suitable boundary condition. Alternatively, we choose a smooth boundary which gradually deforms a TI into a trivial insulator. If we allow time-reversal to be broken at the surface, TI and trivial band insulators can be adiabatically connected, so a smooth boundary is possible. Then we consider the smooth variation of the Hamiltonian across the boundary as a perturbation, and study the response of the system to both the electromagnetic field and the spatial variation of the Hamiltonian.

Consider a general tight-binding Hamiltonian

$$H = \sum_{i,j} \sum_{\alpha\beta} c_\alpha^\dagger(\mathbf{r}_i) h^{\alpha\beta}(\mathbf{r}_i, \mathbf{r}_j, t) c_\beta(\mathbf{r}_j) e^{i A_{ij}(t)} \tag{18}$$

with i, j labeling lattice sites and α, β labeling degrees of freedom (spin and orbitals) at each site. t is the time variable. $\mathbf{r}_i, \mathbf{r}_j$ are the coordinates of the lattice sites i, j, respectively. A_{ij} is the lattice electromagnetic gauge field. Here and below we will take the gauge choice $A_0 = 0$ so that the gauge field only has spatial components. We can write the Hamiltonian in an alternative form:

$$H = \sum_{\mathbf{R},\mathbf{r}} c^\dagger \left(\mathbf{R} + \frac{\mathbf{r}}{2}\right) h(\mathbf{r}; \mathbf{R}, t) c \left(\mathbf{R} - \frac{\mathbf{r}}{2}\right) e^{i\mathbf{A}(\mathbf{R},t)\cdot\mathbf{r}} \tag{19}$$

in which $\mathbf{R}$ and $\mathbf{r}$ denote the center-of-mass position and relative position of the two sites $\mathbf{r}_i, \mathbf{r}_j$ respectively. The internal indices α, β have been omitted, and h is a matrix. We are interested in the Hamiltonians which change smoothly in space-time, so that the $\mathbf{R}, t$ dependence of $h(\mathbf{r}; \mathbf{R}, t)$ is slow and adiabatic. The gauge field A_{ij} is also taken to be smooth and adiabatic, so that it can be replaced by the form $\mathbf{A}(\mathbf{R}, t) \cdot \mathbf{r}$ with $\mathbf{A}(\mathbf{R}, t)$ the vector potentiaal in the continuum limit. For later convenience we further assume the Hamiltonian has the form

$$h(\mathbf{r}; \mathbf{R}, t) = h(\mathbf{r}; \phi(\mathbf{R}, t)) \tag{20}$$

so that the space-time dependence is parameterized by one parameter $\phi(\mathbf{R}, t)$. This assumption is only for the convenience of the derivation, and we will demonstrate later that our results hold for more general space-time dependence of h.

Since $\phi(\mathbf{R}, t)$ and $\mathbf{A}(\mathbf{R}, t)$ vary smoothly in space-time, we can consider them as perturbations, and expand the Hamiltonian in the form of

$$H \simeq H_0 + \sum_{\mathbf{R}} [-\mathbf{j}(\mathbf{R}) \cdot \mathbf{A}(\mathbf{R}, t) + \phi(\mathbf{R}, t)\rho(\mathbf{R})],$$

$$H_0 = \sum_{\mathbf{R},\mathbf{r}} c^\dagger \left(\mathbf{R} + \frac{\mathbf{r}}{2}\right) h(\mathbf{r}; \phi = 0) c \left(\mathbf{R} - \frac{\mathbf{r}}{2}\right),$$

$$\mathbf{j}(\mathbf{R}) = -\sum_{\mathbf{r}} c^\dagger \left(\mathbf{R} + \frac{\mathbf{r}}{2}\right) i\mathbf{r} h(\mathbf{r}; \phi = 0) c \left(\mathbf{R} - \frac{\mathbf{r}}{2}\right),$$

$$\rho(\mathbf{R}) = \sum_{\mathbf{r}} c^\dagger \left(\mathbf{R} + \frac{\mathbf{r}}{2}\right) \partial_\phi h(\mathbf{r}; \phi)|_{\phi=0} c \left(\mathbf{R} - \frac{\mathbf{r}}{2}\right). \tag{21}$$

The operator $\mathbf{j}(\mathbf{R})$ is the charge current which couples to the gauge field, and $\rho(\mathbf{R})$ is the density operator coupled to the ϕ field. It is convenient to write down the action and go to momentum space by Fourier transform:

$$S \simeq S_0 + \int d\Omega \sum_{\mathbf{q}} [-\mathbf{A}(-\mathbf{q},-\Omega) \cdot \mathbf{j}(\mathbf{q},\Omega) + \phi(-\mathbf{q},-\Omega)\rho(\mathbf{q},\Omega)],$$

$$H_0 = \int d\omega \sum_{\mathbf{k}} c_{\mathbf{k}\omega}^\dagger \left(-i\omega + h(\mathbf{k},\phi=0)\right) c_{\mathbf{k}\omega}$$

$$\mathbf{j}(\mathbf{q},\Omega) = \int d\omega \sum_{\mathbf{k}} c_{\mathbf{k}-\mathbf{q}/2,\omega-\Omega/2}^\dagger \nabla_{\mathbf{k}} h(\mathbf{k},\phi=0) c_{\mathbf{k}+\mathbf{q}/2,\omega+\Omega/2},$$

$$\rho(\mathbf{q},\Omega) = \int d\omega \sum_{\mathbf{k}} c_{\mathbf{k}-\mathbf{q}/2,\omega-\Omega/2}^\dagger \left. \partial_\phi h(\mathbf{k},\phi)\right|_{\phi=0} c_{\mathbf{k}+\mathbf{q}/2,\omega+\Omega/2}. \tag{22}$$

Here $c_{\mathbf{k}\omega} = (2\pi N)^{-1/2} \int dt \sum_{\mathbf{r}_i} e^{-i\mathbf{k}\cdot\mathbf{r}+i\omega t} c(\mathbf{r}_i,t), h(\mathbf{k},\phi) = \sum_{\mathbf{r}} h(\mathbf{r},\phi)e^{-i\mathbf{k}\cdot\mathbf{r}}$, and $A(\mathbf{q},\Omega) = (2\pi N)^{-1} \int dt \sum_{\mathbf{R}} A(\mathbf{R},t)e^{-i\mathbf{q}\cdot\mathbf{R}+i\Omega t}$ with N the total number of lattice sites.

With this perturbative expansion, we are ready to calculate the response of the fermion system to fields $\mathbf{A}(\mathbf{q},\Omega)$ and $\phi(\mathbf{q},\Omega)$ by the standard response theory. As was analyzed earlier, there is no interesting topological term if the system is uniform and ϕ is space-time independent. Therefore the only possible terms are those proportional to the space-time gradient $\partial_\mu \phi$. The only candidate for a topological term which is isotropic, uniform, and independent of any continuously tunable parameter (such as the speed of light) is the term of the form

$$S_3 = iG_3 \int d^3x dt \epsilon^{\mu\nu\sigma\tau} \partial_\mu \phi A_\nu \partial_\sigma A_\tau$$

$$= -iG_3 \int \frac{d^4 p d^4 q}{(2\pi)^4} \epsilon^{\mu\nu\sigma\tau} p_\mu \phi(p) A_\nu(-p-q) q_\sigma A_\tau(q). \tag{23}$$

Here $q_\mu = (\mathbf{q},\omega)$ denotes the wavevector and frequency. G_3 in this cubic term is a non-linear response coefficient and is determined by correlation functions of $\mathbf{j}(q)$ and $\rho(q)$:

$$G_3 = \frac{1}{4!} \epsilon^{\mu\nu\sigma\tau} \frac{\partial^2}{\partial p_\mu \partial q_\sigma} \left[\frac{\delta^3 S_3}{\delta\phi(-p)A_\nu(p+q)A_\tau(-q)} \right]\bigg|_{p=q=0}$$

$$= \frac{1}{4!} \epsilon^{\mu\nu\sigma\tau} \frac{\partial^2}{\partial p_\mu \partial q_\sigma} \langle \rho(p) j_\nu(-p-q) j_\tau(q) \rangle \bigg|_{p=q=0}. \tag{24}$$

The correlation function in the second line should be understood as the contribution from the connected Feynman diagram shown in Fig. 2. Using

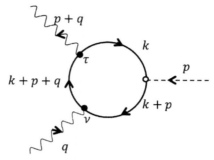

FIGURE 2 The Feynman diagram corresponding to the three-point correlation function in Eq. (24). The solid loop is the electron propagator, the dash line is the external line of ϕ, and the wavy lines are external lines of photon A_μ.

single-particle Green's function $G(k) = [\omega + i\delta - h(\mathbf{k})]^{-1}$ G_3 can be expressed as

$$
\begin{aligned}
G_3 &= \frac{1}{4!} \int \frac{d^4 k}{(2\pi)^4} \epsilon^{\mu\nu\sigma\tau} \frac{\partial^2}{\partial p_\mu \partial q_\sigma} \mathrm{Tr}\left[G_k \frac{\partial h(k + p/2)}{\partial \phi} G(k+p) \frac{\partial h(k+p+q/2)}{\partial k_\nu} \right. \\
&\qquad \left. \cdot G(k+p+q) \frac{\partial h(k+(p+q)/2)}{\partial k_\tau} \right]\Bigg|_{p=q=0} \\
&= -\frac{1}{4!} \int \frac{d^4 k}{(2\pi)^4} \epsilon^{\mu\nu\sigma\tau} \mathrm{Tr}\left[G_k \frac{\partial G^{-1}(k)}{\partial \phi} \frac{\partial G(k)}{\partial k_\mu} \frac{\partial G^{-1}(k)}{\partial k_\nu} \frac{\partial G(k)}{\partial k_\sigma} \frac{\partial G^{-1}(k)}{\partial k_\tau} \right] \\
&= -\frac{1}{4!} \int \frac{d^4 k}{(2\pi)^4} \epsilon^{\mu\nu\sigma\tau} \mathrm{Tr}\left[G\partial_\phi G^{-1} G\partial_\mu G^{-1} G\partial_\nu G^{-1} G\partial_\sigma G^{-1} G\partial_\tau G^{-1} \right]. \quad (25)
\end{aligned}
$$

In this equation we have used $h(\mathbf{k}) = \omega + i\delta - G^{-1}(k)$ to write the correlation function into a more concise form. If we further define a five-dimensional vector $q_a = (q_\mu, \phi)(a = 0, 1, \ldots, 4)$, the derivatives of G^{-1} over ϕ and k_μ can be written as $\partial_a G^{-1}$, and we have

$$
G_3 = -\frac{1}{5!} \int \frac{d^4 k}{(2\pi)^4} \epsilon^{abcde} \mathrm{Tr}\left[G\partial_a G^{-1} G\partial_b G^{-1} G\partial_c G^{-1} G\partial_d G^{-1} G\partial_e G^{-1} \right].
$$
(26)

Response coefficient of similar form has been discussed in the literature for different physical systems [15,49]. G_3 is determined by the Green function G and thus by the single-particle Hamiltonian $h(\mathbf{k}, \phi)$. In a spectral expansion the Hamiltonian can be expressed by its eigenvalues $\epsilon_n(\mathbf{k}, \phi)$ and corresponding eigenstates $|n; \mathbf{k}, \phi\rangle$:

$$
h(\mathbf{k}) = \sum_n \epsilon_n(\mathbf{k}, \phi) |n; \mathbf{k}, \phi\rangle \langle n; \mathbf{k}, \phi|. \quad (27)
$$

Correspondingly, the Green's function is $G(k, \phi) = \sum_n \left(\omega + i\delta - \epsilon_n(\mathbf{k}, \phi) \right)^{-1}$ $|n; \mathbf{k}, \phi\rangle \langle n; \mathbf{k}, \phi|$. In the ground state, all the bands with $\epsilon_n(\mathbf{k}, \phi) > 0$ are unoccupied and those with $\epsilon_n(\mathbf{k},\phi) < 0$ are occupied. For an insulator, there is a finite gap E_g, such that $|\epsilon_n(\mathbf{k},\phi)| \geq E_g$ for all $\mathbf{k},\phi$. Since we are interested in topological properties that should not be sensitive to the details of the energy dispersion, we first evaluate G_3 for special "flat-band" Hamiltonians with $\epsilon_n(\mathbf{k},\phi) = E_g(-E_g)$ for all unoccupied (occupied) bands, respectively. Then the applicability of the result obtained to generic Hamiltonians will be discussed later. In this special case the Green's function is determined by the projection operator to the occupied states $P_-(\mathbf{k},\phi) = \sum_{\epsilon_n<0} |n;\mathbf{k},\phi\rangle \langle n;\mathbf{k},\phi|$ and that to the unoccupied states $P_+(\mathbf{k},\phi) = 1 - P_-(\mathbf{k},\phi)$.

$$G(k,\phi) = \frac{P_+(\mathbf{k},\phi)}{\omega + i\delta - E_g} + \frac{P_-(\mathbf{k},\phi)}{\omega + i\delta + E_g}. \tag{28}$$

Using this simplified expression of the Green's function, G_3 defined in Eq. (26) can be expressed in $P_\pm(\mathbf{k},\phi)$ and the integral over ω can be carried out easily. We will skip the details of the derivation, which can be found in Appendix C3 of Ref. [36], and directly present the result of the calculation:

$$G_3 = \frac{1}{32\pi^3} \int d^3\mathbf{k}\,\epsilon^{ijkl}\mathrm{Tr}\left[P_+\frac{\partial P_-}{\partial k_i}\frac{\partial P_-}{\partial k_j}P_+\frac{\partial P_-}{\partial k_k}\frac{\partial P_-}{\partial k_l}\right]. \tag{29}$$

Here $i,j,k,l = 1,2,3,4$ and $k_4 = \phi$. The projection operators $P_\pm$ are determined by the occupied state wavefunctions, which can be written in a more familiar form by defining the Berry's phase gauge field. For the single-particle states $|\mathbf{k},\phi\rangle$, momentum $\mathbf{k}$ and ϕ are continuous parameters, so that a gauge field in this parameter space can be defined, known as the Berry's phase gauge field [3]:

$$a_i^{nm}(\mathbf{k},\phi) = -i\,\langle n;\mathbf{k},\phi| \frac{\partial}{\partial k_i} |m;\mathbf{k},\phi\rangle. \tag{30}$$

Here n,m runs over all occupied bands. In general for a system with more than one band occupied, the Berry's phase gauge field is non-Abelian. The gauge curvature of the Berry's phase gauge field is related to the projectors $P_\pm$ by

$$f_{ij}^{nm} \equiv \partial_i a_j^{nm} - \partial_j a_i^{nm} + i[a_i,a_j]^{nm}$$
$$= -i\,\langle n;\mathbf{k},\phi| \left(\frac{\partial P_-}{\partial k_i}P_+\frac{\partial P_-}{\partial k_j} - \frac{\partial P_-}{\partial k_j}P_+\frac{\partial P_-}{\partial k_i}\right) |m;\mathbf{k},\phi\rangle. \tag{31}$$

Therefore we obtain the expression for G_3 in terms of the Berry's phase gauge curvature:

$$G_3 = \frac{1}{128\pi^3} \int d^3\mathbf{k}\,\epsilon^{ijkl}\mathrm{Tr}[f_{ij}\,f_{kl}]. \tag{32}$$

By now we have finished the derivation of the response coefficient G_3 corresponding to the term S_3 in Eq. (23), but from this result the relation to the topological term (14) is still not obvious. To clarify that it is helpful to do some more transformations on G_3. The integrand $\epsilon^{ijkl}\mathrm{Tr}[f_{ij}\,f_{kl}]$ is the second

Chern form of the Berry's phase gauge field, which is a topological term if integrated over a closed four-dimensional manifold. The second Chern form can be written as a divergence of the Chern-Simons form [31]:

$$G_3 = \frac{1}{32\pi^3} \int d^3k \partial_i K^i,$$

$$K^i = \epsilon^{ijkl} \text{Tr} \left[a_j \partial_k a_l + \frac{2i}{3} a_j a_k a_l \right]. \tag{33}$$

After the integral, the only non-vanishing term is $\partial_\phi K^\phi$:

$$G_3 = \frac{1}{8\pi^2} \partial_\phi \Theta_{CS}(\phi),$$

$$\Theta_{CS}(\phi) = \frac{1}{4\pi} \int d^3k K^\phi$$

$$= \frac{1}{4\pi} \int d^3k \epsilon^{jkl} \text{Tr} \left[a_j \partial_k a_l + \frac{2i}{3} a_j a_k a_l \right]. \tag{34}$$

In this way, the response coefficient is expressed as the ϕ-derivative of the Chern-Simons term Θ_{CS}. It is important to notice that Θ_{CS} does not contain any derivative over ϕ, and is determined by the occupied states at a given value of ϕ. This is an essential fact since the term (23) containing G_3 is written as

$$S_3 = -\frac{i}{8\pi^2} \int d^3x dt \epsilon^{\mu\nu\sigma\tau} \partial_\phi \Theta_{CS}(\phi) \partial_\mu \phi A_\nu \partial_\sigma A_\tau. \tag{35}$$

For $\phi = \phi(\mathbf{x}, t), \Theta_{CS}(\phi) = \Theta_{CS}(\phi(\mathbf{x}, t))$ acquires space-time dependence via that of ϕ. Using the chain rule we get $\partial_\mu \Theta_{CS} = \partial_\phi \Theta_{CS} \partial_\mu \phi$, so that the ϕ dependence in the action (35) can be removed and we obtain

$$S_3 = -\frac{i}{8\pi^2} \int d^3x dt \epsilon^{\mu\nu\sigma\tau} \partial_\mu \Theta_{CS}(\mathbf{x}, t) A_\nu \partial_\sigma A_\tau$$

$$= \frac{i}{32\pi^2} \int d^3x dt \epsilon^{\mu\nu\sigma\tau} \Theta_{CS}(\mathbf{x}, t) F_{\mu\nu} F_{\sigma\tau}. \tag{36}$$

where we have used integration by parts in the second line. Following the derivation above, one can also see that the assumption (20) we have used so far is not essential. If the space-time dependence of the single-particle Hamiltonian is described not by a single parameter ϕ, but by several parameters $\phi_s, s = 1, 2, \ldots, n$, then the response calculation above can be carried for each ϕ_s, and in the effective action (35) we will obtain $\sum_s \partial_{\phi_s} \Theta_{CS} \partial_\mu \phi_s$ which is still equal to $\partial_\mu \Theta_{CS}$. Therefore the effective action (36) holds for a generic system with an adiabatic space-time dependence of the quadratic Hamiltonian. Comparing Eq. (36) with Eq. (11) we see that the non-linear response calculation indeed leads to the same action as is expected from general considerations, and the value of the coefficient θ in the topological term is determined by Θ_{CS}

in Eq. (34), which is a "topological order parameter" determined by the band structure of a given insulator.

In the last subsection, based on general symmetry consideration of the action (11) we have reached the conclusion that the topological order parameter $\theta = \Theta_{CS}$ must be π (modulo 2π) for a time-reversal invariant system. Therefore the microscopic formula (34) of Θ_{CS} is expected to recover this property. Before discussing the time-reversal property of Θ_{CS}, one should notice that the value of Θ_{CS} is not completely determined by Eq. (34), since the Chern-Simons term is not completely gauge invariant, as has been discussed in Section 1 for the Abelian theory. For N occupied bands labeled by $|n; \mathbf{k}, \phi\rangle$, a unitary $U(N)$ transformation can be defined $|n; \mathbf{k}, \phi\rangle \rightarrow \sum_m U_{mn}(\mathbf{k}, \phi) |m; \mathbf{k}, \phi\rangle$. In such a transformation the gauge field a_i^{nm} transforms as $a_i \rightarrow U^{-1} a_i U - i U^{-1} \partial_i U$. The change of the Chern-Simons term is

$$\Delta \Theta_{CS} = -\frac{i}{12\pi} \int d^3\mathbf{k} \mathrm{Tr} \epsilon^{ijk} \left[U^{-1} \partial_i U U^{-1} \partial_j U U^{-1} \partial_k U \right]. \qquad (37)$$

The right-hand side is a topological invariant which takes discrete values $2\pi n$, with integer n labeling the winding number of $U(N)$ matrix $U(\mathbf{k})$ on the Brillouin zone torus.[2] Therefore we see that Θ_{CS} is only determined modulo 2π. In other words it is $e^{i\Theta_{CS}}$ rather than Θ_{CS} which is completely single-valued and determined by the band structure.

The time-reversal property of Θ_{CS} can be determined from that of the Bloch states $|n; \mathbf{k}\rangle$. From the time-reversal transformation $\Theta\left(|n; \mathbf{k}\rangle\right) = T^{nm}(\mathbf{k}) |m; -\mathbf{k}\rangle$ we obtain the transformation of the gauge potential $a_i(\mathbf{k}) \rightarrow T a_i^*(-\mathbf{k}) T^\dagger - i T \partial_i T^\dagger$. Up to a gauge transformation by $T(\mathbf{k})$, a_i becomes its complex conjugate. Consequently $e^{i\Theta_{CS}} \rightarrow e^{-i\Theta_{CS}}$ under time-reversal, as is expected from the symmetry of the action (36). Therefore the microscopic derivation is consistent with the general discussion in the last subsection, and we obtain the quantization $\Theta_{CS} = 0, \pi$ mod 2π for time-reversal invariant systems. The Z_2 topological invariant $e^{i\Theta_{CS}} = +1(-1)$ corresponds to trivial and topological insulators, respectively. Although derived from a very different approach, the Chern-Simons invariant $e^{i\Theta_{CS}}$ is equivalent to the strong Z_2 topological invariant proposed in the topological band theory [30,12,43] reviewed in Chapter 1 of this volume, which is mathematically proved in Ref. [53].

The derivation above has been made for special "flat-band" Hamiltonians $h(\mathbf{k})$ which have all its eigenvalues equal to $\pm E_g$. For a generic Hamiltonian with energy dispersion $\epsilon_n(\mathbf{k})$, new terms appear in the response coefficient

---

[2] To understand this winding number more intuitively, one can consider the two-band case with $U(\mathbf{k}) = d_0 \sigma_0 + i \mathbf{d} \cdot \boldsymbol{\sigma}$ an $SU(2)$ matrix. Here σ_0 is the identity matrix and $\boldsymbol{\sigma}$ are Pauli matrices. $d_0^2 + \mathbf{d}^2 = 1$. In this case the winding number is reduced to $\epsilon^{abcd} d_a \partial_i d_b \partial_j d_c \partial_k d_d$ ($a = 0,1,2,3$) which is the Jacobian of the mapping $\mathbf{k} \rightarrow d_a(\mathbf{k})$ from Brillouin zone torus T^3 to the unit sphere S^3. Therefore the integral over $\mathbf{k}$ determines the winding number of S^3 on T^3.

G_3 defined in Eq. (26). In general, such dispersion-dependent terms are non-vanishing, leading to corrections to the value of Θ_{CS}. The generic formula of Θ_{CS} has been calculated by a magneto-electric response calculation in Refs. [28, 10], which is also reviewed in Chapter 2. However, for a time-reversal invariant system the quantized value of $\Theta_{CS} = 0, \pi$ mod 2π gets no correction from the energy dispersion, which is guaranteed by its topological quantization. To see that more explicitly one can define a one-parameter family of Hamiltonians

$$h_\lambda(\mathbf{k}) = \sum_n \left[(1 - \lambda)\epsilon_n(\mathbf{k}) + \lambda E_g \frac{\epsilon_n(\mathbf{k})}{|\epsilon_n(\mathbf{k})|} \right] |n; \mathbf{k}\rangle \langle n; \mathbf{k}|, \quad \lambda \in [0,1] \quad (38)$$

$h_\lambda(\mathbf{k})$ defines an interpolation between a generic Hamiltonian $h(\mathbf{k}) = h_0(\mathbf{k})$ and a flat-band Hamiltonian $h_1(\mathbf{k})$. By construction $h_\lambda(\mathbf{k})$ is gapped and time-reversal invariant for each $\lambda \in [0,1]$ (since the eigenstates remain unchanged). Consequently, for each $\lambda \in [0,1]$ the topological order parameter Θ_{CS} must be quantized to 0 or π mod 2π. Since Θ_{CS} should be a continuous function of the Hamiltonian, we conclude that it must remain invariant during the interpolation. In other words, the value of Θ_{CS} for a generic time-reversal invariant Hamiltonian is equal to that of a flat-band Hamiltonian with the same eigenstates, which is given by Eq. (34).

2.3 Physical Consequences of the TFT

In the last two subsections, we have derived the topological term (11) both from the general considerations and from a microscopic derivation in generic non-interacting insulators. In this subsection we will discuss the physical interpretation of this topological response theory, and its observable physical consequences. As has been discussed in Section 2.2, the topological term (11) only affects the electromagnetic response of the system if θ has space-time variations. In other words, although only $\theta = \pi$ describes topological insulators, the topological response of topological insulators can only be observed when time-reversal symmetry breaking perturbations are considered which lead to deviation of θ from the quantized value. For example, consider an interface between a topological insulator with $\theta = \pi$ and a trivial insulator (or vacuum) with $\theta = 0$. Denote the direction perpendicular to the surface as z direction, and the topological insulator (trivial insulator) occupies the $z < 0 (z > 0)$ region, with the interface at $z = 0$. If we break time-reversal symmetry in the neighborhood of the interface, it is possible to open a gap at the surface, and define an adiabatic cross-over between topological and trivial insulators across the surface. In such a system $\theta = \theta(z)$ is well-defined at each spatial location, and continuously interpolates between the limiting values $\theta(z \to \infty) = 2n\pi$ and $\theta(z \to -\infty) = (2m + 1)\pi$ with n, m integers. It should be noted that the bulk values of θ are only defined modulo 2π, so that the integers n, m remain undetermined if there is no interface. However, the spatial gradient $\partial_z \theta$ is a real number that is completely determined by the band structure locally defined around each spatial point. The response equation corresponding to the term (11) is given by Eq. (3) which leads to

$$j_\mu = \frac{1}{4\pi^2} \epsilon^{\mu\nu\sigma\tau} \partial_\nu \theta \partial_\sigma A_\tau. \tag{39}$$

For example with the z-gradient of θ we obtain $j_x = \frac{\partial_z \theta}{4\pi^2} E_y$ which describes a Hall conductivity in the xy plane. If we integrate along z direction we obtain the net two-dimensional current $j_x^{2D} = \sigma_{xy}^{2D} E_y$ with the 2D Hall conductivity of the surface

$$\sigma_{xy}^{2D} = \frac{1}{4\pi^2} \int dz \partial_z \theta = \frac{1}{2\pi} \left((n-m) - \frac{1}{2} \right) \tag{40}$$

Therefore the interface between topological and trivial insulators has a quantized Hall conductivity which is half-integer times the unit $1/2\pi$ (or e^2/h in the physical units). Although the detail of $\theta(z)$ depends on surface conditions, the half-integer quantization of the Hall conductivity is robust and is only determined by different values of θ at the two sides of the interface. Since in Section 1 we have shown that the Hall conductivity of a generic gapped 2D system with no ground state degeneracy should be quantized in unit of $1/2\pi$, the half-integer Hall conductivity of the surface states clearly distinguishes the topological insulator surface state (with time-reversal symmetry breaking gap) from any pure two-dimensional system [11,36].

The half-quantized Hall effect of the topological surface states is a direct consequence of the topological response theory. When one considers experimental measurements of this surface Hall conductivity, it is important to realize that the half-quantized Hall conductivity cannot be directly measured in a standard transport measurement. In ordinary 2D quantum Hall experiments, contacts can be made to the edges of the quantum Hall droplet, but the surface of a topological insulator is always closed and does not have an edge. Therefore, the topological response of a topological insulator should be measured by non-contact measurements, which detects electromagnetic response rather than charge current. The electromagnetic properties of the system are described by the topological term together with the ordinary Maxwell term:

$$S_{\text{eff}} = \int d^3x dt \left[\frac{i\theta}{4\pi^2} \mathbf{E} \cdot \mathbf{B} + \frac{1}{8\pi\alpha} \left(\epsilon \mathbf{E}^2 + \frac{1}{\mu} \mathbf{B}^2 \right) \right]. \tag{41}$$

Here $\alpha = e^2/\hbar c$ (in Gauss unit) is the dimensionless fine structure constant which controls the strength of electromagnetic interaction. The equations of motion of this effective action can be written as the standard Maxwell equations

$$\nabla \cdot \mathbf{D} = 4\pi\rho, \nabla \cdot \mathbf{B} = 0,$$

$$\nabla \times \mathbf{E} = -\frac{1}{c} \frac{\partial \mathbf{B}}{\partial t}, \nabla \times \mathbf{H} = \frac{1}{c} \frac{\partial \mathbf{D}}{\partial t} + \frac{4\pi}{c} \mathbf{j} \tag{42}$$

with the modified constituent equations

$$\mathbf{H} = \mathbf{B} - 4\pi \mathbf{M} + \frac{\theta \alpha}{\pi} \mathbf{E},$$

$$\mathbf{D} = \mathbf{E} + 4\pi \mathbf{P} - \frac{\theta \alpha}{\pi} \mathbf{B}, \qquad (43)$$

For simplicity we have written the action (41) with an isotropic dielectric constant and permittivity, but the equations of motion (42) and (43) apply to more generic systems with ϵ and μ tensors. In this description, the topological response current, such as the Hall current at the surface, is included as a correction to the bound current density, and ρ and $\mathbf{j}$ in Eq. (42) remain the ordinary "free charge and current density." The constituent equations (43) provide a clear physical illustration of the topological response property described by the θ term. The charge polarization obtains a new contribution proportional to $\theta \mathbf{B}$ and the magnetization obtains a contribution proportional to $\theta \mathbf{E}$. In other words, $\frac{\theta}{\pi}$ is a magneto-electric coefficient, which is quantized to $\alpha(n + 1/2)$ for topological insulators. This effect is called the topological magneto-electric (TME) effect [36]. As a direct and generic consequence of the topological response theory, the topological magnetoelectric effect serves as a defining property of 3D topological insulators, just like the role of quantized Hall conductivity in the quantum Hall insulators. Various different physical consequences of this effect have been studied, which we will review in the following.

2.3.1 Image Magnetic Monopole Effect

One of the most direct consequences of the TME effect is the image magnetic monopole effect [38]. A probe charge close to the surface of an ordinary insulator will induce charge polarization of the insulator which can be described by an image charge below the surface. When the same thing is done with a topological insulator (with surface states gapped by time-reversal symmetry breaking), due to the TME effect a magnetization is induced by the electric field of the external charge, which can be described by an image magnetic monopole below the surface, as is illustrated in Fig. 3.

The position and the magnetic flux of the image monopole can be determined straightforwardly from solving the modified Maxwell equations with a point-like source charge and $\theta(z) = 0$ for $z > 0$ and $\theta(z) = \theta_0 = (2n + 1)\pi$ for $z < 0$. [3] For a probe charge at a distance d to the surface, the image monopole stays at the mirror position below the surface, the same as the image charge (which also exists for topological insulators, just like for ordinary insulators).

---

[3] As is discussed earlier, θ should be a smooth function of z. The step function form is considered as an approximation of a smooth function, which does not affect the properties of the image monopole to the leading order.

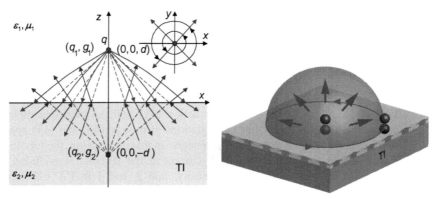

FIGURE 3 (Left) The image charge and image monopole induced by an external charge. The red and blue arrows show the distribution of the electric field of the probe charge and the magnetic field characterized by the image monopole. The inset is a top-down view showing the in-plane component of the electric field on the surface (red arrows) and the circulating surface current (black circles). (Right) Illustration of the fractional statistics induced by the image monopole effect. From [38]. (For interpretation of the references to color in this figure legend, the reader is referred to the web version of this book.)

The image charge q_1 and monopole g_1 are given by

$$q_1 = \frac{1}{\epsilon_1} \frac{(\epsilon_1 - \epsilon_2)(1/\mu_1 + 1/\mu_2) - \alpha^2 \theta_0^2/\pi^2}{(\epsilon_1 + \epsilon_2)(1/\mu_1 + 1/\mu_2) + \alpha^2 \theta_0^2/\pi^2} q,$$

$$g_1 = -\frac{2\alpha\theta_0/\pi}{(\epsilon_1 + \epsilon_2)(1/\mu_1 + 1/\mu_2) + \alpha^2 \theta_0^2/\pi^2} q \qquad (44)$$

in which q is the probe charge, and $\epsilon_{1,2}, \mu_{1,2}$ the dielectric constant and permittivity of the ordinary and topological region, respectively. Similarly, an observer staying in the topological insulator region will observe an image charge $q_2 = q_1$ and an image monopole $g_2 = -g_1$ at the same position as the external charge. It should be noted that similar effect also occurs in other insulators with a surface Hall effect or equivalently a bulk magneto-electric effect, such as a quantum Hall state at the interface [17], and multiferroic insulators.

When we consider multiple external charges and their motion, the image monopole effect leads to an interesting fractional statistics. As is illustrated in Fig. 3, when two external charges wind around each other, each of them moves together with its image monopole, which makes it a bound state of charge and monopole, known as a "dyon" discussed in high energy physics [61]. Consequently each charge sees a magnetic flux from the image monopole of the other charge. When the distance R between two charges is much larger than the distance d of each charge to the surface, the flux is $g/2$. If the external charge is an electron, we have $q = e$ and the statistical angle is given by

$$\varphi = \frac{g_1 q}{2\hbar c} = \frac{\alpha^2 \theta_0/\pi}{(\epsilon_1 + \epsilon_2)(1/\mu_1 + 1/\mu_2) + \alpha^2 \theta_0^2/\pi^2}. \tag{45}$$

Such an image monopole attachment is a new mechanism of statistic transmutation. In general, the statistical angle obtained is irrational. The image monopole can be detected directly by local probes sensitive to small magnetic fields, such as scanning superconducting quantum interference devices (scanning SQUID) and scanning magnetic force microscopy (scanning MFM) [38]. It is interesting to note that the generalization of the image monopole effect has been considered for an imperfect surface with finite longitudinal conductivity [34], in which case the image monopole decays in time.

2.3.2 Topological Kerr and Faraday Rotation

The TME effect also has consequences for the propagation of polarized light. Consider a linearly polarized light injected perpendicularly to the interface between topological and trivial insulators. Due to the TME effect, electric field generates magnetization in the topological insulator region, so that the polarization plane of the transmitted light will be rotated. Similar rotation occurs to the reflected light. Such polarization plane rotation of linearly polarized light is known as Faraday effect (for transmitted light) and Kerr effect (for reflected light). For a single surface between semi-infinite topological insulator and trivial insulator (or vacuum), the Faraday and Kerr rotation angles can be calculated by solving the modified Maxwell equations, which leads to [36,21] (See the left panel of Fig. 4).

$$\tan \theta_K = \frac{2\alpha \theta_0/\pi \sqrt{\epsilon_1/\mu_1}}{\epsilon_2/\mu_2 - \epsilon_1/\mu_1 + \alpha^2 \theta_0^2/\pi^2}, \tag{46}$$

$$\tan \theta_F = \frac{\alpha \theta_0/\pi}{\sqrt{\epsilon_1/\mu_1} + \sqrt{\epsilon_2/\mu_2}}. \tag{47}$$

Here $\epsilon_{1,2}, \mu_{1,2}$, and θ_0 has the same meaning as in the discussion of image monopole effect. From this result we see that the topological contribution is of order $\alpha \sim 1/137$, which is observable by the current experimental techniques. However, it should be noted that the effective action (41) only describes response properties in the long wavelength limit, so that the topological Faraday/Kerr effects can only be observed for photons with frequency $\hbar\omega \ll E_g$. Here E_g is the minimal energy gap of the system (including the surface region).

Similar to the image monopole effect, the Faraday/Kerr angles depend on material parameters $\epsilon_{1,2}, \mu_{1,2}$, and thus cannot be used to directly probe the topologically quantized value of θ. However, it is possible to cancel the ϵ, μ dependence by considering light transmission and reflection at both surfaces of a TI film [57,27]. By measuring the Faraday/Kerr angles at two

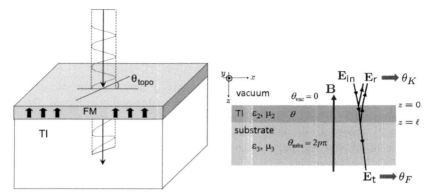

FIGURE 4 (Left) Illustration of the Faraday rotation of a linearly polarized light on a T-breaking surface of a topological insulator. Adapted from [36]. (Right) Kerr and Faraday rotation on two surfaces of a topological insulator film. By measuring the Kerr and Faraday rotation at two frequencies corresponding to the maxima and minima of reflectivity, the quantized topological order parameter θ of the topological insulator (green region) can be measured. (For interpretation of the references to color in this figure legend, the reader is referred to the web version of this book.) Adapted from [27].

special frequencies corresponding to reflectivity maxima and minima, enough information is obtained so that the ϵ and μ dependence can be canceled and the half-quantized surface Hall conductivity of the two surfaces can be independently determined [27] (See the right panel of Fig. 4).

2.3.3 Charge of a Magnetic Monopole (Witten Effect)

If there is a real magnetic monopole in a topological insulator, according to the TME effect the magnetic field generates charge polarization, so that the monopole obtains an electric charge $q = e\frac{\theta}{2\pi}\frac{g}{\phi_0}$ with $\phi_0 = hc/e$ the flux quanta [36]. In high energy physics this effect is known as the Witten effect [61] which transmutes a monopole into a dyon. In principle, topological insulators can be used to detect magnetic monopoles through this effect [41]. Physically, this effect can be understood by considering a monopole gradually moving into a topological insulator. The charge polarization is proportional to the magnetic field and thus increases gradually, which leads to a charge pumping from the boundary of the system to the monopole location [42]. It is interesting to note that the effect is not symmetric between electric and magnetic field—a monopole carries electric charge due to the θ term, but a charge does not obtain a monopole flux.

Although the Witten effect is difficult to verify due to the absence of magnetic monopoles, there are proposals which may realize an effective magnetic monopole. In Ref. [47] the Coulomb interaction between electrons at the two surfaces of a TI film is considered. Such interaction may lead to an "exciton condensation" of the surface states, which means a coherent hopping between top and bottom surfaces forms spontaneously, with a phase $e^{i\phi}$. Such a phase in tunneling is equivalent to a lattice gauge field, which couples to the TI in the same way as the electromagnetic field. When the phase has a vortex, it is equivalent to a magnetic monopole, which thus carries a half charge.

2.3.4 Other Related Effects

In all the effects discussed above, the boundary condition at the surface is considered static, and the change of θ across the surface is fixed. If the time-reversal symmetry breaking at the surface is induced by a magnetic film at the surface, the magnetic moment of the film can be dynamical. The flip of the magnetic moment direction reverses the surface Hall conductivity. If there are different magnetic domains, chiral edge states are trapped along the domain walls since the Hall conductivity changes by integer across the domain wall. Therefore the TME effect leads to nontrivial coupling between the dynamics of the surface magnetic film and the electromagnetic and charge transport properties of the topological insulator. Such couplings have been studied in various situations [65,29,64,13,66,33]. Both charge current and charge density can be coupled to the magnetic structures at the surface through this effect, which provides an approach to the electric control of magnetic moment [13,66,33]. Another related idea is to consider the dynamics of θ not only at the boundary but also in the bulk. If there is a bulk anti-ferromagnetic long range order developed in an interacting topological insulator, the topological order parameter θ can deviate from the quantized value and depend on the anti-ferromagnetic Neel vector (as long as time-reversal and inversion symmetries are both broken). Therefore θ becomes a dynamical field, which is in general a gapped component of the spin wave mode [25]. In such a system the term $\theta \mathbf{E} \cdot \mathbf{B}$ becomes a nonlinear coupling between spin fluctuation and electromagnetic field, which can be measured experimentally. For example, in a uniform background magnetic field there is a linear coupling between θ and $\mathbf{E}$ which leads to a hybridization of spin wave and photon modes, named as the "axionic polariton" mode.

3 TFT DESCRIPTION OF MORE GENERAL TOPOLOGICAL STATES

3.1 Phase Space Chern-Simons Theories

The procedure of deriving topological terms from the response theory presented in Section 2.2 can be straightforwardly generalized to insulators in other spatial dimensions and symmetry classes. In generic dimensions, an insulator coupled to electromagnetic field can be described by the Hamiltonian (19) if its space-time inhomogeneity is smooth compared to the lattice constant scale. In Section 2.2 it has been shown that the result of the response theory is independent of the form of the space-time dependence parameterized by $\phi(\mathbf{R},t)$. Therefore we do not introduce such parametrization and directly expand the Hamiltonian (19) in the gauge field and the gradient of the single-particle Hamiltonian:

$$H \simeq H_0 + \sum_{\mathbf{R}} \left[-\mathbf{j}(\mathbf{R}) \cdot \mathbf{A}(\mathbf{R},t) + R^{\mu} F_{\mu}(\mathbf{R}) \right]. \tag{48}$$

Here $\mathbf{j}(\mathbf{R})$ is the same as that defined in Eq. (21), and

$$F_{\mu}(\mathbf{R}) = \sum_{\mathbf{r}} c^{\dagger} \left(\mathbf{R} + \frac{\mathbf{r}}{2} \right) \left. \frac{\partial}{\partial R^{\mu}} h(\mathbf{r}; \mathbf{R},t) \right|_{\mathbf{R}=t=0} c \left(\mathbf{R} - \frac{\mathbf{r}}{2} \right) \tag{49}$$

with $\mu = 0,1,2,\ldots,D$ for D spatial dimensions, and R^{μ} the space-time vector with $R^0 = t$. The action can be written in momentum space as

$$S \simeq S_0 + \int d\Omega \sum_{\mathbf{q}} \left[-\mathbf{j}(-\mathbf{q},-\mathbf{\Omega}) \cdot \mathbf{A}(\mathbf{q},\Omega) \right] + i \left. \frac{\partial}{\partial q^{\mu}} F^{\mu}(\mathbf{q},\Omega) \right|_{q^{\mu}=0}$$

$$S_0 = \int d\Omega \sum_{\mathbf{k}} c^{\dagger}_{\mathbf{k}\Omega} \left(i\Omega + h(\mathbf{k}; R^{\mu} = 0) \right) c_{\mathbf{k}\Omega}$$

$$\mathbf{j}(\mathbf{q},\Omega) = \int d\omega \sum_{\mathbf{k}} c^{\dagger}_{\mathbf{k}-\mathbf{q}/2,\omega-\Omega/2} \nabla_{\mathbf{k}} h(\mathbf{k}; R^{\mu} = 0) c_{\mathbf{k}+\mathbf{q}/2,\omega+\Omega/2}$$

$$F_{\mu}(\mathbf{q}) = \int d\omega \sum_{\mathbf{k}} c^{\dagger}_{\mathbf{k}-\mathbf{q}/2,\omega-\Omega/2} \left. \frac{\partial}{\partial R^{\mu}} h(\mathbf{k}; \mathbf{R},t) \right|_{R^{\mu}=0} c_{\mathbf{k}+\mathbf{q}/2,\omega+\Omega/2} \tag{50}$$

From this expression one can see the similarity between $\mathbf{j}(\mathbf{q})$ and $F_{\mu}(\mathbf{q})$ which corresponds to gradient of $h(\mathbf{k},\mathbf{R})$ in momentum and real space coordinates, respectively. The response theory can be calculated in each order of $A_{\mu}(\mathbf{q})$ and $F_{\mu}(\mathbf{q})$. If we define the $2D + 2$ dimensional phase space coordinate $q^A = (\mathbf{R},t;\mathbf{k},\omega)$, the kernel in $\mathbf{j}(\mathbf{q})$ and $F_{\mu}(\mathbf{q})$ can be written in a unified form $\partial_A h(q^A) = \partial_A G^{-1}(q_A)$. Following similar derivation as that of G_3 in Section 2.2, one can extract the topological terms in the response theory which are anti-symmetric terms in the gauge curvature. Multiple topological terms can be defined for a given system, which comes from different powers of A_{μ} in the perturbative expansion. In spatial dimension D a generic topological response term has the form

$$S_n = i \int d^D x dt \Omega_n^{\mu_1 \ldots \mu_{2n-1}} A_{\mu_1} \partial_{\mu_2} A_{\mu_3} \ldots \partial_{\mu_{2n-2}} A_{\mu_{2n-1}}, \tag{51}$$

$$\Omega_n^{A_1 A_2 \ldots A_{2n-1}} = \frac{1}{(2D - 2n + 3)!} \int \frac{d^D k d\omega}{(2\pi)^{D+1}} \epsilon^{A_1 A_2 \ldots A_{2n-1} A_{2n} \ldots A_{2D+2}}$$

$$\cdot \mathrm{Tr} \left(G\partial_{A_{2n}} G^{-1} G\partial_{A_{2n+1}} G^{-1} \ldots G\partial_{A_{2D+2}} G^{-1} \right).$$

Here $\Omega_n^{A_1 A_2 \ldots A_{2n-1}}$ is a rank-$(2n - 1)$ anti-symmetric tensor in the $2D + 2$ dimensional phase space, and $\Omega_n^{\mu_1 \ldots \mu_{2n-1}}$ is obtained by restricting A_i to

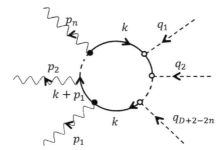

FIGURE 5   The Feynman diagram corresponding to the topological term in Eq. (51). The vertices connected to dash lines and wavy lines stand for F^μ and j^μ, respectively. The circle is the electron Green's function.

space-time coordinates μ_i. The corresponding Feynman diagram is shown in Fig. 5. For example, the 3D topological insulator case discussed in last section corresponds to taking $2n - 1 = 3, D = 3$, and $\Omega_n^{\mu_1\mu_2\mu_3}$ involves five powers of $G\partial_A G^{-1}$, as is shown in Eq. (26). The relation to the notation in the last section is that $\Omega_n^{\mu_1\mu_2\mu_3} = \frac{1}{8\pi^2}\epsilon^{\mu_1\mu_2\mu_3\mu_4}\partial_{\mu_4}\Theta_{CS}$. Similar to the discussion in Section 2.2, if one ignores the energy dispersion contribution and consider flat-band Hamiltonians, the Green functions have the form (28), and the "topological order parameter" $\Omega_n^{A_1A_2...A_{2n-1}}$ can be reduced to a Chern form of the Berry's phase gauge field after taking the integral over ω:

$$\Omega_n^{a_1a_2...a_{2n-1}} = \frac{1}{2^{D-n+1}n!(D+1-n)!}\int \frac{d^Dk}{(2\pi)^D}\epsilon^{a_1a_2...a_{2D+1}}$$
$$\cdot \mathrm{Tr}\left[f_{a_{2n}a_{2n+1}}f_{a_{2n+2}a_{2n+3}}\cdots f_{a_{2D}a_{2D+1}}\right]. \quad (52)$$

Here each a_i labels the $2D + 1$ dimensional phase space of $(\mathbf{R}, t, \mathbf{k})$, with ω index already integrated out. Ω_n is the $(D - n + 1)$th Chern form of the phase space gauge field integrated over the D-dimensional Brillouin zone. The power n in Eq. (51) must satisfy $2n - 1 \le D + 1$. For example for $D = 3$, we have $n \le 5/2$ so that $n = 0, 1, 2$. It is interesting to discuss the special case of $n = 0$, which corresponds to the term with only F^μ and no electromagnetic field. The topological term in this case becomes a Wess-Zumino-Witten (WZW) term [60,62] of the Green function $G(q^A)$ in the phase space:

$$S_0 = \frac{1}{(2D+2)!}\int d^Dx dt \frac{d^Dk d\omega}{(2\pi)^{D+1}}\int_0^1 du\,\epsilon^{A_1A_2...A_{2D+2}}$$
$$\cdot \mathrm{Tr}\left(G\partial_{A_1}G^{-1}G\partial_{A_2}G^{-1}\ldots G\partial_{A_{2D+2}}G^{-1}G\partial_u G^{-1}\right). \quad (53)$$

Here u is an interpolation parameter which is defined such that $G(q^A, 1) = G(q^A)$ corresponds to the physical system studied, and $G(q^A, 0) = G_0(\omega)$ a trivial reference system with no momentum and space-time dependence. Such an interpolation parameter is necessary to write down the WZW term in a local form. After integrating over ω, S_0 can be written as the $2D + 1$ dimensional Chern-Simons term of the Berry's connection a_b defined in phase space:

$$S_0 = \frac{1}{(D+1)!} \int d^D x dt \frac{d^D k}{(2\pi)^D} \epsilon^{b_1 b_2 \ldots b_{2D+1}} \mathrm{Tr} \left[a_{b_1} \partial_{b_2} a_{b_3} \ldots \partial_{b_{2D}} a_{b_{2D+1}} \right] + \ldots \tag{54}$$

Here ... stand for additional terms determined by gauge invariance [31], such as the $\frac{i}{3} a_i a_j a_k$ term in Eq. (34). If the adiabatic deformation of the Hamiltonian is parameterized by some order parameter field such as the mass vector of a Dirac fermion, this phase space WZW term can reduce to a Wess-Zumino-Witten term of the corresponding non-linear σ-model in space-time [63].

For a given system the topological terms S_n for different n describe different topological response properties. The total topological response theory is given by a sum of these terms, which interestingly is simply the Chern-Simons term of the combined gauge field $A_b + a_b$:

$$S_{\text{topo}} = \frac{1}{(D+1)!} \int d^D x dt \frac{d^D k}{(2\pi)^D} \epsilon^{b_1 b_2 \ldots b_{2D+1}} \mathrm{Tr} \left[(A_{b_1} + a_{b_1}) \partial_{b_2} (A_{b_3} + a_{b_3}) \right.$$
$$\left. \ldots \partial_{b_{2D}} (A_{b_{2D+1}} + a_{b_{2D+1}}) \right] + \ldots \tag{55}$$

Notice that the electromagnetic gauge field has only space-time components A_μ and other components are zero. The phase space Chern-Simons action (55) provides a unified description of topological electromagnetic response properties of insulators in generic dimensions. It should be noticed that not every term S_n corresponds to a class of topological insulators. As has been seen in the example of 3D topological insulators ($D = 3$, $n = 2$), the response coefficient (Θ_{CS}) is not quantized unless time-reversal symmetry is imposed. The topological classification of topological insulators is determined by the topological response theory (55) together with the discrete symmetry requirement. The universal form of the topological response theory (55) in different dimensions also naturally describes the periodicity of the classification of topological states in spatial dimensions. In Ref. [36] it was pointed out that the topological insulators with time-reversal or particle-hole symmetries have the same topological classification in dimensions D and $D + 8$, which is related to the Bott periodicity [4]. When the discussion is generalized to more generic symmetry classes with both time-reversal and particle-hole symmetries [46, 22], the classification of topological insulators and superconductors is shown to satisfy the Bott periodicity in both spatial dimensions and symmetry classes.

As an example of the phase space Chern-Simons theory we can consider the $D = 2$ case, where there are three terms S_0, S_1, S_2. S_2 is the familiar Chern-Simons term describing quantum Hall effect, with the coefficient $\Omega_2^{\mu\nu\tau} = \frac{1}{16\pi^2} \epsilon^{\mu\nu\tau} \int d^2 k \epsilon^{ij} \mathrm{Tr}[f_{ij}] = \frac{1}{4\pi} \epsilon^{\mu\nu\tau} C_1$. Here $C_1 = \frac{1}{2\pi} \int d^2 k \mathrm{Tr}[f_{xy}]$ is the first Chern number in the momentum space. S_1 has the form of $\Omega_1^\mu A_\mu$, which thus describes a ground state current $j^\mu = \Omega_1^\mu$. Ω_1^μ is defined in Eq. (52) which is determined by the phase space second Chern form. In Ref. [36] it was shown that such a current is a topological response property of a 2D topological insulator. As is discussed in Chapter 1, the 2D topological insulator has helical edge states with odd number of channels moving in each direction along the edge. Consider the simplest edge state with a linear dispersion, described by the effective Hamiltonian $H = \int dx \psi^\dagger(x)(-iv\partial_x)\sigma_z\psi(x)$. Here $\psi(x)$ has two components describing the left and right movers with opposite spins. When a time-reversal symmetry breaking field such as a magnetic field is applied to the edge, the edge state can obtain two possible mass terms $H_M = \int dx m\psi^\dagger(x)\left(\cos\theta(x)\sigma_x + \sin\theta(x)\sigma_y\right)\psi(x)$. The parameter θ can be controlled by the direction of the external magnetic field. When θ is space-time dependent, it is known that a charge density and current is introduced, which is determined by the Goldstone-Wilczek formula [19,14]: $j_\mu = \frac{1}{2\pi}\epsilon_{\mu\nu}\partial^\nu\theta$. If one describes the 2D topological insulator with the boundary magnetic field by the bulk theory rather than a boundary effective theory, and calculate the response of the system to electromagnetic field and the boundary T-breaking field, one obtains $\Omega_1^\mu = \frac{1}{2\pi}\epsilon^{\mu\nu\tau}\partial_\nu\theta\hat{n}_\tau$, with $\hat{n}$ the normal direction of the boundary. In particular, when θ changes by π around a domain wall, which corresponds to a flip of the magnetic field direction, there is a half charge at the domain wall determined by $Q = \int \rho(x)dx = \frac{1}{2\pi}\int \partial_x\theta dx$, which is a signature that distinguishes 2D topological insulator from trivial insulators [35]. Finally, S_0 is a second Chern-Simons term of the internal gauge field a_b, which describes a pure Berry's phase obtained by the system during an adiabatic evolution, in absence of electromagnetic field.

3.2 Further Investigation on Interacting Topological States

In this chapter we have presented two approaches to obtain the topological field theory description. One is the "bottom-up," or phenomenological, approach taken in Sections 1 and 2.1, and the other is the "top-down," or microscopic, approach for non-interacting fermions discussed in Sections 2.2 and 3.1. For interacting systems, it is generally difficult to obtain topological terms from an explicit response theory calculation, and the bottom-up approach has to be applied. The discussion in Section 1 on Chern-Simons term and that in Section 2.1 on θ-term can be easily generalized to other space-time dimensions. For all even spatial dimensions $D = 2n$, there is a Chern-Simons term of A_μ with an integer-quantized coefficient, so that the insulators in these dimensions are classified by an integer, if no additional symmetry other than charge conservation is imposed. When discrete symmetries are considered, one can determine whether the topological classification survives in the symmetry requirement by studying the symmetry property of the Chern-Simons term. Under time-reversal, $A_0 \rightarrow A_0$ and $A_i \rightarrow -A_i$ (with A_i the spatial components), so that the Chern-Simons terms in $D = 4n - 2$ dimensions are time-reversal odd, and those in $D = 4n$ dimensions are time-reversal even. Therefore time-reversal invariant topological insulators classified by an integer are defined in $D = 4n$ dimensions, but not in $D = 4n - 2$ since the Chern-Simons term is required to vanish by symmetry. Similarly if one considers particle-hole symmetry which transforms $A_\mu \rightarrow -A_\mu$, the situation is the reverse, with the Chern-Simons terms in $D = 4n - 2$ even and those in $D = 4n$ odd. Therefore particle-hole invariant topological insulators classified by an integer are defined in $D = 4n - 2$. Similar symmetry considerations on the θ terms lead to the conclusion that Z_2 invariant topological insulators exist in $D = 4n - 1$ for time-reversal invariant states, and in $D = 4n - 3$ for particle-hole invariant states.

In interacting systems the topological order parameters—integer or Z_2 coefficients of the topological terms—cannot be directly calculated in response theory, since interaction introduces additional Feynman diagrams which may modify the response. The single-particle formula for the topological order parameter such as Θ_{CS} in Eq. (34) cannot be directly applied to interacting case, since the single-particle occupied states are not well-defined, nor do the Berry's phase gauge field. However, single-particle Green's function $G(\mathbf{k}, \omega)$ is still well-defined, as long as the system has translation symmetry. (Even for disordered system without translation symmetry, the ensemble average over impurity configurations can define a translation invariant Green's function.) The topological order parameters can be defined by using Green's functions [54,16], as is shown in Eq. (26), and more generally in Eq. (51). For example, for the 4D quantum Hall insulator the integer-valued topological invariant is

$$C_2 = \frac{\pi^2}{15} \int \frac{d^4 k d\omega}{(2\pi)^5} \epsilon^{\mu\nu\sigma\tau\pi} \mathrm{Tr}\left[G\partial_\mu G^{-1} G\partial_\nu G^{-1} G\partial_\sigma G^{-1} G\partial_\tau G^{-1} G\partial_\pi G^{-1} \right].$$

(56)

For the 3D time-reversal invariant topological insulator, the Z_2 topological invariant Θ_{CS} can be written as an integral of G_3 in Eq. (26):

$$\Theta_{CS} = \frac{\pi}{6} \int_0^1 du \int \frac{d^3 k d\omega}{(2\pi)^4} \epsilon^{\mu\nu\sigma\tau}$$
$$\cdot \mathrm{Tr}\left[G\partial_\mu G^{-1} G\partial_\nu G^{-1} G\partial_\sigma G^{-1} G\partial_\tau G^{-1} G\partial_u G^{-1} \right] \quad (57)$$

Here u is an interpolation parameter so that $G(\mathbf{k}, \omega; u = 0) = (\omega + i\delta - h_0)^{-1}$ a trivial reference function, and $G(\mathbf{k}, \omega; u = 1)$ is the Green's function of the topological insulator. The details of this interpolation process can only change Θ_{CS} by integer multiple of 2π, so that $e^{i\Theta_{CS}}$ is well-defined and takes the value of ± 1 when time-reversal symmetry is preserved. Although for an interacting system there are generically corrections to the response coefficients, the coefficient of the topological terms, such as the θ term for a time-reversal invariant system, is quantized and has to remain constant unless the system undergoes a phase transition. Since the Green's function formulas are topological invariant, and hold for non-interacting topological states, they also apply to any interacting topological state that is adiabatically connected to a non-interacting one. The Green's function approach has been generalized to other symmetry classes [8,56].

So far we have been discussing the topological response of topological insulators to an external electromagnetic field. In a generic topological state of matter, it is not always sufficient to consider electromagnetic response. For example, topological insulators have been generalized to topological superconductors. In the mean-field approximation, the quasi-particles of superconductors are described by a Bogoliubov-de-Gennes Hamiltonian which has the same form as an insulator Hamiltonian with an additional particle-hole symmetry. However, the coupling of superconductors to electromagnetic field is drastically different from that of insulators, due to Meissner effect. The topological terms above do not apply to topological superconductors. To generalize the topological field theory approach to more generic topological states such as superconductors, other external fields can be considered. One natural candidate in the superconductor case is the gravitational field, which does not require any conserved charge, and can be coupled to any states of matter with energy-momentum conservation. For example, it was proposed that the 3D time-reversal invariant topological superconductors [46,37,44] are described by a gravitational counterpart of the θ term, where the second Chern form of electromagnetic field in action (11) is replaced by the Pontryagin invariant of the Riemann curvature [55,45]: $S_\theta = -\frac{\theta}{1536\pi^2} \int d^4x \epsilon^{\mu\nu\rho\sigma} R^\alpha_{\beta\mu\nu} R^\beta_{\alpha\rho\sigma}$. Reference [45] also discussed such topological response to gravitational field and electromagnetic field, and the mixed response in different dimensions and symmetry classes, which linked different topological states of matter to different types of "quantum anomalies." In a more recent work [39], a more complete topological field theory description of the 3D topological superconductor is derived, which contains the θ terms of both electromagnetic field and gravitational field, and the phase θ is not a topological order parameter but a dynamical field determined by the superconducting pairing order parameter. Since the superconducting order parameter is a bosonic field with dynamics, the topological field theory in Ref. [39] is not a topological response theory but a dynamical topological field theory, similar to the Chern-Simons theory of fractional quantum Hall effect that we discussed in Section 1.

Beside the topological insulators and topological superconductors, more exotic topological states of matter have been studied in topological field theory approach and other approaches. Without discussing further details, we will refer the reader to the main works along this direction. Similar to the generalization of integer to fractional quantum Hall effect, 3D topological insulators can also be generalized to fractional topological insulators, which can be described by a non-Abelian θ term of the emergent gauge field [26,48]. When time-reversal symmetry is broken at the surface of fractional topological insulators, the surface state behaves like half of a fractional quantum Hall state, which is a fractionalized version of the half quantum Hall effect for the ordinary (weakly interacting) topological insulators. A different topological field theory description to fractional topological insulators is the BF theory with a topological term [7]. From a more mathematical approach, a generic framework of $(3 + 1)$-dimensional topological field theory is developed [52], which may be related to the fractional topological insulators described by non-Abelian θ terms. Recently, spin models that can realize the Walker-Wang theory have been constructed [50]. For topological states without fractionalization, a generic classification in general dimension and symmetry conditions has been developed based on group cohomology theory [6,5]. From the viewpoint of topological field theory, this approach can be roughly understood as a classification of topological non-linear σ-models for a given symmetry group.

ACKNOWLEDGMENT

This work is supported by the National Science Foundation through the grant No. DMR-1151786.

REFERENCES

[1] Alvarez-Gaume L, Witten E. Gravitational anomalies. Nucl Phys B 1984;234(2):269–330.
[2] Bernevig BA, Hughes TL, Zhang SC. Quantum spin Hall effect and topological phase transition in HgTe quantum wells. Science 2006;314:1757.
[3] Berry MV. Quantal phase factors accompanying adiabatic changes. Proc R Soc London, A Math Phys Sci 1984;392(1802):45–57.
[4] Bott R. An application of the morse theory to the topology of lie groups. Bull Soc Math France 1956;84:251–81.
[5] Chen X, Liu ZX, Wen XG. Two-dimensional symmetry-protected topological orders and their protected gapless edge excitations. Phys Rev B 2011;84(23):235141.
[6] Chen Xie, Gu Zheng-Cheng, Liu Zheng-Xin, Wen Xiao-Gang. Symmetry protected topological orders and the cohomology class of their symmetry group. 2011. e-print arxiv:1106.4772.
[7] Cho GY, Moore JE. Topological bf field theory description of topological insulators. Ann Phys 2011.
[8] Essin Andrew M, Gurarie Victor. Bulk-boundary correspondence of topological insulators from their respective green's functions. Phys Rev B 2011;84:125132.
[9] Essin Andrew M, Moore Joel E, Vanderbilt David. Magnetoelectric polarizability and axion electrodynamics in crystalline insulators. Phys Rev Lett 2009;102:146805.
[10] Essin Andrew M, Turner Ari M, Moore Joel E. Vanderbilt David. Orbital magnetoelectric coupling in band insulators. Phys Rev B 2010;81:205104.
[11] Fu L, Kane CL. Topological insulators with inversion symmetry. Phys Rev B 2007;76:045302.
[12] Fu L, Kane CL, Mele EJ. Topological insulators in three dimensions. Phys Rev Lett 2007;98:106803.
[13] Garate Ion, Franz M. Inverse spin-galvanic effect in the interface between a topological insulator and a ferromagnet. Phys Rev Lett 2010;104:146802.
[14] Goldstone J, Wilczek F. Phys Rev Lett 1981;47:986.
[15] Golterman Marten FL, Jansen Karl, Kaplan David B. Phys Lett B 1993;301:219.
[16] Gurarie V. Single-particle green's functions and interacting topological insulators. Phys Rev B 2011;83:085426.
[17] Haldane, FDM, Liang Chen. Magnetic flux of vortices on the two-dimensional hall surface. Phys Rev Lett 1984;53:2591.

[18] Jackiw R. Fractional charge and zero modes for planar systems in a magnetic field. Phys Rev D 1984;29:2375–7.
[19] Jackiw R, Rebbi C. Solitons with fermion number 1/2. Phys Rev D 1976;13:3398.
[20] Kaluza T. Sitzungsber Preuss Akad Wiss Berlin (Math Phys) 1921:966.
[21] Karch A. Electric-magnetic duality and topological insulators. Phys Rev Lett 2009;103:171601.
[22] Kitaev A. Periodic table for topological insulators and superconductors. AIP Conf Proc 2009;1134:22.
[23] Klein O. Z Phys 1926;37:895.
[24] Laughlin RB. Quantized hall conductivity in two dimensions. Phys Rev B 1981;23:5632.
[25] Rundong Li, Jing Wang, Qi XL, Zhang SC. Dynamical axion field in topological magnetic insulators. Nat Phys 2010;6:284.
[26] Maciejko J, Qi XL, Karch A, Zhang SC. Fractional topological insulators in three dimensions. Phys Rev Lett 2010;105:246809.
[27] Joseph Maciejko, Xiao-Liang Qi, Dennis Drew H, Zhang Shou-Cheng. Topological quantization in units of the fine structure constant. Phys Rev Lett 2010;105:166803.
[28] Andrei Malashevich, Ivo Souza, Sinisa Coh, David Vanderbilt. Theory of orbital magnetoelectric response. New J Phys 2010;12:053032.
[29] Mondal S, Sen D, Sengupta K, Shankar R. Tuning the conductance of dirac fermions on the surface of a topological insulator. Phys Rev Lett 2010:104.
[30] Moore JE, Balents L. Topological invariants of time-reversal-invariant band structures. Phys Rev B 2007;75:121306.
[31] Nakahara Mikio. Geometry, topology, and physics. A. Hilger; 1990.
[32] Niemi AJ, Semenoff GW. Axial-anomaly-induced fermion fractionization and effective gauge-theory actions in odd-dimensional space-times. Phys Rev Lett 1983;51:2077–80.
[33] Nomura Kentaro, Nagaosa Naoto. Electric charging of magnetic textures on the surface of a topological insulator. Phys Rev B 2010;82:161401(R).
[34] Pesin DA, MacDonald AH. Topological magneto-electric effect decay. 2012. e-print arxiv:1207.4444.
[35] Xiao-Liang Qi, Taylor Hughes, Shou-Cheng Zhang. Fractional charge and quantized current in the quantum spin Hall state. Nat Phys 2008;4:273.
[36] Xiao-Liang Qi, Taylor Hughes, Shou-Cheng Zhang. Topological Field Theory of Time-Reversal Invariant Insulators. Phys Rev B 2008;78:195424.
[37] Xiao-Liang Qi, Hughes Taylor L, Raghu S, Shou-Cheng Zhang. Time-reversal-invariant topological superconductors and superfluids in two and three dimensions. Phys Rev Lett 2009;102:187001.
[38] Xiao-Liang Qi, Rundong Li, Jiadong Zang, Shou-Cheng Zhang. Inducing a magnetic monopole with topological surface states. Science 2009;323:1184.
[39] Qi Xiao-Liang, Witten Edward, Zhang Shou-Cheng. Axion topological field theory of topological superconductors. 2012. e-print arxiv:1206.1407.
[40] Xiao-Liang Qi, Shou-Cheng Zhang. Topological insulators and superconductors. Rev Mod Phys 2011;83:1057–110.
[41] Rosenberg G, Franz M. Witten effect in a crystalline topological insulator. Phys Rev B 2010;82:035105.
[42] Rosenberg G, Guo HM, Franz M. Wormhole effect in a strong topological insulator. Phys Rev B 2010;82:041104(R).
[43] Roy R. Z2 classification of quantum spin hall systems: an approach using time-reversal invariance. Phys Rev B 2009;79:195321.
[44] Roy Rahul. 2008. e-print arxiv:0803.2868.
[45] Ryu Shinsei, Moore Joel E, Ludwig Andreas WW. Electromagnetic and gravitational responses and anomalies in topological insulators and superconductors. Phys Rev B 2012;85:045104.
[46] Schnyder Andreas P, Ryu Shinsei, Furusaki Akira, Ludwig Andreas WW. Classification of topological insulators and superconductors in three spatial dimensions. Phys Rev B 2008;78:195125.
[47] Seradjeh B, Moore JE, Franz M. Exciton condensation and charge fractionalization in a topological insulator film. Phys Rev Lett 2009;103:066402.

[48] Swingle B, Barkeshli M, McGreevy J, Senthil T. Correlated topological insulators and the fractional magnetoelectric effect. Phys Rev B 2011;83:195139.
[49] Volovik GE. JETP Lett 2002;75:63.
[50] von Keyserlingk CW, Burnell FJ, Simon SH. Three-dimensional topological lattice models with surface anyons. 2012. Arxiv preprint arxiv:1208.5128.
[51] von Klitzing K, Dorda G, Pepper M. Phys Rev Lett 1980;45:494.
[52] Walker K, Wang Z. (3+ 1)-tqfts and topological insulators. Front Phys 2011:1–10.
[53] Zhong Wang, Xiao-Liang Qi, Shou-Cheng Zhang. Equivalent topological invariants of topological insulators. New J Phys 2010;12:065007.
[54] Zhong Wang, Xiao-Liang Qi, Shou-Cheng Zhang. Topological order parameters for interacting topological insulators. Phys Rev Lett 2010;105:256803.
[55] Zhong Wang, Xiao-Liang Qi, Shou-Cheng Zhang. Topological field theory and thermal responses of interacting topological superconductors. Phys Rev B 2011;84:014527.
[56] Zhong Wang, Xiao-Liang Qi, Shou-Cheng Zhang. Topological invariants for interacting topological insulators with inversion symmetry. Phys Rev B 2012;85:165126.
[57] MacDonald AH. Tse Wang-Kong. Giant magneto-optical kerr effect and universal faraday effect in thin-film topological insulators, Phys Rev Lett 2010;105:057401.
[58] Wen X-G. Phys Rev B 1991;44:5708.
[59] Wen XG, Niu Q. Ground-state degeneracy of the fractional quantum hall states in the presence of a random potential and on high-genus riemann surfaces. Phys Rev B 1990;41:9377–96.
[60] Wess J, Zumino B. Consequences of anomalous ward identities. Phys Lett B 1971;37(1):95–7.
[61] Witten E. Phys Lett B 1979;86:283.
[62] Witten E. Global aspects of current algebra. Nucl Phys B 1983;223(2):422–32.
[63] Hong Yao, Dung-Hai Lee. Topological insulators and topological nonlinear σ models. Phys Rev B 2010;82:245117.
[64] Yazyev Oleg V, Moore Joel E, Louie Steven G. Spin polarization and transport of surface states in the topological insulators bi_2se_3 and bi_2te_3 from first principles. Phys Rev Lett 2010;105(26):266806.
[65] Takehito Yokoyama, Yukio Tanaka, Naoto Nagaosa. Anomalous magnetoresistance of a two-dimensional ferromagnet/ferromagnet junction on the surface of a topological insulator. Phys Rev B 2010;81(12):121401.
[66] Takehito Yokoyama, Jiadong Zang, Naoto Nagaosa. Theoretical study of the dynamics of magnetization on the topological surface. Phys Rev B 2010;81(24):241410.
[67] Zhang SC, Int J. Mod Phys B 1992;6:25.
[68] Zhang SC, Hu JP. A four-dimensional generalization of the quantum Hall effect. Science 2001;294:823.

Part II

Experimental Discoveries

Quantum Spin Hall State in HgTe

C. Brüne, H. Buhmann and L.W. Molenkamp
Physikalisches Institut, Experimentelle Physik 3, Universität Würzburg, Würzburg, Germany

Chapter Outline Head

1 HgTe QUANTUM WELLS

First we focus on the bandstructure of bulk HgTe and HgTe quantum wells, which distinguishes these systems from most other semiconductors. HgTe is a zincblende II–VI material. The bonds in the material are formed between the 6s electrons from the Hg atoms and 5p electrons from Te. Consequently the bands in the crystal which are close to the Fermi energy will evolve from these energy levels. This combination of s- and p-states is common to most conventional zincblende semiconductors. HgTe however is special in terms of the energetic position of the resulting energy bands. This is because both Hg and Te are relatively heavy atoms, so that relativistic corrections to the positions of the energy levels become very important. Figure 1 schematically shows how the positions of the energy bands develop in HgTe (on the left) compared to those in CdTe (on the right) when applying these relativistic corrections [1]. The corrections to the unperturbed Hamiltonian H_0 are visualized in the following order from left to right: Darwin term H_D, the relativistic mass velocity

Topological Insulators. http://dx.doi.org/10.1016/B978-0-444-63314-9.00005-6

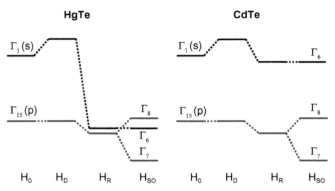

FIGURE 1 The evolution of the main energy bands in HgTe and CdTe. The impact of relativistic corrections onto the band positions is visualized. From left to right the influence of the Darwin term (H_D), the mass velocity correction (H_R), and finally the spin-orbit interaction (H_{SO}) is shown. The very strong mass velocity correction for the Hg s-states leads to a band inversion in HgTe. Adapted from [1].

correction H_R, and finally the spin-orbit coupling correction H_{SO}. While the Darwin correction is qualitatively similar for both compounds the relativistic mass velocity correction is quite different. This is caused by the difference in atomic masses and core charges in Hg and Cd. The mass velocity correction for HgTe is so strong that the energy position of the Γ_6 state (originating from the Hg s-states) is lowered to nearly the same level as those of the Te p-states. Finally, the spin-orbit interaction will split the p-states into the Γ_8 and Γ_7 states. As a result, the Γ_8 band is lifted above the Γ_6 state and we end up with the inverted band structure that distinguishes HgTe from most other materials. The spin-orbit splitting in CdTe is similar to that in HgTe, since it occurs in the Te p-states, which are the same for both materials. But since H_R in CdTe is much smaller than in HgTe, CdTe exhibits a normal band ordering with the Γ_6 state being the first conduction band state and the Γ_8 marking the first valence band state. The inversion is therefore a result of the interplay between H_R and H_{SO} in HgTe.

While the band inversion makes HgTe a topological material, it also has another consequence: Since the Γ_8 state consists of the degenerate light- and heavy-hole bands while the Γ_6 state only forms a single band, bulk HgTe has a semimetallic band structure, see Fig. 2 on the left [2,3]. The Γ_6 band now is a valence band at an energy ≈ 300 meV below the Γ_8 light- and heavy-hole bands. Since the filling of the Γ_6 states only compensates one of the Γ_8 bands, the heavy-hole band remains occupied and thereby a valence band, while the light-hole band now becomes the conduction band.

Bulk HgTe thus is a topological semimetal. To turn it into a topological insulator it is therefore necessary to lift the Γ_8 degeneracy and open up a band gap. In general this can be achieved by lowering the point group symmetry.

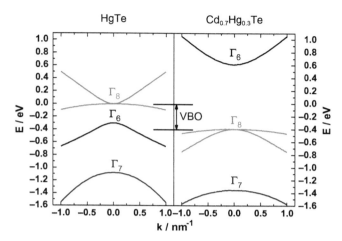

FIGURE 2   Band structure around the Γ-point for HgTe (left) and $Cd_{0.7}Hg_{0.3}Te$ on the right.

For thin, 2-dimensional layers, this is easily realized by growing quantum well structures. In three dimensions, the degeneracy can be lifted by utilizing growth induced strain [4].

For experiments on 2-dimensional topological insulators we use $Cd_{0.7}Hg_{0.3}Te/HgTe/Cd_{0.7}Hg_{0.3}Te$ quantum well heterostructures. The band structure for $Cd_{0.7}Hg_{0.3}Te$ and HgTe at the Γ-point is shown in Fig. 2. $Cd_{0.7}Hg_{0.3}Te$ exhibits a conventional band structure with Γ_6 above Γ_8 and a band gap of ≈ 1 eV. The resulting quantum wells are so-called type III quantum wells with a characteristic band edge profile as shown in the inset of Fig. 3 [5]. In these quantum wells the band structure can be tuned over a wide range by changing the confinement strength through the well thickness. In Fig. 3 the evolution of the subband energy versus the well thickness is shown. Subbands depicted in red act as conduction bands in the corresponding region while those in blue are valence bands. For wide quantum wells, the confinement energy is small and the subband ordering retains the band inversion of bulk HgTe with the H_1 subband (originating from the Γ_8 states) being the first conduction band and the E_1 subband (originating from the Γ_6 band) being one of the valence bands. For thin quantum wells however E_1 becomes the first conduction band and H_1 the first valence band [6,7]. These two regimes correspond to the topologically trivial case (thin quantum wells) and the topological insulator case (thick quantum wells). Since the topology changes between these regimes, they have to be connected by a state with a vanishing band gap. For a critical thickness of 6.3 nm the band gap collapses and a zero-gap state is formed [7–9]. These characteristics will enable us to test the 2-dimensional topological insulator regime by comparing its characteristics with those of the normal regime. This will be the content of the next section.

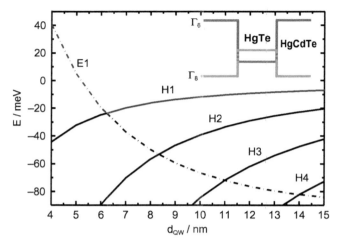

FIGURE 3   Subband energy against quantum well thickness for HgTe quantum wells. Subbands originating from Γ_6 states are labeled as E_i (dashed) and those evolving from the Γ_8 bands with H_i (straight lines). The colors indicate regions in which the subbands act as conduction band (red) and valence band (blue). HgTe quantum wells are so-called type III quantum wells. The inset shows the schematic band edge profile for such a system. (For interpretation of the references to color in this figure legend, the reader is referred to the web version of this book.)

2 THE QUANTUM SPIN HALL EFFECT

The quantum spin Hall effect is the signature state of a 2-dimensional topological insulator and it describes the existence of edge states on the sides of a 2-dimensional topological insulator system. These edge states are supposed to consist of two counterpropagating oppositely spin polarized edge channels in the band gap of the material [2,10,11]. Figure 4 shows a cartoon picture of the quantum spin Hall effect in a HgTe quantum well system.

To find evidence for these edge states we perform transport measurements on HgTe quantum well devices [12,13]. The $Cd_{0.7}Hg_{0.3}Te$/HgTe quantum well structures are fabricated by molecular beam epitaxy on $Cd_{0.96}Zn_{0.04}Te$ substrates [14]. The quantum well structures are modulation doped (by Iodine doping of the barriers) such that the initial carrier densities are in the low $10^{11}\,cm^{-2}$ regime. Figure 5 shows a schematic drawing of the layer sequence of a symmetric quantum well structure with typical layer thicknesses. The electron mobilities are typically of the order of $3 \times 10^5\,cm^2\,V^{-1}\,s^{-1}$ or higher. The measurements were carried out on 6-terminal Hallbar devices; a device schematic is shown as inset in Fig. 6. These were patterned with low temperature optical and e-beam lithography processes onto the heterostructures. The samples are fitted with top gate structures consisting of a Ti/Au layer. Since HgTe, like all II–VI semiconductors, exhibits a leaky Schottky barrier when connected to a metal, the gates are insulated from the quantum well

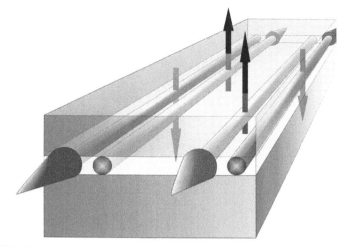

FIGURE 4   Cartoon picture of the quantum spin Hall effect in HgTe quantum wells.

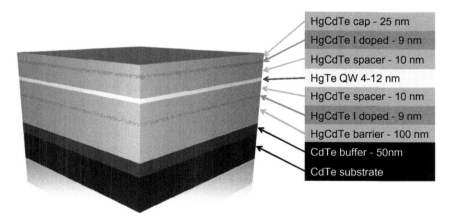

HgCdTe cap - 25 nm

HgCdTe I doped - 9 nm

HgCdTe spacer - 10 nm

HgTe QW 4-12 nm

HgCdTe spacer - 10 nm

HgCdTe I doped - 9 nm

HgCdTe barrier - 100 nm

CdTe buffer - 50nm

CdTe substrate

FIGURE 5   Schematic layer sequence of a $Cd_{0.7}Hg_{0.3}Te/HgTe/Cd_{0.7}Hg_{0.3}Te$ quantum well structure used in the experiments.

structure by a SiO_2/Si_3N_4 multilayer insulator film. The gate electrodes can be used to tune the carrier density continuously from n-conductance through the band gap to p-conductance and to modify the electric field across the well. Transport measurements are done at temperatures of 30 mK in a $^3He/^4He$-dilution refrigerator fitted with a 18 T magnet and at 1.8 K in a ^4He cryostat fitted with a 10 T magnet.

To find evidence for the existence of the quantum spin Hall effect we measure the longitudinal resistance of the Hallbar while shifting the Fermi

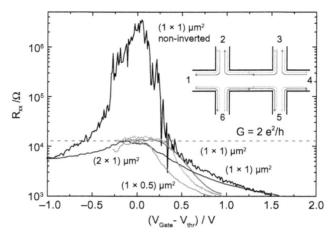

FIGURE 6 Four-terminal longitudinal resistance of HgTe quantum well structures in the inverted and normal regime. The black trace was taken on a sample in the normal regime while the red blue and green curves correspond to samples in the inverted state. The dimensions of the Hallbar devices are given by the labels in their corresponding colors. The inset shows a schematic drawing of a Hallbar structure with the quantum spin Hall edge states.

level through the band gap by using the top gate. In a 2-dimensional topological insulator system one expects to detect a quantized conductance/resistance when the Fermi energy is located inside the band gap. In a conventional system the conductance should drop to zero/ the resistance should rise to very large values in this situation. The measurements have been performed on small-sized Hallbars with dimensions of some few μm. This size is chosen to ensure that the system dimensions are below the inelastic mean free path since inelastic scattering events could introduce backscattering in the helical edge states [12, 15]. The measurements are performed in a typical 4-terminal geometry with the current being applied between contacts 1 and 4 and the longitudinal voltage detected between the contacts 2 and 3 or 5 and 6, respectively.

The resulting data is shown in Fig. 6[1]. Curves for the following samples are displayed: The black curve has been obtained from a Hallbar sample with dimensions of $L \times W = (1 \times 1) \mu m^2$ patterned from a 5.5 nm thick quantum well. The results for a 7.3 nm thick quantum well are shown in green, using a Hallbar with $L \times W = (1 \times 0.5) \mu m^2$, and in red and orange for $L \times W = (1 \times 1) \mu m^2$. Finally, the blue curve represents the results obtained on a sample with a length to width ratio of $(2 \times 1) \mu m^2$ fabricated from a 7.5 nm thick quantum well. In a control experiment, we observe a high resistance for the sample with

---

[1] Please note that Figs. 6, 10 will appear in B/W in print and color in the web version. Based on this, please approve the footnote 1 which explains this.

conventional band ordering (black). In this sample, the resistance rises above $10^6\Omega$ when the Fermi energy crosses the band gap indicating a conventional insulator behavior. For samples with an inverted band structure, however, the resistance stays finite in the band gap with values close to $h/2e^2$. This is the expected value for transport through a helical edge state system in the measured configuration (a detailed explanation will follow in the next section). These results thus provide first evidence for the existence of the quantum spin Hall effect in inverted HgTe quantum wells.

Apart from the observation of a quantized resistance the experiment additionally provides an indication that the observed effect is caused by edge channel transport. As one can see from the measurements in Fig. 6, the value of $h/2e^2$ is reached independent of the sample geometry of the Hallbars. This is typical for edge transport and cannot be explained by normal diffusive transport behavior. While this result provides the first evidence for the existence of edge state transport in inverted HgTe quantum well structures, we so far did not present evidence for the proposed helical nature of these edge channels. To do so we performed a set of non-local transport experiments in the quantum spin Hall regime. These experiments will be described in the following section.

3 NON-LOCAL TRANSPORT IN THE QUANTUM SPIN HALL STATE

The measurements presented in the last section have been performed in a standard 4-terminal geometry and we observe a conductance of $2e^2/h$ when measuring the longitudinal resistance. At first glance, when compared to similar measurements on quantum Hall systems this is a somewhat surprising observation. For chiral quantum Hall edge states, a 4-terminal longitudinal measurement will yield a vanishing longitudinal resistance. This difference can be understood when applying the Landauer-Büttiker quantum transport formalism [16,17] to the helical edge state system of the quantum spin Hall effect. In the Landauer-Büttiker formalism, the relation between the current and the voltage is described as:

$$I_i = \frac{e^2}{h} \sum_j (T_{ji}V_i - T_{ij}V_j). \tag{1}$$

In this equation, I_i denotes the current flowing out of the ith contact into the sample region, V_i is the voltage on the ith contact, and T_{ij} is the transmission probability between contact i and j. To ensure that the total current is conserved, one demands that $\sum_i I_i = 0$; the voltage leads can be defined by setting the net current on the corresponding contact to zero.

Solving this equation for a general 2-dimensional sample can be complicated since the number of conduction channels scales with the sample width, which will lead to a complex and non-universal transmission matrix T_{ij}. For edge channel transport, however, the equation is significantly simplified. For example, in the $\nu = 1$ quantum Hall state only the elements $T_{i+1,i} = 1$ remain while all other elements vanish. This is due to the chiral nature of the quantum Hall edge states allowing transport only in one direction along one edge of the sample. Edge states for electrons moving in the opposite direction are located on the opposite sides of the sample. If we now apply this formula to the quantum Hall case and calculate the expected resistance for a 4-terminal measurement of the longitudinal resistance on a 6-terminal Hallbar we indeed get $R_{14,23} = 0$.

In the case of the quantum spin Hall effect, however, the result will be different, since there are two counterpropagating channels on each side of the sample (as schematically shown in the inset of Fig. 6). The transmission of each edge state is still perfect, as in the quantum Hall case, since the edge states are protected against backscattering by time reversal symmetry. But since there is a forward and backward moving edge channel on each side of the sample, the non-vanishing edge transmission matrix elements are now given by

$$T_{i+1,i} = T_{i,i+1} = 1. \tag{2}$$

The result for the 4-terminal resistance in such a system is $R_{14,23} = h/2e^2$ which corresponds to the values measured in the experiments shown in the previous section. This result also implies that in the quantum spin Hall regime, all contacts to the mesa act as a source for dissipation. (This is in contrast to the situation in the quantum Hall regime where contacts downstream from the voltage probes do not influence the potentials at the voltage probes themselves.)

Contacts are metallic regions with a quasi-infinite number of low energy degrees of freedom available with which the quantum spin Hall states can interact. This will introduce irreversibility on the macroscopic level, break the time reversal symmetry, and destroy the phase coherence. This effectively broken time reversal symmetry inside the contacts lifts the protection against dissipation of the quantum spin Hall states [15].

The expected differences between chiral and helical edge states can be further tested in non-local transport experiments [15]. These experiments are conducted on samples which have been specially designed to observe non-local signals (schematically shown in the insets of Figs. 7 and 8).

Figure 7 shows a measurement in a fully non-local configuration on a device which for obvious reasons we refer to as an "H-bar." In this experiment the current is applied along one part of the sample between contacts 1 and 4, while the voltage is measured between contacts 2 and 3 on the other leg. The measured non-local signal reaches ≈ 6.5 kΩ when the Fermi energy is located inside the band gap. This result agrees very well with the expected conductance value of $4e^2/h$ or 6.45 kΩ in resistance from Landauer-Büttiker calculations. This data

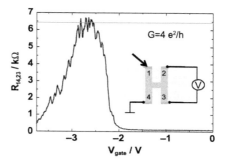

FIGURE 7 Non-local measurement on a 4-terminal sample. The green line indicates the expected conductance value of $4e^2/h$. (For interpretation of the references to color in this figure legend, the reader is referred to the web version of this book.)

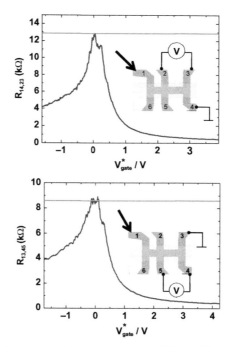

FIGURE 8 Non-local measurements in the quantum spin Hall state. The schematic sample layout and the measured contact configurations are displayed as insets. The green lines indicate the expected quantized conductance values. (For interpretation of the references to color in this figure legend, the reader is referred to the web version of this book.)

provides direct evidence for the edge channel transport; neither diffusive nor ballistic transport will show such a large non-local signal.

A striking difference between chiral (quantum Hall) and helical (quantum spin Hall) edge states arises when one uses the two measurement configurations shown in Fig. 8. In the upper panel a measurement in a typical longitudinal configuration is presented (same data as the blue trace in Fig. 6). The current is applied between the contacts 1 and 4 while the voltage is probed between contacts 2 and 3. This measurement yields a conductance of $2\,e^2/h$ as expected from the Landauer-Büttiker model. If one now applies the current between contacts 1 and 3 and measures the voltage between contacts 4 and 5 (bottom inset) we measure a different value of $3\,e^2/h$. This result is consistent with calculations for helical edge states and distinguishes the quantum spin Hall from the quantum hall system. For chiral edge channels (as well as in diffusive transport) there is only one longitudinal resistivity and these two configurations should deliver the same resistance values. From these insights and observations we can understand the importance of contacts and contact configurations for resistance measurements in the quantum spin Hall state.

The Landauer-Büttiker picture can also be used to explain why relatively small samples are needed to observe these effects. For this we can consider the influence of potential fluctuations on the quantum spin Hall state. In our samples a main source for potential fluctuations are interface states between the gate oxide and the sample surface. These interface states screen the gate potential locally and thereby lead to potential fluctuations in the 2-dimensional electron gas. A large enough potential fluctuation will shift the conduction (valence) band edge below (above) the Fermi energy locally and a metallic puddle will form. If this puddle is large enough and has sufficient spin-orbit coupling it can lead to inelastic scattering and loss of spin information thus effectively acting in the same way as a contact. A cartoon picture showing the influence of a potential fluctuation on the quantum spin Hall state is shown in Fig. 9.

The three panels show the impact of a metallic puddle (gray)—growing in size from top to bottom—on the transport properties of the quantum spin Hall state. The upper panel shows the situation for small metallic puddles. If these puddles are small enough they will not allow for inelastic scattering and the loss of phase coherence. Thus transport will only happen in the two edge channels connecting the macroscopic contacts in the direction of the applied voltage and the transport will be completely phase coherent (denoted by solid lines). Larger puddles will allow for partial dephasing and backscattering (middle panel). Incoherent transport is visualized as dotted lines. Even larger metallic puddles (lower panel) will lead to completely incoherent transport thus leading to a fully dissipative behavior similar to the situation in macroscopic contacts. Additional in-depth discussion and examples can be found in the supplementary material of reference [15].

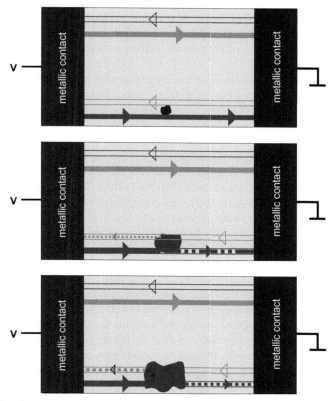

FIGURE 9   Cartoon picture showing the impact of potential fluctuations on the quantum spin Hall effect. For large enough potential fluctuations a metallic puddle is formed (gray). The three panels show the influence of a metallic puddle (growing in size from top to bottom) on the edge channels.

4 SPIN POLARIZATION OF THE QUANTUM SPIN HALL EDGE STATES

So far we obtained evidence for the existence of the quantum spin Hall effect and its edge state nature. The spin polarization of the edge states however still needs confirmation. This will be the topic of the present section.

The direct detection of a magnetic field generated by the spin polarized carriers inside a helical edge state is challenging. The magnetic field generated by a current flowing through the edge state exceeds the component originating from the spin polarized electrons [18]. This problem can be circumvented by converting the magnetic into an electric signal. This can be achieved by utilizing the spin Hall effect in a degenerate (metallic) semiconductor and its counterpart, the inverse spin Hall effect. The spin Hall effect is the appearance of a spin current flowing perpendicular to a charge current in systems where spin-orbit interaction is present [19,20]. In the case of the inverse spin Hall effect, the Onsager counterpart of the spin Hall effect, a spin current is transformed into a charge current.

HgTe quantum wells feature strong spin-orbit interaction effects due to their narrow band gaps and the large atomic spin-orbit coupling. This leads to a large Rashba splitting in HgTe quantum wells when the Fermi energy is located in the conduction or valence band and a perpendicular electric field is applied across the structure [21]. Such strong Rashba effects enable the observation of the ballistic intrinsic spin Hall effect [22–24]. We have performed experiments on metallic H-bars in which we were able to measure non-local transport signals that originate from the intrinsic spin Hall effect. These experiments show spin Hall related signals when the Fermi energy is located in the valence band while the response is an order of magnitude lower for transport occurring in the conduction band [25]. This behavior is expected and due to the difference in Rashba splitting in conduction and valence band [5].

For an all electrical detection of the spin polarization of the quantum spin Hall states we now utilize this metallic spin Hall effect [26]. The device design is once again similar to the H-bar design used to detect non-local signals shown in the previous section (Fig. 8). The H-bar is now fitted with two top gates, one for each leg of the structure, so that we can tune the carrier density independently in both legs. Figure 10 shows an electron micrograph of the split gate structure used in the experiments, the sample dimensions are displayed in yellow.

The experiment follows two main approaches: The first approach is to use the quantum spin Hall channels to detect the spin polarized currents from the spin Hall effect. The measurement configuration for this approach is schematically drawn in Fig. 11. The upper leg between contacts 1 and 2 will be tuned into the quantum spin Hall state. In the lower leg the Fermi energy is located in the metallic regime of either the conduction or valence band. If we now apply a current in the lower leg between contacts 3 and 4, this current will lead to a spin polarized current flowing perpendicular to the charge current. This spin current causes a difference in chemical potential for the two spin states and the quantum spin Hall states will couple selectively to the potential of the matching edge channel. The potential difference will thereby be transferred to contacts 1 and 2 and should lead to a non-local voltage signal that can be detected in the experiment. Non-local signals can only be observed when the metallic regions exhibit a spin Hall effect and the quantum spin Hall states are spin polarized or otherwise the non-local voltage will be zero. Note that the above experiment is a direct demonstration of the magnetoelectric effect of the topological surface state in a 2-dimensional system [27]. The detection of non-local signals is therefore evidence for the spin polarization of the quantum spin Hall states. We can furthermore test if the spin Hall effect signal strength differs in the n- and p-conducting regime as one would expect from the different spin-orbit splitting strengths. Finally we can check if we are able to observe a strong non-local

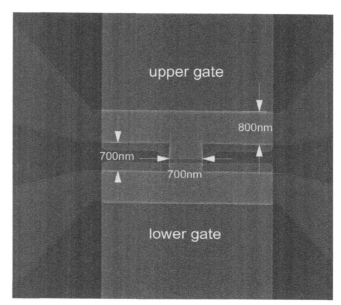

FIGURE 10 Electron micrograph of a split gate H-structure used in the experiments. Dimensions of the sample are shown in yellow.

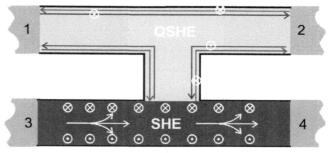

FIGURE 11 Concept for utilizing the quantum spin Hall state as spin polarization detector. Here the spin Hall effect, induced by a current flowing in the lower leg, leads to a spin accumulation on the edges of the leg which can be transformed into a non-local signal by the spin polarized quantum spin Hall states.

signal if we tune the Fermi level into the gap in both legs of the structure, similar to the experiments we described in the last section.

During the measurement, the gate voltage on the lower leg is held constant while the upper gate is swept such that the Fermi level changes from n-type metallic through the band gap to p-type metallic. This means the spin injection properties from the lower leg (where the current is applied) remain unchanged while we vary the upper gate voltage to investigate the spin detection capabilities of the different regimes.

Figure 12 shows the resulting non-local resistance signals for this measurement configuration. The measurement was performed for three different, fixed, gate voltages on the lower gate. Panel (a) shows the non-local resistance for metallic p-type transport and panel (b) for metallic n-type transport in the lower leg. These measurements show a significant increase in the non-local signal and the occurrence of a pronounced maximum when the gate voltage is swept through the quantum spin Hall regime. The maximum non-local resistance that can be observed is an order of magnitude higher for injection from the p-conducting regime. This is again in agreement with the difference in spin Hall effect strength between valence and conduction band as mentioned above. The existence of these large non-local signals when the detection occurs through quantum spin Hall states is evidence for the spin polarization of the quantum spin Hall edge states. These signals can only arise if the metallic region exhibits a spin Hall effect which can be detected by spin polarized edge states.

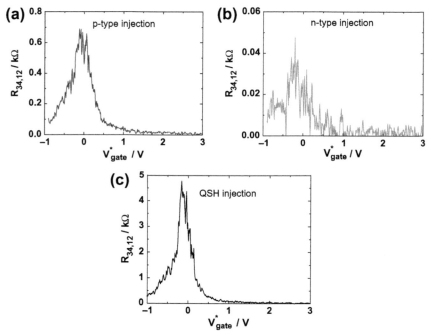

FIGURE 12 Non-local resistance data for measurement configuration 1. Current is applied on the lower leg while the non-local voltage is measured on the upper leg. Panel (a) shows the signal for a p-type lower leg. In (b) the lower leg is n-conducting while (c) corresponds to a Fermi level inside the band gap.

As a further test we can measure the non-local signals that develop when both injection and detection happen in the quantum spin Hall regime. The result is shown in Fig. 12(c). Since this measurement configuration is similar to the non-local resistance measurements presented in the previous section, one would expect to observe a strong non-local resistance in this measurement as a signature of edge channel transport in the whole sample. As can be seen in panel (c) this is indeed the case and the non-local resistance reaches values of several kΩ. This value is slightly lower than the value one would expect for perfect quantization, which is not surprising due to the use of two separate gates in this experiment. In the region between the two gates (see also 10) the gating effect will be weak and metallic puddles can form. These will act as dephasing centers and lead to additional backscattering as described in the previous section.

The second experimental approach focuses on using the quantum spin Hall channels to inject a spin polarized current. A schematic layout for this experiment is shown in Fig. 13. Here the current is applied in the upper leg between the contacts 1 and 2 and the edge channels will inject a spin polarized current into the lower leg. This spin polarized current will lead to a different chemical potential for spin up and spin down electrons. The spin polarized current will lead to a voltage difference between contacts 3 and 4 due to the inverse spin Hall effect [23]. Due to the Onsager-Casimir symmetry relations for non-local resistances this setup should yield similar results to the first approach [16,28]. The Onsager-Casimir symmetry relation for a 4-terminal device is

$$R_{mn,kl}(\mathbf{B}) = R_{kl,mn}(-\mathbf{B}). \tag{3}$$

Current probes are denoted by mn and voltage probes by kl while B refers to the applied magnetic field. Since the magnetic field is zero in our experiment we expect $R_{34,12} = R_{12,34}$.

In the experiment we apply the gate voltages in the same way as before. This means we can now modify the injection properties by scanning the upper gate voltage while the detection of non-local signals happens with a fixed gate

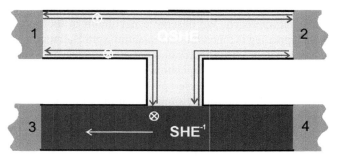

FIGURE 13 Concept for using the quantum spin Hall state as spin injector. If a current is injected in the upper leg the spin polarized edge states will induce a spin current in the lower leg. This spin current will be transformed into a non-local voltage by the inverse spin Hall effect.

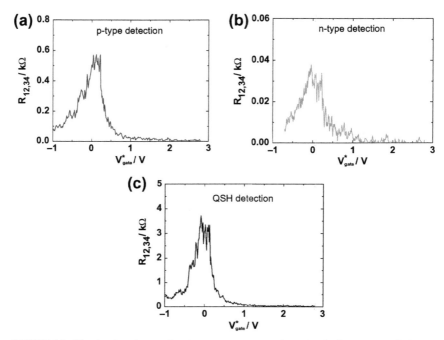

FIGURE 14 Non-local resistance data for measurement configuration 1. Current is applied on the upper leg while the non-local voltage is measured on the lower leg. Panel (a) shows the signal for a p-type lower leg. In (b) the lower leg is n-conducting while (c) corresponds to a Fermi level inside the band gap.

voltage on the lower gate. The results for this setup are shown in Fig. 14. Panel (a) and panel (b) show the results for p-type and n-type detection, respectively. The non-local response for detection using the quantum spin Hall state is displayed in panel (c).

As expected we obtain results that resemble the observations in the first experimental approach. We observe a maximum in the non-local resistance when the upper leg is tuned into the quantum spin Hall regime while the lower leg remains metallic. This observation serves as evidence for an injection of spin polarized electrons from the quantum spin Hall edge states which are detected in the lower leg via the inverse spin Hall effect. We again observe an order of magnitude larger non-local signal when the lower leg is tuned into the p-conducting regime compared to the n-conducting case. Additionally, when the lower leg is tuned into the quantum spin Hall regime we can again observe the largest non-local resistance signals since we reach the fully edge state dominated transport.

In summary these two experiments show that transport in the quantum spin Hall states is indeed spin polarized. This provides the last missing experimental evidence that the picture of the quantum spin Hall state as counterpropagating spin polarized edge channels is correct.

5 CONCLUSION

This chapter has focused on the experimental verification of the quantum spin Hall effect in 2-dimensional topological insulators and its properties. We use HgTe quantum wells as a model system for a 2-dimensional topological insulator. In this system the band inversion of the Γ_6 and Γ_8 states leads to a topologically non-trivial band structure. Transport studies on these quantum well structures give experimental evidence for the existence of the quantum spin Hall effect. Furthermore non-local transport experiments confirm that the quantum spin Hall state consists of two counterpropagating oppositely spin polarized edge states.

REFERENCES

[1] Chu J, Sher A. Physics and properties of narrow gap semiconductors. New York: Springer; 2008.
[2] Bernevig BA, Zhang SC. Phys Rev Lett 2006;96:106802.
[3] Chadi D, Walter JP, Cohen ML. Phys Rev B 1972;5:3058–64.
[4] Fu L, Kane CL. Phys Rev B 2007;76:045302.
[5] Novik EG, Pfeuffer-Jeschke A, Jungwirth T, Latussek V, Becker CR, Landwehr G, et al. Phys Rev B 2005;72:035321.
[6] Pfeuffer-Jeschke A. Ph.D. thesis. Universität Würzburg; 2000.
[7] Bernevig BA, Hughes TL, Zhang SC. Science 2006;314:1757.
[8] Dai X, Hughes TL, Qi X-L, Fang Z, Zhang S-C. Phys Rev B 2008;77:125319.
[9] Büttner B, Liu CX, Tkachov G, Novik EG, Brüne C, Buhmann H, et al. Nature Phys 2011;7:418.
[10] Kane CL, Mele EJ. Phys Rev Lett 2005;95:226801.
[11] Kane CL, Mele EJ. Phys Rev Lett 2005;95:146802.
[12] König M, Wiedmann S, Brüne C, Roth A, Buhmann H, Molenkamp LW, et al. Science 2007;318:766.
[13] König M, Buhmann H, Molenkamp LW, Hughes TL, Liu CX, Qi XL, et al. J Phys Soc Jpn 2008;77:031007.
[14] Becker C, Brüne C, Schäfer M, Roth A, Buhmann H, Molenkamp LW. Phys Stat Sol 2007. http://dx.doi.org/10.1002/pssc.200775402.
[15] Roth A, Brüne C, Buhmann H, Molenkamp LW, Maciejko J, Qi X-L, et al. Science 2009;325:294.
[16] Büttiker M. Phys Rev Lett 1986;57:1761.
[17] Büttiker M. Phys Rev B 1988;38:9375.
[18] Nowack KC, Spanton EM, Baenninger M, König M, Kirtley JR, Kalisky B, et al. Imaging currents in HgTe quantum wells in the quantum spin Hall regime; 2012. Available from: arXiv:1212.2203. [cond-mat.mes-hall].
[19] Dyakonov MI, Perel VI. Phys Lett A 1971;35:459.
[20] Hirsch JE. Phys Rev Lett 1999;83:1834.
[21] Gui YS, Becker CR, Dai N, Liu J, Qiu ZJ, Novik EG, et al. Phys Rev B 2004;70:115328.
[22] Sinova J, Culcer D, Niu Q, Sinitsyn NA, Jungwirth T, MacDonald AH. Phys Rev Lett 2004;92:126603.
[23] Hankiewicz EM, Molenkamp LW, Jungwirth T, Sinova J. Phys Rev B 2004;70:241301.
[24] Nikolić BK, Sourma S, Zârbo L, Sinova J. Phys Rev Lett 2005;95:046601.

[25] Brüne C, Roth A, Buhmann H, Hankiewicz EM, Molenkamp LW, Maciejko J, et al. Nature
      Phys 2010;6:448.
[26] Brüne C, Roth A, Novik EG, König M, Buhmann H, Hankiewicz EM, et al. Nature Phys
      2012;8:485.
[27] Essin AM, Moore JE, Vanderbilt D. Phys Rev Lett 2009;102:146805.
[28] Büttiker M. IBM J Res Dev 1988;32:317.

Chapter 6

Topological Surface States: A New Type of 2D Electron Systems

M. Zahid Hasan[*,†], Su-Yang Xu[*,‡], David Hsieh[*,§], L. Andrew Wray[*,‡,¶] and Yuqi Xia[*]

[*]Joseph Henry Laboratories, Department of Physics, Princeton University, Princeton, NJ 08544, USA
[†]Princeton Institute for the Science and Technology of Materials, School of Engineering and Applied Science, Princeton University, Princeton, NJ 08544, USA
[‡]Advanced Light Source, Lawrence Berkeley National Laboratory, Berkeley, California, 94305, USA
[§]Department of Physics, California Institute of Technology, Pasadena, CA 91125, USA
[¶]SLAC National Accelerator Laboratory, Stanford, CA 94305, USA

Chapter Outline Head

Topological Insulators. http://dx.doi.org/10.1016/B978-0-444-63314-9.00006-8

143

1 INTRODUCTION

The three-dimensional topological insulator is the first example in nature of a topologically ordered electronic phase existing in bulk solids. Their topological order can be realized at room temperatures without magnetic fields and they can be turned into magnets and exotic superconductors leading to worldwide interest and activity in topological insulators [1–7]. All of the 2D topological insulator examples (Integer Quantum Hall (IQH), Quantum Spin Hall (QSH)) including the fractional one (FQH) involving Coulomb interaction are understood in the standard picture of quantized electron orbits in a spin-independent or spin-dependent magnetic field, the 3D topological insulator defies such description and is a novel type of topological order which cannot be reduced to multiple copies of quantum-Hall-like states. In fact, the 3D topological insulator exists not only in zero magnetic field but differs from the 2D variety in three very important aspects: (1) they possess topologically protected 2D metallic surfaces (Topological Surface States, a new type of 2DEG) rather than the 1D edges, (2) they can work at room temperature (300 K and beyond) rather than cryogenic (sub-K) temperatures required for the QSH effects, and (3) they occur in standard bulk semiconductors rather than at buried interfaces of ultraclean semiconductor heterostructures and thus tolerate stronger disorder than the IQH-like states. One of the major challenges in going from quantum-Hall-like 2D states to 3D topological insulators is to employ new experimental approaches/methods to precisely probe this novel form of topological-order since the standard tools and settings that work for IQH-states also work for QSH-states. The method to probe 2D topological order is exclusively with charge transport. In a 3D topological insulator, the boundary itself supports a 2DEG and transport is not (Z_2) topologically quantized hence *cannot* directly probe the topological invariants ν_o nor the topological quantum numbers analogous to the Chern numbers of the IQH systems. This is *unrelated* to the fact that the present materials have some extrinsic or residual/impurity conductivity in their naturally grown bulk. In this paper, we review the birth of momentum- and spin-resolved spectroscopy as a new experimental approach and as a highly boundary sensitive method to study and prove topological-order via the direct measurements of the topological invariants ν_o that are associated with the Z_2 topology of the spin-orbit band structure and opposite parity band inversions. These experimental methods led to the experimental discovery of the first 3D topological insulator in Bi-Sb semiconductors which further led to the discovery of the Bi_2Se_3 class—the most widely researched topological insulator to this date. We discuss the fundamental properties of the novel topologically spin-momentum locked half Dirac metal on the surfaces of the topological insulators and how they emerge from topological phase transitions due to increasing spin-orbit coupling in the bulk. These methods and their derivatives are now being applied by others worldwide for further finer investigations of topological-order and for discovering new topological insulator states as well as exotic topological quantum phenomena. We also review how spectroscopic methods are leading to the identification of spin-orbit superconductors that may work as Majorana platforms and can be used to identify topological superconductors—yet another class of new states of matter.

2 THE BIRTH OF MOMENTUM-RESOLVED SPECTROSCOPY AS A DIRECT EXPERIMENTAL PROBE OF Z_2 TOPOLOGICAL-ORDER

Ordered phases of matter such as a superfluid or a ferromagnet are usually associated with the breaking of a symmetry and are characterized by a local order parameter [8]. The typical experimental probes of these systems are sensitive to order parameters. In the 1980s, two new phases of matter were realized by subjecting 2D electron gases at buried interfaces of semiconductor heterostructures to large magnetic fields. These new phases of matter, the 2D integer and 2D fractional quantum Hall states, exhibited a new and rare type of order that is derived from an organized collective quantum entangled motion of electrons [9–12]. These so-called "2D topologically ordered insulators" do not exhibit any symmetry breaking and are characterized by a topological number [13] as opposed to a local order parameter. The most striking manifestation of this 2D topological order is the existence of one-way propagating 1D metallic states confined to their edges, which lead to remarkable quantized charge transport phenomena. To date the experimental probe of their topological quantum numbers is based on charge transport, where measurements of the quantization of transverse magneto-conductivity $\sigma_{xy} = ne^2/h$ (where e is the electric charge and h is Planck's constant) reveal the value of the topological number n that characterizes these quantum Hall states [14].

Recently, a third type of 2D topological insulator, the spin quantum Hall insulator, was theoretically predicted [15,16] and then experimentally discovered [17]. This class of quantum-Hall-like topological phases can exist in spin-orbit materials without external magnetic fields, and can be described as an ordinary quantum Hall state in a spin-dependent magnetic field. Their topological order gives rise to counterpropagating 1D edge states that carry opposite spin polarization, often described as a superposition of a spin-up and spin-down quantum Hall edge state (Fig. 1). Like conventional quantum Hall systems, the 2D spin quantum Hall insulator (QSH) is realized at a buried solid interface. The first realization of this phase was made in (Hg,Cd)Te quantum wells using *charge transport* by measuring a longitudinal conductance of about $2e^2/h$ at mK temperatures [17]. The quantum spin Hall state can be thought of as two copies of integer quantum Hall states (IQH) and are protected by a Z_2 invariant.

It was also realized that a fundamentally new type of genuinely three-dimensional topological order might be realized in bulk crystals without need for an external magnetic field [6,18,19]. Such a 3D topological insulator cannot be reduced to multiple copies of the IQH and such phases would be only the

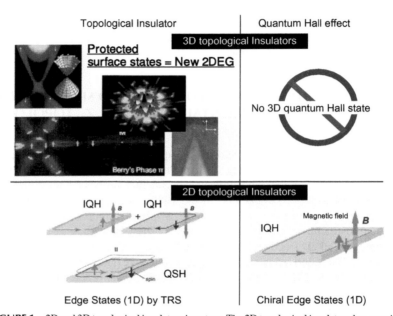

FIGURE 1 2D and 3D topological insulators in nature. The 2D topological insulator class consists of the 2D quantum Hall states (IQH) and 2D quantum spin Hall state (QSH). The latter is constructed from two copies of the former. On the other hand, a 3D quantum Hall state is forbidden in nature, so the 3D topological insulator represents a new type of topologically ordered phase. The protected surface states form a novel type of topological metal (half Dirac metal) where electron's spin is locked to its momentum but exhibit no spin quantum Hall effect. Images are adapted from Refs. [1,3,7,20–22,24,45].

fourth type of topologically ordered phase to be discovered in nature, and the first type to fall outside the quantum-Hall-like 2D topological states (IQH, FQH, QSH). Instead of having quantum-Hall type 1D edge states, these so-called 3D topological insulators would possess unconventional metallic 2D topological surface states called spin-textured helical metals, a type of 2D electron gas long thought to be impossible to realize. However, it was recognized that 3D topological insulators would NOT necessarily exhibit a topologically (Z_2) quantized charge transport by themselves as carried out in the conventional transport settings of all quantum-Hall-like measurements. Therefore, their 3D topological quantum numbers (Z_2), the analogs of n (Chern numbers), could not be measured via the charge transport based methods even if a complete isolation of surface charge transport becomes routinely possible. Owing to the 2D nature of the two surface conduction channels (top and bottom surfaces of a typical thin sample) that contribute together in a 3D topological insulator case, it was theoretically recognized that it would not be possible to measure the topological invariants due to the lack of a quantized transport response of the 2D surface that measures the Z_2 topological invariants [19].

Here we review the development of spin- and angle-resolved photoemission spectroscopy (spin-ARPES) as the new approach/method to probe 3D topological order [20–35], which today constitutes the experimental standard for identifying topological order in bulk solids. 3D topological insulators are also experimentally studied by many others worldwide using various techniques such as ARPES [36–44], scanning tunneling spectroscopies (STM) [45–52], transport [53–64], optical methods [65–72], and even on nano-structured samples [38,55,56]. Here, we will review the procedures for (i) separating intrinsic bulk bands from surface electronic structures using incident photon energy modulated ARPES, (ii) mapping the surface electronic structure across the Kramers momenta to establish the topologically non-trivial nature of the surface states, (iii) using spin-ARPES to map the spin texture of the surface states to reveal topological quantum numbers and Berry's phases, and (iv) measuring the topological parent compounds to establish the microscopic origins of 3D topological order. These will be discussed in the context of $Bi_{1-x}Sb_x$, which was the first 3D topological insulator to be experimentally discovered in nature and a textbook example of how this method is applied. The confluence of three factors, having a detailed spectroscopic procedure to measure 3D topological order, their discovery in fairly simple bulk semiconductors, and being able to work at room temperatures, has led to worldwide efforts to study 3D topological physics and led to over 100 compounds being identified as 3D topological insulators to date.

3 SEPARATION OF INSULATING BULK FROM METALLIC SURFACE STATES USING INCIDENT PHOTON ENERGY MODULATED ARPES

Three-dimensional topological order is predicted to occur in semiconductors with an inverted-band gap, therefore 3D topological insulators are often searched for in systems where a band gap inversion is known to take place as a function of some control parameter. The experimental signature of being in the vicinity of a bulk band inversion is that the bulk band dispersion should be described by the massive Dirac equation rather than Schrödinger equation, since the system must be described by the massless Dirac equation exactly at the bulk band inversion point.

The early theoretical treatments [18,73] focused on the strongly spin-orbit coupled, band-inverted $Bi_{1-x}Sb_x$ series as a possible realization of 3D topological order for the following reason. Bismuth is a semimetal with strong spin-orbit interactions. Its band structure is believed to feature an indirect negative gap between the valence band maximum at the T point of the bulk Brillouin zone (BZ) and the conduction band minima at three equivalent L points [74,75] (here we generally refer to these as a single point, L). The valence and conduction bands at L are derived from antisymmetric (L_a) and symmetric

(L_s) p-type orbitals, respectively, and the effective low-energy Hamiltonian at this point is described by the (3+1)-dimensional relativistic Dirac equation [76–78]. The resulting dispersion relation, $E(\vec{k}) = \pm\sqrt{(\vec{v})^2(\vec{k})^2 + \Delta^2} \approx \vec{v}\cdot\vec{k}$, is highly linear owing to the combination of an unusually large band velocity $\vec{v}$ and a small gap Δ (such that $|\Delta/|\vec{v}|| \approx 5 \times 10^{-3}$ Å^{-1}) and has been used to explain various peculiar properties of bismuth [76–78]. Substituting bismuth with antimony is believed to change the critical energies of the band structure as follows (see Fig. 2). At an Sb concentration of $x \approx 4\%$, the gap Δ between L_a and L_s closes and a massless three-dimensional (3D) Dirac point is realized. As x is further increased this gap re-opens with inverted symmetry ordering, which leads to a change in sign of Δ at each of the three equivalent L points in the BZ. For concentrations greater than $x \approx 7\%$ there is no overlap between the valence band at T and the conduction band at L, and the material becomes an inverted-band insulator. Once the band at T drops below the valence band at L, at $x \approx 8\%$, the system evolves into a direct-gap insulator whose low-energy physics is dominated by the spin-orbit coupled Dirac particles at L [18,74].

High-momentum-resolution angle-resolved photoemission spectroscopy [Fig. 2(a) and (b)] performed with varying incident photon energy (IPEM-ARPES) allows for measurement of electronic band dispersion along various momentum-space ($\vec{k}$-space) trajectories in the 3D bulk BZ. ARPES spectra taken along two orthogonal cuts through the L point of the bulk BZ of Bi$_{0.9}$Sb$_{0.1}$ are shown in Fig. 2(c) and (e). A Λ-shaped dispersion whose tip lies less than 50 meV below the Fermi energy (E_F) can be seen along both directions. Additional features originating from surface states that do not disperse with incident photon energy are also seen. Owing to the finite intensity between the bulk and surface states, the exact binding energy (E_B) where the tip of the Λ-shaped band dispersion lies is unresolved. The linearity of the bulk Λ-shaped bands is observed by locating the peak positions at higher E_B in the momentum distribution curves (MDCs), and the energy at which these peaks merge is obtained by extrapolating linear fits to the MDCs. Therefore 50 meV represents a lower bound on the energy gap Δ between L_a and L_s. The magnitude of the extracted band velocities along the k_x and k_y directions are $7.9 \pm 0.5 \times 10^4$ and $10.0 \pm 0.5 \times 10^5$ ms^{-1}, respectively, which are similar to the tight binding values 7.6×10^4 and 9.1×10^5 ms^{-1} calculated for the L_a band of bismuth [75]. Our data are consistent with the extremely small effective mass of $0.002m_e$ (where m_e is the electron mass) observed in magneto-reflection measurements on samples with $x = 11\%$ [79]. The Dirac point in graphene, co-incidentally, has a band velocity ($|v_F| \approx 10^6$ ms^{-1}) [80] comparable to what we observe for Bi$_{0.9}$Sb$_{0.1}$, but its spin-orbit coupling is several orders of magnitude weaker, and the only known method of inducing a gap in the Dirac spectrum of graphene is by coupling to an external chemical substrate (which is a trivial gap) [81]. The Bi$_{1-x}$Sb$_x$ series thus provides a rare opportunity to study relativistic Dirac Hamiltonian physics in a 3D condensed matter system where the intrinsic (rest) mass gap can be easily tuned.

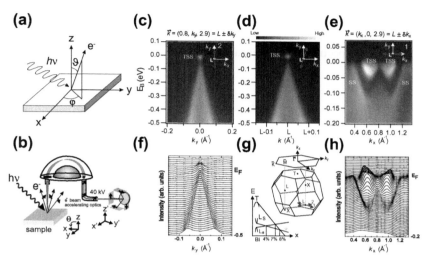

FIGURE 2 The first 3D topological insulator (2007): Dirac-like dispersion signaling band inversion via spin-orbit interaction. (a) Schematic of an ARPES experimental geometry. The kinetic energy of photoelectrons and their angles of emission (θ, ϕ) determine its electronic structure. (b) Energy and momentum analysis takes place through a hemispherical analyzer and the spin analysis is performed using a Mott detector. Selected ARPES intensity maps of $Bi_{0.9}Sb_{0.1}$ are shown along three $\vec{k}$-space cuts through the L point of the bulk 3D Brillouin zone (BZ). The presented data are taken in the third BZ with $L_z = 2.9$ Å^{-1} with a photon energy of 29 eV. The cuts are along (c) the k_y direction, (d) a direction rotated by approximately 10° from the k_y direction, and (e) the k_x direction. Each cut shows a Λ-shaped bulk band whose tip lies below the Fermi level signalling a bulk gap. The (topological) surface states are denoted (T)SS and are all identified in Fig. 3. (f) Momentum distribution curves (MDCs) corresponding to the intensity map in (c). (h) Log scale plot of the MDCs corresponding to the intensity map in (e). The red lines are guides to the eye for the bulk features in the MDCs. (g) Schematic of the bulk 3D BZ of $Bi_{1-x}Sb_x$ and the 2D BZ of the projected (1 1 1) surface. The high symmetry points $\bar{\Gamma}, \bar{M}$, and $\bar{K}$ of the surface BZ are labeled. Schematic evolution of bulk band energies as a function of x is shown. The L band inversion transition occurs at $x \approx 0.04$, where a 3D gapless Dirac point is realized, and the composition we study here (for which $x = 0.1$) is indicated by the green arrow. A more detailed phase diagram based on our experiments is shown in Fig. 4(c). (Adapted from [7,20]).

Studying the band dispersion perpendicular to the sample surface provides a way to differentiate bulk states from surface states in a 3D material. To visualize the near-E_F dispersion along the 3D L-X cut (X is a point that is displaced from L by a k_z distance of $3\pi/c$, where c is the lattice constant), in Fig. 3(a) we plot energy distribution curves (EDCs), taken such that electrons at E_F have fixed in-plane momentum $(k_x, k_y) = (L_x, L_y) = (0.8, 0.0$ Å$^{-1})$, as a function of photon energy ($h\nu$). There are three prominent features in the EDCs: a non-dispersing, k_z independent, peak centered just below E_F at about -0.02 eV; a broad non-dispersing hump centered near -0.3 eV; and a strongly dispersing hump that coincides with the latter near $h\nu = 29$ eV. To understand which bands these features originate from, we show ARPES intensity maps along an in-plane cut $\bar{K}\bar{M}\bar{K}$ (parallel to the k_y direction) taken using $h\nu$ values of 22,

29, and 35 eV, which correspond to approximate k_z values of $L_z - 0.3 \text{ Å}^{-1}$, L_z, and $L_z + 0.3 \text{ Å}^{-1}$, respectively [Fig. 3(b)]. At $h\nu = 29$ eV, the low-energy ARPES spectral weight reveals a clear Λ-shaped band close to E_F. As the

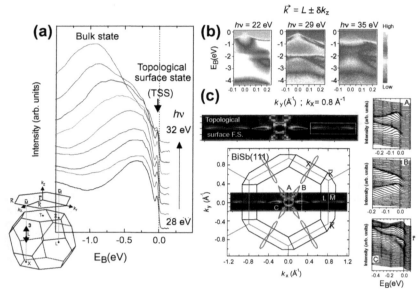

FIGURE 3 The first 3D topological insulator: Topological Surface States and electronic band dispersion along the k_z direction in momentum space. Surface states are experimentally identified by studying their out-of-plane momentum dispersion through the systematic variation of incident photon energy. (a) Energy distribution curves (EDCs) of $Bi_{0.9}Sb_{0.1}$ with electrons at the Fermi level (E_F) maintained at a fixed in-plane momentum of $(k_x = 0.8 \text{ Å}^{-1}, k_y = 0.0 \text{ Å}^{-1})$ are obtained as a function of incident photon energy to identify states that exhibit no dispersion perpendicular to the (1 1 1)-plane along the direction shown by the double-headed arrow labeled "3" in the inset. Selected EDC data sets with photon energies of 28–32 eV in steps of 0.5 eV are shown for clarity. The non-energy dispersive (k_z independent) peaks near E_F are the topological surface states (TSS). (b) ARPES intensity maps along cuts parallel to k_y taken with electrons at E_F fixed at $k_x = 0.8 \text{ Å}^{-1}$ with respective photon energies of $h\nu = 22, 29$, and 35 eV. The faint Λ-shaped band at $h\nu = 22$ eV and $h\nu = 35$ eV shows some overlap with the bulk valence band at L ($h\nu = 29$ eV), suggesting that it is a resonant surface state degenerate with the bulk state in some limited k-range near E_F. The flat band of intensity centered at about -2 eV in the $h\nu = 22$ eV scan originates from Bi $5d$ core level emission from second-order light. (c) Projection of the bulk BZ (black lines) onto the (1 1 1) surface BZ (green lines). Overlay (enlarged in inset) shows the high resolution Fermi surface (FS) of the metallic SS mode, which was obtained by integrating the ARPES intensity (taken with $h\nu = 20$ eV) from -15 to 10 meV relative to E_F. The six tear-drop shaped lobes of the surface FS close to $\bar{\Gamma}$ (center of BZ) show some intensity variation between them that is due to the relative orientation between the axes of the lobes and the axis of the detector slit. The sixfold symmetry was however confirmed by rotating the sample in the $k_x - k_y$ plane. EDCs corresponding to the cuts A, B, and C are also shown; these confirm the gapless character of the surface states in bulk insulating $Bi_{0.9}Sb_{0.1}$. (Adapted from [7,20]). (For interpretation of the references to color in this figure legend, the reader is referred to the web version of this book.)

photon energy is either increased or decreased from 29 eV, this intensity shifts to higher binding energies as the spectral weight evolves from the Λ-shaped into a ∪-shaped band. Therefore the dispersive peak in Fig. 2(a) comes from the bulk valence band, and for $h\nu = 29$ eV the high symmetry point $L = (0.8, 0, 2.9)$ appears in the third bulk BZ. In the maps of Fig. 3(b) with respective $h\nu$ values of 22 and 35 eV, overall weak features near E_F that vary in intensity remain even as the bulk valence band moves far below E_F. The survival of these weak features over a large photon energy range (17–55 eV) supports their surface origin. The non-dispersing feature centered near -0.3 eV in Fig. 3(a) comes from the higher binding energy (valence band) part of the full spectrum of surface states, and the weak non-dispersing peak at -0.02 eV reflects the low-energy part of the surface states that cross E_F away from the $\bar{M}$ point and forms the surface Fermi surface [Fig. 3(c)].

4 WINDING NUMBER COUNT: COUNTING OF SURFACE FERMI SURFACES ENCLOSING KRAMERS POINTS TO IDENTIFY TOPOLOGICALLY NON-TRIVIAL SURFACE SPIN-TEXTURED STATES

Having established the existence of an energy gap in the bulk state of $Bi_{0.9}Sb_{0.1}$ (Figs. 2 and 3) and observed linearly dispersive bulk bands uniquely consistent with strong spin-orbit coupling model calculations [75–78], we now discuss the topological character of its surface states, which are found to be gapless [Fig. 3(c)]. In general, the states at the surface of spin-orbit coupled compounds are allowed to be spin split owing to the loss of space inversion symmetry $[E(k,\uparrow) = E(-k,\uparrow)]$. However, as required by Kramers' theorem, this splitting must go to zero at the four time-reversal invariant momenta (TRIM) in the 2D surface BZ. As discussed in [18,19], along a path connecting two TRIM in the same BZ, the Fermi energy inside the bulk gap will intersect these singly degenerate surface states either an even or an odd number of times. When there are an even number of surface state crossings, the surface states are topologically Z_2 trivial because disorder or correlations can remove *pairs* of such crossings by pushing the surface bands entirely above or below E_F. When there are an odd number of crossings, however, at least one surface state must remain gapless, which makes it non-trivial [18,19,73]. The existence of such topologically non-trivial surface states can be theoretically predicted on the basis of the *bulk* band structure only, using the Z_2 invariant that is related to the quantum Hall Chern number [15]. Materials with band structures with $Z_2 = +1 (\nu_0 = 0)$ are ordinary Bloch band insulators that are topologically equivalent to the filled shell atomic insulator, and are predicted to exhibit an even number (including zero) of surface state crossings. Materials with bulk band structures with $Z_2 = -1 (\nu_0 = 1)$ on the other hand, which are expected

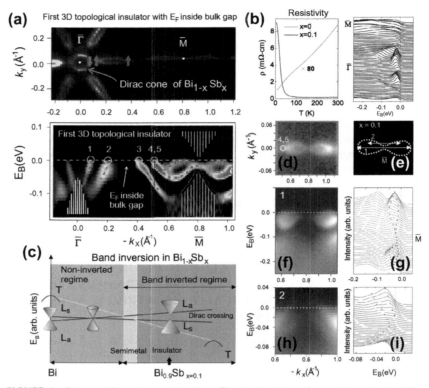

FIGURE 4 The first 3D topological insulator: The topological gapless surface states in bulk insulating (observed via the bulk band dispersion gap) $Bi_{0.9}Sb_{0.1}$. (a) The surface Fermi surface and surface band dispersion second derivative image (SDI) of $Bi_{0.9}Sb_{0.1}$ along $\bar{\Gamma} - \bar{M}$. The shaded white area shows the projection of the bulk bands based on ARPES data, as well as a rigid shift of the tight binding bands to sketch the unoccupied bands above the Fermi level. To maintain high-momentum resolution, data were collected in two segments of momentum space, then the intensities were normalized using background level above the Fermi level. A non-intrinsic flat band of intensity near E_F generated by the SDI analysis was rejected to isolate the intrinsic dispersion. The Fermi crossings of the surface state are denoted by yellow circles, with the band near $-k_x \approx 0.5$ Å^{-1} counted twice owing to double degeneracy. The red lines are guides to the eye. An in-plane rotation of the sample by 60° produced the same surface state dispersion. The EDCs along $\bar{\Gamma} - \bar{M}$ are shown to the right. There are a total of five crossings from $\bar{\Gamma} - \bar{M}$ which indicates that these surface states are topologically non-trivial. The number of surface state crossings in a material (with an odd number of Dirac points) is related to the topological Z_2 invariant (see text). (b) The resistivity curves of Bi and $Bi_{0.9}Sb_{0.1}$ reflect the contrasting transport behaviors. The presented resistivity curve for pure bismuth has been multiplied by a factor of 80 for clarity. (c) Schematic variation of bulk band energies of $Bi_{1-x}Sb_x$ as a function of x (based on band calculations and on [18,74]). $Bi_{0.9}Sb_{0.1}$ is a direct gap bulk Dirac point insulator well inside the inverted-band regime, and its surface forms a "topological metal"—the 2D analog of the 1D edge states in quantum spin Hall systems. (d) ARPES intensity integrated within ±10 meV of E_F originating solely from the surface state crossings. The image was plotted by stacking along the negative k_x direction a series of scans taken parallel to the k_y direction. (e) Outline of $Bi_{0.9}Sb_{0.1}$ surface state ARPES intensity near E_F measured in (d). White lines show scan directions "1" and "2." (f) Surface band dispersion

to exist in rare systems with strong spin-orbit coupling acting as an internal quantizing magnetic field on the electron system [82], and inverted bands at an odd number of high symmetry points in their bulk 3D BZs, are predicted to exhibit an odd number of surface state crossings, precluding their adiabatic continuation to the atomic insulator [6,16–19,73]. Such "topological quantum Hall metals" [6,19] cannot be realized in a purely 2D electron gas system such as the one realized at the interface of GaAs/GaAlAs systems.

In our experimental case, namely the (1 1 1) surface of $Bi_{0.9}Sb_{0.1}$, the four TRIM are located at $\bar{\Gamma}$ and three $\bar{M}$ points that are rotated by 60° relative to one another. Owing to the threefold crystal symmetry (A7 bulk structure) and the observed mirror symmetry of the surface Fermi surface across $k_x = 0$ (Fig. 3), these three $\bar{M}$ points are equivalent (and we henceforth refer to them as a single point, $\bar{M}$). The mirror symmetry $[E(k_y) = E(-k_y)]$ is also expected in this system. The complete details of the surface state dispersion observed in our experiments along a path connecting $\bar{\Gamma}$ and $\bar{M}$ are shown in Fig. 4(a); finding this information is made possible by our experimental separation of surface states from bulk states. As for bismuth (Bi), two surface bands emerge from the bulk band continuum near $\bar{\Gamma}$ to form a central electron pocket and an adjacent hole lobe [83–85]. It has been established that these two bands result from the spin-splitting of a surface state and are thus singly degenerate [85,86]. On the other hand, the surface band that crosses E_F at $-k_x \approx 0.5$ Å^{-1}, and forms the narrow electron pocket around $\bar{M}$, is clearly doubly degenerate, as far as we can determine within our experimental resolution. This is indicated by its splitting below E_F between $-k_x \approx 0.55$ Å^{-1} and $\bar{M}$, as well as the fact that this splitting goes to zero at $\bar{M}$ in accordance with Kramers' theorem. In semimetallic single crystal bismuth, only a single surface band is observed to form the electron pocket around $\bar{M}$ [87,88]. Moreover, this surface state overlaps, hence becomes degenerate with, the bulk conduction band at L (L projects to the surface $\bar{M}$ point) owing to the semimetallic character of Bi [Fig. 4(b)]. In $Bi_{0.9}Sb_{0.1}$ on the other hand, the states near $\bar{M}$ fall completely inside the bulk energy gap preserving their purely surface character at $\bar{M}$ [Fig. 4(a)]. The surface Kramers doublet point can thus be defined in the bulk insulator (unlike in Bi [83–88]) and is experimentally located in $Bi_{0.9}Sb_{0.1}$ samples to lie approximately 15 ± 5 meV below E_F at $\vec{k} = \bar{M}$ [Fig. 4(a)]. For the precise

---

FIGURE 4 (*Continued*) along direction "1" taken with $h\nu = 28$ eV and the corresponding EDCs (g). The surface Kramers degenerate point, critical in determining the topological Z_2 class of a band insulator, is clearly seen at $\bar{M}$, approximately 15 ± 5 meV below E_F. (We note that the scans are taken along the negative k_x direction, away from the bulk L point.) (h) Surface band dispersion along direction "2" taken with $h\nu = 28$ eV and the corresponding EDCs (i). This scan no longer passes through the $\bar{M}$ point, and the observation of two well-separated bands indicates the absence of Kramers degeneracy as expected, which cross-checks the result in (a). (Adapted from [7,20].) (For interpretation of the references to color in this figure legend, the reader is referred to the web version of this book.)

location of this Kramers point, it is important to demonstrate that our alignment is strictly along the $\bar{\Gamma} - \bar{M}$ line. To do so, we contrast high resolution ARPES measurements taken along the $\bar{\Gamma} - \bar{M}$ line with those that are slightly offset from it [Fig. 4(e)]. Fig. 4(f)–(i) shows that with k_y offset from the Kramers point at $\bar{M}$ by less than 0.02 Å^{-1}, the degeneracy is lifted and only one band crosses E_F to form part of the bow-shaped electron distribution [Fig. 4(d)]. Our finding of five surface state crossings (an odd rather than an even number) between $\bar{\Gamma}$ and $\bar{M}$ [Fig. 4(a)], confirmed by our observation of the Kramers degenerate point at the TRIM, indicates that these gapless surface states are topologically non-trivial. This corroborates our bulk electronic structure result that $Bi_{0.9}Sb_{0.1}$ is in the insulating band-inverted ($Z_2 = -1$ ($\nu_0 = 1$)) regime [Fig. 4(c)], which contains an odd number of bulk (gapped) Dirac points.

These experimental results taken collectively strongly suggest that $Bi_{0.9}Sb_{0.1}$ is quite distinct from graphene [80,89] and represents a novel state of quantum matter: a strongly spin-orbit coupled insulator with an odd number of Dirac points with a negative Z_2 topological Hall phase, which realizes the "parity anomaly without Fermion doubling." These works further demonstrate a general methodology for possible future investigations of *novel topological orders* in exotic quantum matter.

5 SPIN-RESOLVING THE SURFACE STATES TO IDENTIFY THE NON-TRIVIAL TOPOLOGICAL PHASE AND ESTABLISH A 2D HELICAL METAL PROTECTED FROM BACKSCATTERING

Strong topological materials are distinguished from ordinary materials such as gold by a topological quantum number, $\nu_0 = 1$ or 0, respectively [6,15]. For $Bi_{1-x}Sb_x$, theory has shown that ν_0 is determined solely by the character of the bulk electronic wavefunctions at the L point in the three-dimensional (3D) Brillouin zone (BZ). When the lowest energy conduction band state is composed of an antisymmetric combination of atomic p-type orbitals (L_a) and the highest energy valence band state is composed of a symmetric combination (L_s), then $\nu_0 = 1$, and vice versa for $\nu_0 = 0$ [18]. Although the bonding nature (parity) of the states at L is not revealed in a measurement of the bulk band structure, the value of ν_0 can be determined from the spin textures of the surface bands that form when the bulk is terminated. In particular, a $\nu_0 = 1$ topology requires the terminated surface to have a Fermi surface (FS) that supports a non-zero Berry's phase (odd as opposed to even multiple of π), which is not realizable in an ordinary spin-orbit material.

In a general inversion symmetric spin-orbit insulator, the bulk states are spin degenerate because of a combination of space inversion symmetry [$E(\vec{k},\uparrow) = E(-\vec{k},\uparrow)$] and time-reversal symmetry [$E(\vec{k},\uparrow) = E(-\vec{k},\downarrow)$]. Because space inversion symmetry is broken at the terminated surface, the spin degeneracy of surface bands can be lifted by the spin-orbit interaction [19–21]. However,

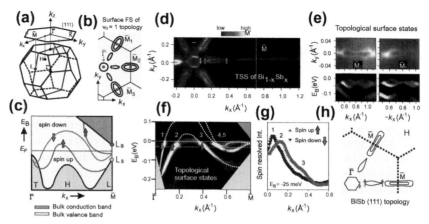

FIGURE 5 Spin texture of a topological insulator encodes Z_2 topological order of the bulk. (a) Schematic sketches of the bulk Brillouin zone (BZ) and (1 1 1) surface BZ of the $Bi_{1-x}Sb_x$ crystal series. The high symmetry points $(L, H, T, \Gamma, \bar{\Gamma}, \bar{M}, \bar{K})$ are identified. (b) Schematic of Fermi surface pockets formed by the surface states (SS) of a topological insulator that carries a Berry's phase. (c) Partner switching band structure topology: Schematic of spin-polarized SS dispersion and connectivity between $\bar{\Gamma}$ and $\bar{M}$ required to realize the FS pockets shown in panel-(b). L_a and L_s label bulk states at L that are antisymmetric and symmetric, respectively, under a parity transformation (see text). (d) Spin-integrated ARPES intensity map of the SS of $Bi_{0.91}Sb_{0.09}$ at E_F. Arrows point in the measured direction of the spin. (e) High resolution ARPES intensity map of the SS at E_F that enclose the $\bar{M}_1$ and $\bar{M}_2$ points. Corresponding band dispersion (second derivative images) are shown below. The left-right asymmetry of the band dispersions is due to the slight offset of the alignment from the $\bar{\Gamma} - \bar{M}_1 (\bar{M}_2)$ direction. (f) Surface band dispersion image along the $\bar{\Gamma} - \bar{M}$ direction showing five Fermi level crossings. The intensity of bands 4,5 is scaled up for clarity (the dashed white lines are guides to the eye). The schematic projection of the bulk valence and conduction bands are shown in shaded blue and purple areas. (g) Spin-resolved momentum distribution curves presented at $E_B = -25$ meV showing single spin degeneracy of bands at 1, 2, and 3. Spin-up and -down correspond to spin pointing along the $+\hat{y}$ and $-\hat{y}$ directions, respectively. (h) Schematic of the spin-polarized surface FS observed in our experiments. It is consistent with a $\nu_0 = 1$ topology (compare (b) and (h)). (Adapted from [21]).

according to Kramers' theorem [16], they must remain spin degenerate at four special time-reversal invariant momenta $(\vec{k}_T = \bar{\Gamma}, \bar{M})$ in the surface BZ [11], which for the (1 1 1) surface of $Bi_{1-x}Sb_x$ are located at $\bar{\Gamma}$ and three equivalent $\bar{M}$ points [see Fig. 5(a)].

Depending on whether ν_0 equals 0 or 1, the Fermi surface pockets formed by the surface bands will enclose the four $\vec{k}_T$ an even or odd number of times, respectively. If a Fermi surface pocket does not enclose $\vec{k}_T (= \bar{\Gamma}, \bar{M})$, it is irrelevant for the Z_2 topology [18,90]. Because the wavefunction of a single electron spin acquires a geometric phase factor of π [91] as it evolves by 360° in momentum space along a Fermi contour enclosing a $\vec{k}_T$, an odd number of Fermi pockets enclosing $\vec{k}_T$ in total imply a π geometrical (Berry's) phase [18]. In order to realize a π Berry's phase the surface bands must be spin-polarized and exhibit a partner switching [18] dispersion behavior between a pair of $\vec{k}_T$.

This means that any pair of spin-polarized surface bands that are degenerate at $\bar{\Gamma}$ must not re-connect at $\bar{M}$, or must separately connect to the bulk valence and conduction bands in between $\bar{\Gamma}$ and $\bar{M}$. The partner switching behavior is realized in Fig. 5(c) because the spin-down band connects to and is degenerate with different spin up bands at $\bar{\Gamma}$ and $\bar{M}$.

We first investigate the spin properties of the topological insulator phase [21]. Spin-integrated ARPES [92] intensity maps of the (1 1 1) surface states of insulating $Bi_{1-x}Sb_x$ taken at the Fermi level (E_F) [Fig. 5(d) and (e)] show that a hexagonal FS encloses $\bar{\Gamma}$, while dumbbell-shaped FS pockets that are much weaker in intensity enclose $\bar{M}$. By examining the surface band dispersion below the Fermi level [Fig. 5(f)] it is clear that the central hexagonal FS is formed by a single band (Fermi crossing 1) whereas the dumbbell-shaped FSs are formed by the merger of two bands (Fermi crossings 4 and 5) [20].

This band dispersion resembles the partner switching dispersion behavior characteristic of topological insulators. To check this scenario and determine the topological index ν_0, we have carried out spin-resolved photoemission spectroscopy. Figure 5(g) shows a spin-resolved momentum distribution curve taken along the $\bar{\Gamma} - \bar{M}$ direction at a binding energy $E_B = -25$ meV [Fig. 5(g)]. The data reveal a clear difference between the spin-up and spin-down intensities of bands 1, 2, and 3, and show that bands 1 and 2 have opposite spin whereas bands 2 and 3 have the same spin (detailed analysis discussed later in text). The former observation confirms that bands 1 and 2 form a spin-orbit split pair, and the latter observation suggests that bands 2 and 3 (as opposed to bands 1 and 3) are connected above the Fermi level and form one band. This is further confirmed by directly imaging the bands through raising the chemical potential via doping. Irrelevance of bands 2 and 3 to the topology is consistent with the fact that the Fermi surface pocket they form does not enclose any $\vec{k}_T$. Because of a dramatic intrinsic weakening of signal intensity near crossings 4 and 5, and the small energy and momentum splitting of bands 4 and 5 lying at the resolution limit of modern spin-resolved ARPES spectrometers, no conclusive spin information about these two bands can be drawn from the methods employed in obtaining the data sets in Fig. 5(g) and (h). However, whether bands 4 and 5 are both singly or doubly degenerate does not change the fact that an odd number of spin-polarized FSs enclose the $\vec{k}_T$, which provides evidence that $Bi_{1-x}Sb_x$ has $\nu_0 = 1$ and that its surface supports a non-trivial Berry's phase. This directly implies an absence of backscattering in electronic transport along the surface (Fig. 6), which has been re-confirmed by numerous scanning tunneling microscopy studies that show quasi-particle interference patterns that can only be modeled assuming an absence of backscattering [45–47].

Shortly after the discovery of $Bi_{1-x}Sb_x$, physicists sought to find a simpler version of a 3D topological insulator consisting of a single surface state instead of five. This is because the surface structure of $Bi_{1-x}Sb_x$ was rather complicated and the band gap was rather small. This motivated a search for topological insulators with a larger band gap and simpler surface spectrum. A second

Helical spin texture directly implies absence of backscattering

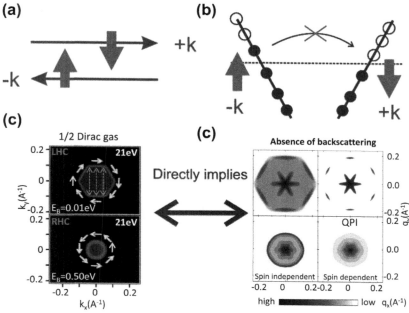

FIGURE 6 Helical spin texture naturally leads to absence of elastic backscattering for surface transport: No "U" turn on a 3D topological insulator surface. (a) Our measurement of a helical spin texture in both $Bi_{1-x}Sb_x$ and in Bi_2Se_3 directly shows that there is (b) an absence of backscattering. (c) ARPES measured FSs are shown with spin directions based on polarization measurements. L(R)HC stands for left(right)-handed chirality. (d) Spin-independent and spin-dependent scattering profiles on FSs in (c) relevant for surface quasi-particle transport are shown which is sampled by the quasi-particle interference (QPI) modes. (Adapted from [32]).

generation of 3D topological insulator materials, especially Bi_2Se_3, offers the potential for topologically protected behavior in ordinary crystals at room temperature and zero magnetic field. Starting in 2008, work by the Princeton group used spin-ARPES and first-principles calculations to study the surface band structure of Bi_2Se_3 and observe the characteristic signature of a topological insulator in the form of a single Dirac cone that is spin-polarized (Figs. 7 and 8) such that it also carries a non-trivial Berry's phase [22,24]. Concurrent theoretical work by [23] used electronic structure methods to show that Bi_2Se_3 is just one of several new large band-gap topological insulators. These other materials were soon after also identified using this ARPES technique we describe [25,26].

The Bi_2Se_3 surface state is found from spin-ARPES and theory to be a nearly idealized single Dirac cone as seen from the experimental data in Figs. 7 and 16. An added advantage is that Bi_2Se_3 is stoichiometric (i.e., a pure compound rather than an alloy such as $Bi_{1-x}Sb_x$) and hence can be prepared, in principle,

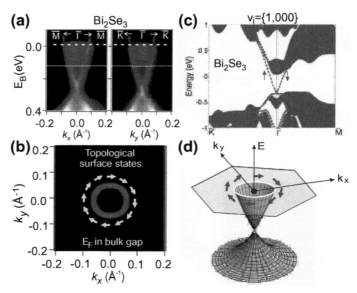

FIGURE 7 Spin-momentum locked helical fermions reveal Z_2 topological order of the bulk: spin-momentum locked helical surface Dirac fermions are hallmark signatures of topological insulators. (a) ARPES data for Bi_2Se_3 reveals surface electronic states with a single spin-polarized Dirac cone. The surface Fermi surface (b) exhibits a chiral left-handed spin texture. Data is from a sample doped to a Fermi level [thin line in (a)] that is in the bulk gap. ARPES measurements are carried out in P-polarizations mode which couples strongly to the (dominant) p_z orbital component of the surface state wavefunction. (c) Surface electronic structure of Bi_2Se_3 computed in the local density approximation. The shaded regions describe bulk states, and the red lines are surface states. (d) Schematic of the spin-polarized surface state dispersion in the Bi_2X_3 topological insulators. (Adapted from [22]).

at higher purity. While the topological insulator phase is predicted to be quite robust to disorder, many experimental probes of the phase, including ARPES of the surface band structure, are clearer in high-purity samples. Finally and perhaps most important for applications, Bi_2Se_3 has a large band gap of around 0.3 eV (3600 K). This indicates that in its high-purity form Bi_2Se_3 can exhibit topological insulator behavior at room temperature and greatly increases the potential for applications, which we discuss in greater depth later in the review.

6 IDENTIFYING THE ORIGIN OF 3D TOPOLOGICAL ORDER VIA A BULK BAND GAP INVERSION TRANSITION

We investigated the quantum origin of topological order in the $Bi_{1-x}Sb_x$ and Bi_2Se_3 classes of materials. It has been theoretically speculated that the novel topological order in $Bi_{1-x}Sb_x$ originates from the parities of the electrons in pure Sb and not Bi [18,74]. It was also noted [90] that the origin of the topological effects can only be tested by measuring the spin texture of the

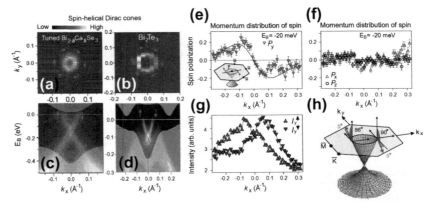

FIGURE 8 Direct detection of Topological Order: spin-momentum locking of spin-helical Dirac electrons in Bi_2Se_3 and Bi_2Te_3 using spin-resolved ARPES. (a) ARPES intensity map at E_F of the (1 1 1) surface of tuned $Bi_{2-\delta}Ca_\delta Se_3$ (see text) and (b) the (1 1 1) surface of Bi_2Te_3. Red arrows denote the direction of spin around the Fermi surface. (c) ARPES dispersion of tuned $Bi_{2-\delta}Ca_\delta Se_3$ and (d) Bi_2Te_3 along the k_x cut. The dotted red lines are guides to the eye. The shaded regions in (c) and (d) are our calculated projections of the bulk bands of pure Bi_2Se_3 and Bi_2Te_3 onto the (1 1 1) surface, respectively. (e) Measured y component of spin polarization along the $\bar{\Gamma} - \bar{M}$ direction at $E_B = -20$ meV, which only cuts through the surface states. Inset shows a schematic of the cut direction. (f) Measured x (red triangles) and z (black circles) components of spin polarization along the $\bar{\Gamma} - \bar{M}$ direction at $E_B = -20$ meV. Error bars in (e) and (f) denote the standard deviation of $P_{x,y,z}$, where typical detector counts reach 5×10^5; Solid lines are numerical fits [93]. (g) Spin-resolved spectra obtained from the y component spin-polarization data. The non-Lorentzian lineshape of the $I_y^\uparrow$ and $I_y^\downarrow$ curves and their non-exact merger at large $|k_x|$ is due to the time evolution of the surface band dispersion, which is the dominant source of statistical uncertainty. a.u., arbitrary units. (h) Fitted values of the spin-polarization vector $\mathbf{P} = (S_x, S_y, S_z)$ are $(\sin(90°)\cos(-95°), (\sin 90°)\sin(-95°), \cos(90°))$ for electrons with $+k_x$ and $(\sin(86°)\cos(85°), \sin(86°)\sin(85°), \cos(86°))$ for electrons with $-k_x$, which demonstrates the topological helicity of the spin-Dirac cone. ARPES measurements are carried out in P-polarizations mode which couples strongly to the (dominant) p_z orbital component of the surface state wavefunction. The angular uncertainties are of order $\pm 10°$ and the magnitude uncertainty is of order ± 0.15. (Adapted from [24]).

Sb surface, which has not been measured. Based on quantum oscillation and magneto-optical studies, the bulk band structure of Sb is known to evolve from that of insulating $Bi_{1-x}Sb_x$ through the hole-like band at H rising above E_F and the electron-like band at L sinking below E_F [74]. The relative energy ordering of the L_a and L_s states in Sb again determines whether the surface state pair emerging from $\bar{\Gamma}$ switches partners [Fig. 9(a)] or not [Fig. 9(b)] between $\bar{\Gamma}$ and $\bar{M}$, and in turn determines whether they support a non-zero Berry's phase.

In a conventional spin-orbit metal such as gold, a free-electron like surface state is split into two parabolic spin-polarized sub-bands that are shifted in $\vec{k}$-space relative to each other [94]. Two concentric spin-polarized Fermi surfaces are created, one having an opposite sense of in-plane spin rotation from

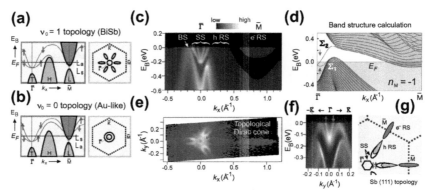

FIGURE 9 Topological character of parent compound revealed on the (1 1 1) surface states. Schematic of the bulk band structure (shaded areas) and surface band structure (red and blue lines) of Sb near E_F for a (a) topologically non-trivial and (b) topological trivial (gold-like) case, together with their corresponding surface Fermi surfaces, are shown. (c) Spin-integrated ARPES spectrum of Sb(1 1 1) along the $\bar{\Gamma} - \bar{M}$ direction. The surface states are denoted by SS, bulk states by BS, and the hole-like resonance states and electron-like resonance states by h RS and e⁻ RS, respectively. (d) Calculated surface state band structure of Sb(1 1 1) based on the methods in [75,90]. The continuum bulk energy bands are represented with pink shaded regions, and the lines show the discrete bands of a 100 layer slab. The red and blue single bands, denoted Σ_1 and Σ_2, are the surface states bands with spin polarization $\langle \vec{P} \rangle \propto +\hat{y}$ and $\langle \vec{P} \rangle \propto -\hat{y}$, respectively. (e) ARPES intensity map of Sb(1 1 1) at E_F in the $k_x - k_y$ plane. The only one FS encircling $\bar{\Gamma}$ seen in the data is formed by the inner V-shaped SS band seen in panel-(c) and (f). The outer V-shaped band bends back toward the bulk band best seen in data in panel-(f). (f) ARPES spectrum of Sb(1 1 1) along the $\bar{\Gamma} - \bar{K}$ direction shows that the outer V-shaped SS band merges with the bulk band. (g) Schematic of the surface FS of Sb(1 1 1) showing the pockets formed by the surface states (unfilled) and the resonant states (blue and purple). The purely surface state Fermi pocket encloses only one Kramers degenerate point ($\vec{k}_T$), namely, $\bar{\Gamma}(=\vec{k}_T)$, therefore consistent with the $\nu_0 = 1$ topological classification of Sb which is different from Au (compare (b) and (g)). As discussed in the text, the hRS and e⁻ RS count trivially. (Adapted from [21]). (For interpretation of the references to color in this figure legend, the reader is referred to the web version of this book.)

the other, that enclose $\bar{\Gamma}$. Such a Fermi surface arrangement, like the schematic shown in Fig. 9(b), does not support a non-zero Berry's phase because the $\vec{k}_T$ are enclosed an even number of times (2 for most known materials).

However, for Sb, this is not the case. Figure 9(c) shows a spin-integrated ARPES intensity map of Sb(1 1 1) from $\bar{\Gamma}$ to $\bar{M}$. By performing a systematic incident photon energy dependence study of such spectra, previously unavailable with He lamp sources [95], it is possible to identify two V-shaped surface states (SS) centered at $\bar{\Gamma}$, a bulk state located near $k_x = -0.25$ Å^{-1} and resonance states centered at about $k_x = 0.25$ Å^{-1} and $\bar{M}$ that are hybrid states formed by surface and bulk states [92]. An examination of the ARPES intensity map of the Sb(1 1 1) surface and resonance states at E_F [Fig. 9(e)] reveals that the central surface FS enclosing $\bar{\Gamma}$ is formed by the inner V-shaped SS only. The outer V-shaped SS on the other hand forms part of a tear-drop shaped FS that does *not* enclose $\bar{\Gamma}$, unlike the case in gold. This tear-drop shaped FS is formed partly by the outer V-shaped SS and partly by the hole-like resonance state. The electron-like resonance state FS enclosing $\bar{M}$ does not affect the determination of ν_0 because it must be doubly spin degenerate.

Such a FS geometry [Fig. 9(g)] suggests that the V-shaped SS pair may undergo a partner switching behavior expected in Fig. 9(a). This behavior is most clearly seen in a cut taken along the $\bar{\Gamma} - \bar{K}$ direction since the top of the bulk valence band is well below E_F [Fig. 9(f)] showing only the inner V-shaped SS crossing E_F while the outer V-shaped SS bends back toward the bulk valence band near $k_x = 0.1$ Å^{-1} before reaching E_F. The additional support for this band dispersion behavior comes from tight binding surface calculations on Sb [Fig. 9(d)], which closely match with experimental data below E_F. Our observation of a single surface band forming a FS enclosing $\bar{\Gamma}$ suggests that pure Sb is likely described by $\nu_0 = 1$, and that its surface may support a Berry's phase.

Confirmation of a surface π Berry's phase rests critically on a measurement of the relative spin orientations (up or down) of the SS bands near $\bar{\Gamma}$ so that the partner switching is indeed realized, which cannot be done without spin resolution. Spin resolution was achieved using a Mott polarimeter that measures two orthogonal spin components of a photoemitted electron [97,98]. These two components are along the y' and z' directions of the Mott coordinate frame, which lie predominantly in and out of the sample (1 1 1) plane, respectively. Each of these two directions represents a normal to a scattering plane defined by the photoelectron incidence direction on a gold foil and two electron detectors mounted on either side (left and right) [Fig. 10(a)]. Strong spin-orbit coupling of atomic gold is known to create an asymmetry in the scattering of a photoelectron off the gold foil that depends on its spin component normal to the scattering plane [98]. This leads to an asymmetry between the left intensity ($I^L_{y',z'}$) and right intensity ($I^R_{y',z'}$) given by $A_{y',z'} = (I^L_{y',z'} - I^R_{y',z'})/(I^L_{y',z'} + I^R_{y',z'})$, which is related to the spin polarization $P_{y',z'} = (1/S_{eff}) \times A_{y',z'}$ through the Sherman function $S_{eff} = 0.085$ [97,98]. Spin-resolved momentum distribution curve data sets of the SS bands along the $-\bar{M} - \bar{\Gamma} - \bar{M}$ cut at $E_B = -30$ meV [Fig. 10(b)] are shown for maximal intensity. Figure 10(d) displays both y' and z' polarization components along this cut, showing clear evidence that the bands are spin polarized, with spins pointing largely in the (1 1 1) plane. In order to estimate the full 3D spin-polarization vectors from a two-component measurement (which is not required to prove the partner switching or the Berry's phase), we fit a model polarization curve to our data following the recent demonstration in Ref. [99], which takes the polarization directions associated with each momentum distribution curve peak [Fig. 10(c)] as input parameters, with the constraint that each polarization vector has length one (in angular momentum units of $\hbar/2$). Our fitted polarization vectors are displayed in the sample (x, y, z) coordinate frame [Fig. 10(f)], from which we derive the spin-resolved momentum distribution

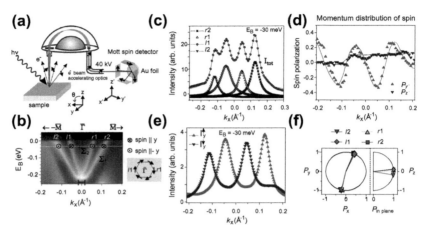

FIGURE 10 Spin texture of topological surface states and topological chirality. (a) Experimental geometry of the spin-resolved ARPES study. At normal emission ($\theta = 0°$), the sensitive y'-axis of the Mott detector is rotated by 45° from the sample $\bar{\Gamma}$ to $-\bar{M}(\| - \hat{x})$ direction, and the sensitive z'-axis of the Mott detector is parallel to the sample normal ($\| \hat{z}$). (b) Spin-integrated ARPES spectrum of Sb(1 1 1) along the $-\bar{M} - \bar{\Gamma} - \bar{M}$ direction. The momentum splitting between the band minima is indicated by the black bar and is approximately 0.03 Å^{-1}. A schematic of the spin chirality of the central FS based on the spin-resolved ARPES results is shown on the right. (c) Momentum distribution curve of the spin averaged spectrum at $E_B = -30$ meV (shown in (b) by white line), together with the Lorentzian peaks of the fit. (d) Measured spin-polarization curves (symbols) for the detector y' and z' components together with the fitted lines using the two-step fitting routine [99]. (e) Spin-resolved spectra for the sample y component based on the fitted spin-polarization curves shown in (d). Up (down) triangles represent a spin direction along the $+(-)\hat{y}$ direction. (f) The in-plane and out-of-plane spin polarization components in the sample coordinate frame obtained from the spin-polarization fit. Overall spin-resolved data and the fact that the surface band that forms the central electron pocket has $\langle \vec{P} \rangle \propto -\hat{y}$ along the $+k_x$ direction, as in (e), suggest a left-handed chirality (schematic in (b) and see text for details). (Adapted from [21]).

curves for the spin components parallel ($I_y^\uparrow$) and anti-parallel ($I_y^\downarrow$) to the y direction as shown in Fig. 10(e). There is a clear difference in $I_y^\uparrow$ and $I_y^\downarrow$ at each of the four momentum distribution curve peaks indicating that the surface state bands are spin polarized [Fig. 10(e)], which is possible to conclude even without a full 3D fitting. Each of the pairs $l2/l1$ and $r1/r2$ has opposite spin, consistent with the behavior of a spin split pair, and the spin polarization of these bands is reversed on either side of $\bar{\Gamma}$ in accordance with the system being time-reversal symmetric [$E(\vec{k},\uparrow) = E(-\vec{k},\downarrow)$] [Fig. 10(f)]. The measured spin texture of the Sb(1 1 1) surface states (Fig. 10), together with the connectivity of the surface bands (Fig. 9), uniquely determines its belonging to the $\nu_0 = 1$ class. Therefore the surface of Sb carries a non-zero (π) Berry's phase via the inner V-shaped band and pure Sb can be regarded as the parent metal of the $Bi_{1-x}Sb_x$ topological insulator class, in other words, the topological order

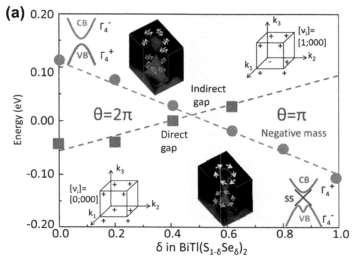

FIGURE 11 Observation of bulk band inversion across a topological quantum phase transition. Energy levels of Γ_4^-. (blue circles) and Γ_4^+ (green squares) bands are obtained from ARPES measurements as a function of composition δ. CB: conduction band; VB: valence band. Parity eigenvalues (+ or −) of Bloch states are shown. The topological invariants, ν_i, obtained from the parity eigenvalues are presented as $[\theta/\pi = \nu_0; \nu_1, \nu_2, \nu_3]$ where $\theta = \pi\nu_0$ is the axion angle [96] and ν_0 is the strong invariant. ARPES measurements are carried out in P-polarizations mode which couples strongly to the (dominant) p_z orbital component of the surface state wavefunction. (Adapted from [32]).

originates from the Sb wavefunctions. A recent work [32] has demonstrated a topological quantum phase transition as a function of chemical composition from a non-inverted to an inverted semiconductor as a clear example of the origin of topological order (Fig. 11).

Our spin-polarized measurement methods (Fig. 5 and 10) uncover a new type of topological quantum number n_M, which is not related to the time-reversal symmetry, but a consequence of the mirror symmetries of a crystalline system [90]. Topological band theory suggests that the bulk electronic states in the mirror ($k_y = 0$) plane can be classified in terms of a number n_M (=integer) [90], which defines both the number and the chirality of the spin-helical edge states propagating along the mirror planes of a crystal (see Fig. 12).

For example, we now determine the value of n_M of antimony surface states from our data. From Fig. 5, it is seen that a single (one) surface band, which switches partners at $\bar{M}$, connects the bulk valence and conduction bands, so $|n_M| = 1$. The sign of n_M is related to the direction of the spin polarization $\langle \vec{P} \rangle$ of this band [90], which is constrained by mirror symmetry to point along $\pm\hat{y}$. Since the central electron-like FS enclosing $\bar{\Gamma}$ intersects six mirror invariant points [see Fig. 10(b)], the sign of n_M distinguishes two distinct types of

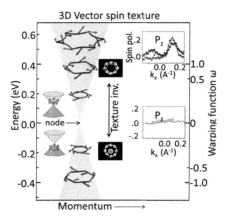

FIGURE 12 Spin-texture evolution of topological surface bands as a function of energy away from the Dirac node (left axis) and geometrical warping factor ω (right axis). The warping factor is defined as $\omega = \frac{k_F(\bar{\Gamma}-\bar{M})-k_F(\bar{\Gamma}-\bar{K})}{k_F(\bar{\Gamma}-\bar{M})+k_F(\bar{\Gamma}-\bar{K})} \times \frac{2+\sqrt{3}}{2-\sqrt{3}}$ where $\omega = 0$, $\omega = 1$, and $\omega > 1$ implies circular, hexagonal, and snowflake-shaped FSs, respectively. The sign of ω indicates texture chirality for LHC (+) or RHC (−). Insets: Out-of-plane 3D spin-polarization measurements at corresponding FSs. ARPES measurements are carried out in P-polarizations mode which couples strongly to the (dominant) p_z orbital component of the surface state wavefunction. (Adapted from [32]).

handedness for this spin-polarized FS. Figures 5(f) and 10 show that for both $Bi_{1-x}Sb_x$ and Sb, the surface band that forms this electron pocket has $\langle \vec{P} \rangle \propto -\hat{y}$ along the k_x direction, suggesting a left-handed rotation sense for the spins around this central FS thus $n_M = -1$. We note that similar analysis regarding the mirror symmetry and mirror eigenvalues $n_M = -1$ can be applied to the single Dirac cone surface states in the Bi_2Se_3 material class as well as the recently discovered topological crystalline insulator phase in $Pb_{1-x}Sn_xTe$ material Ref. [36, 101].

As a matter of fact, a non-zero (non-trivial) topological mirror number does not require a non-zero Z_2 topological number. Or in other words, there is no necessary correlation between a mirror symmetry protected topological order and a time-reversal symmetry protected topological order. However, since most of the Z_2 topological insulators ($Bi_{1-x}Sb_x$, Bi_2Se_3, Bi_2Te_3, etc.) also possess mirror symmetries in their crystalline form, topological mirror order n_M typically coexists with the strong Z_2 topological order, and only manifests itself as the chirality or the handness of the spin texture. One possible way to isolate the mirror topological order from the Z_2 order is to work with systems that feature an even number of bulk band inversions. This approach naturally excludes a non-trivial Z_2 order which strictly requires an odd number of band inversions. More importantly, if the locations of the band inversions coincide with the mirror planes in momentum space, it will lead to a topologically non-trivial phase protected by the mirror symmetries of the crystalline system that is irrelevant to the time-reversal symmetry protection and the Z_2 (Kane-Mele) topological order. Such exotic new phase of topological order, noted as topological crystalline insulator [100] protected by space group mirror symmetries, has very recently been theoretically predicted and experimentally identified in the $Pb_{1-x}Sn_x$ Te(Se) alloy systems [35–37,101]. An anomalous $n_M = -2$ topological mirror number in $Pb_{1-x}Sn_x$ Te, distinct from the $n_M = -1$ case observed in the Z_2 topological insulators, has also been experimentally determined using spin-resolved ARPES measurements as shown in Ref. [36]. Moreover, the mirror symmetry can be generalized to other space group symmetries, leading to a large number of distinct topological crystalline insulators awaited to be discovered, some of which are predicted to exhibit non-trivial crystalline order even without spin-orbit coupling as well as topological crystalline surface states in non-Dirac (e.g., quadratic) fermion forms [100].

In 3D Z_2 (Kane-Mele) TIs, the spin-resolved experimental measurements shown above reveal an intimate and straightforward connection between the topological numbers (v_0, n_M) and the non-trivial Berry's phase. The v_0 determines whether the surface electrons support a non-trivial Berry's phase. The 2D Berry's phase is a critical signature of topological order and is not realizable in isolated 2D electron systems, nor on the surfaces of conventional spin-orbit or exchange coupled magnetic materials. A non-zero Berry's phase is known, theoretically, to protect an electron system against the almost universal weak-localization behavior in their low temperature transport [18,102] and is expected to form the key element for fault-tolerant computation schemes [102,103], because the Berry's phase is a geometrical agent or mechanism for protection against quantum decoherence [104]. Its remarkable realization on the $Bi_{1-x}Sb_x$ surface represents an unprecedented example of a 2D π Berry's phase, and opens the possibility for building realistic prototype systems to test quantum computing modules. In general, our results demonstrate that spin-ARPES is a powerful probe of 3D topological order, which opens up a new search front for topological materials for novel spin-devices and fault-tolerant quantum computing.

7 TOPOLOGICAL PROTECTION AND TUNABILITY OF THE SURFACE STATES OF A 3D TOPOLOGICAL INSULATOR

The predicted topological protection of the surface states of Sb implies that their metallicity cannot be destroyed by weak time-reversal symmetric perturbations. In order to test the robustness of the measured gapless surface states of Sb, we introduce such a perturbation by randomly substituting Bi into the Sb crystal matrix. Another motivation for performing such an experiment is that the formalism developed by Fu and Kane [18] to calculate the Z_2 topological invariants relies on inversion symmetry being present in the bulk crystal, which they assumed to hold true even in the random alloy $Bi_{1-x}Sb_x$. However, this formalism is simply a device for simplifying the calculation and the non-trivial

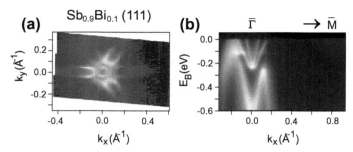

FIGURE 13 Robustness against disorder. Spin split surface states survive alloying disorder in
$Sb_{0.9}Bi_{0.1}$. (a) ARPES intensity map at E_F of single crystal $Sb_{0.9}Bi_{0.1}$ (1 1 1) in the $k_x - k_y$
plane taken using 20 eV photons. (b) ARPES intensity map of $Sb_{0.9}Bi_{0.1}$ (1 1 1) along the $\bar{\Gamma} - \bar{M}$
direction taken with $h\nu = 22$ eV photons. The band dispersion is not symmetric about $\bar{\Gamma}$ because
of the threefold rotational symmetry of the bulk states about the $\langle 111 \rangle$ axis. (Adapted from [21]).

$\nu_0 = 1$ topological class of $Bi_{1-x}Sb_x$ is predicted to hold true even in the
absence of inversion symmetry in the bulk crystal [18]. Therefore introducing
light Bi substitutional disorder into the Sb matrix is also a method to examine the
effects of alloying disorder and possible breakdown of bulk inversion symmetry
on the surface states of Sb(1 1 1). We have performed spin-integrated ARPES
measurements on single crystals of the random alloy $Sb_{0.9}Bi_{0.1}$. Figure 13
shows that both the surface band dispersion along $\bar{\Gamma} - \bar{M}$ as well as the surface
state Fermi surface retain the same form as that observed in Sb(1 1 1), and
therefore the "topological metal" surface state of Sb(1 1 1) fully survives the
alloy disorder. Since Bi alloying is seen to only affect the band structure of
Sb weakly, it is reasonable to assume that the topological order is preserved
between Sb and $Bi_{0.91}Sb_{0.09}$ as we observed.

In a simpler fashion compared to $Bi_{1-x}Sb_x$, the topological insulator
behavior in Bi_2Se_3 is associated with a single band inversion at the surface
Brillouin zone center. Owing to its larger band gap compared with $Bi_{1-x}Sb_x$,
ARPES has shown that its topological properties are preserved at room
temperature [24]. Two defining properties of topological insulators, spin-
momentum locking of surface states and π Berry phase, can be clearly
demonstrated in the Bi_2Se_3 series. The surface states are protected by time-
reversal symmetry, which implies that the surface Dirac node should be robust
in the presence of nonmagnetic disorder. On the other hand, a gap at the
Dirac point is theoretically expected to open in the presence of magnetism
along the out-of-plane direction perpendicular to the sample surface. Here
magnetism is required to break time-reversal symmetry whereas its out-of-plane
direction is critical to break additional protections by crystalline-symmetries
[100]. Unlike in theory, magnetism in real topological insulator materials
exhibits complex phenomenology [31, 39, 105–110]. Experimental studies of a
magnetic gap especially on the quantitative level are challenging because many
other physical or chemical changes are found to also lead to extrinsic gap-like

feature at the Dirac point [43,51,109,110]. We utilize spin-resolved angle-resolved photoemission spectroscopy to measure the momentum space spin configurations in systematically magnetically (Mn) doped, non-magnetically (Zn) doped, and ultrathin quantum coherent topological insulator films [34]. Figure 14(b) shows the out-of-plane spin-polarization (P_z) measurements of the electronic states in the vicinity of the Dirac point gap of a Mn(2.5%)-

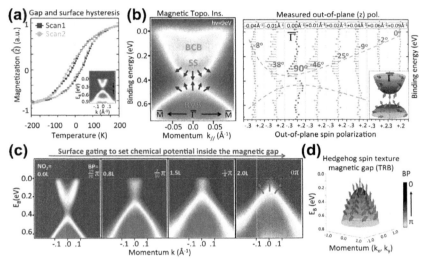

FIGURE 14 Hedgehog spin texture and Berry's phase tuning in a magnetic topological insulator. (a) Magnetization measurements using magnetic circular dichroism show out-of-plane ferromagnetic character of the Mn-Bi$_2$Se$_3$ MBE film surface through the observed hysteretic response. The inset shows the ARPES observed gap at the Dirac point in the Mn(2.5%)-Bi$_2$Se$_3$ film sample. (b) Spin-integrated and spin-resolved measurements on a representative piece of Mn(2.5%)-Bi$_2$Se$_3$ film sample using 9 eV photons. Left: Spin-integrated ARPES dispersion map. The blue arrows represent the spin-texture configuration in close vicinity of the gap revealed by our spin-resolved measurements. Right, Measured out-of-plane spin polarization as a function of binding energy at different momentum values. The momentum value of each spin-polarization curve is noted on the top. The polar angles (θ) of the spin-polarization vectors obtained from these measurements are also noted. The 90° polar angle observed at $\bar{\Gamma}$ point suggests that the spin vector at $\bar{\Gamma}$ is along the vertical direction. The spin behavior at $\bar{\Gamma}$ and its surrounding momentum space reveals a hedgehog-like spin configuration for each Dirac band separated by the gap. Inset shows a schematic of the revealed hedgehog-like spin texture. (c) Measured surface state dispersion upon *in situ* NO$_2$ surface adsorption on the Mn-Bi$_2$Se$_3$ surface. The NO$_2$ dosage in the unit of Langmuir (1 L = 1 × 10^{-6} torr s) and the tunable Berry's phase (BP) associated with the topological surface state are noted on the top-left and top-right corners of the panels, respectively. The red arrows depict the time-reversal breaking out-of-plane spin texture at the gap edge based on the experimental data. (d) The time-reversal breaking spin texture features a singular hedgehog-like configuration when the chemical potential is tuned to lie within the magnetic gap, corresponding to the experimental condition presented in the last panel in panel (c). In this work, the Mn distribution on the sample surface was found to be inhomogeneous, which can be sample preparation dependent. (Adapted from [34]).

Bi_2Se_3 sample. The surface electrons at the time-reversal invariant $\bar{\Gamma}$ point [red curve in Fig. 14(b)] are clearly observed to be spin-polarized in the out-of-plane direction. The opposite sign of P_z for the upper and lower Dirac band shows that the Dirac point spin degeneracy is indeed lifted up ($E(k_{//} = 0, \uparrow) \neq E(k_{//} = 0, \downarrow)$), which manifestly breaks the time-reversal symmetry on the surface of our $Mn(2.5\%)-Bi_2Se_3$ samples. Systematic spin-resolved measurements as a function of binding energy and momentum reveal a Hedgehog-like spin texture (inset of Fig. 14(b). Such exotic spin groundstate in a magnetic topological insulator enables a tunable Berry's phase on the magnetized topological surface [34], as experimentally demonstrated by our chemical gating via NO_2 surface adsorption method shown in Figs. 14(c) and (d).

Many of the interesting theoretical proposals that utilize topological insulator surfaces require the chemical potential to lie at or near the surface Dirac point. This is similar to the case in graphene, where the chemistry of carbon atoms naturally locates the Fermi level at the Dirac point. This makes its density of carriers highly tunable by an applied electrical field and enables applications of graphene to both basic science and microelectronics. The surface Fermi level of a topological insulator depends on the detailed electrostatics of the surface and is not necessarily at the Dirac point. Moreover, for naturally grown Bi_2Se_3 the bulk Fermi energy is not even in the gap. The observed n-type behavior is believed to have caused Se vacancies. By appropriate chemical modifications, however, the Fermi energy of both the bulk and the surface can be controlled. This allowed [24] to reach the sweet spot in which the surface Fermi energy is tuned to the Dirac point (Fig. 15). This was achieved by doping bulk with a small concentration of Ca, which compensates the Se vacancies, to place the Fermi level within the bulk band gap. The surface was the hole doped by exposing the surface to NO_2 gas to place the Fermi level at the Dirac point, and has been shown to be effective even at room temperature (Fig. 16). These results collectively show how ARPES can be used to study the topological protection and tunability properties of the 2D surface of a 3D topological insulator.

These techniques to identify, characterize, and manipulate the topological bulk and surface states of 3D topological insulators have opened the way for performing surface-sensitive transport measurements on bulk crystals [53,54], exfoliated nano-devices [56,57] and thin films [59–61] as well as surface sensitive optical measurements [66,72].

8 FUTURE DIRECTIONS: TOPOLOGICAL SUPERCONDUCTORS AND TOPOLOGICAL CRYSTALLINE INSULATORS

Recent measurements [27] show that surface instabilities cause the spin-helical topological insulator band structure of Bi_2Se_3 to remain well defined and non-degenerate with bulk electronic states at the Fermi level of optimally

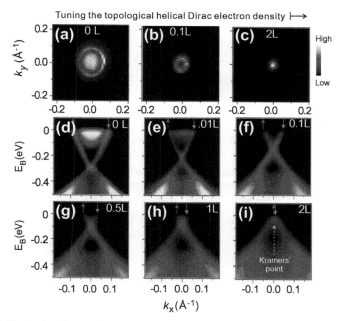

FIGURE 15 Surface Gating : Tuning the density of helical Dirac electrons to the spin-degenerate Kramers point and topological transport regime. (a) A high resolution ARPES mapping of the surface Fermi surface (FS) near $\bar{\Gamma}$ of $Bi_{2-\delta}Ca_{\delta}Se_3$ (1 1 1). The diffuse intensity within the ring originates from the bulk-surface resonance state [6]. (b) The FS after 0.1 Langmuir (L) of NO_2 is dosed, showing that the resonance state is removed. (c) The FS after a 2 L dosage, which achieves the Dirac charge neutrality point. (d) High resolution ARPES surface band dispersions through after an NO_2 dosage of 0 L, (e) 0.01 L, (f) 0.1 L, (g) 0.5 L, (h) 1 L, and (i) 2 L. The arrows denote the spin polarization of the bands. We note that due to an increasing level of surface disorder with NO_2 adsorption, the measured spectra become progressively more diffuse and the total photoemission intensity from the buried $Bi_{2-\delta}Ca_{\delta}Se_3$ surface is gradually reduced. (Adapted from [24]).

doped superconducting $Cu_{0.12}Bi_2Se_3$, and that this is also likely to be the case for superconducting variants of p-type Bi_2Te_3. These surface states provide a highly unusual physical setting in which superconductivity cannot take a conventional form, and is expected to realize one of two novel states that have not been identified elsewhere in nature. If superconducting pairing has even parity, as is nearly universal among the known superconducting materials, the surface electrons will achieve a 2D non-Abelian superconductor state with non-commutative Majorana fermion vortices that can potentially be manipulated to store quantum information. Surface vortices will be found at the end of bulk vortex lines as drawn in Fig. 17. If superconducting pairing is odd, the resulting state is a novel state of matter known as a "topological superconductor" with Bogoliubov surface quasi-particles present below the superconducting critical temperature of 3.8 K. As drawn in Fig. 17(c), these low temperature surface states would be gapless, likely making it impossible to adiabatically manipulate

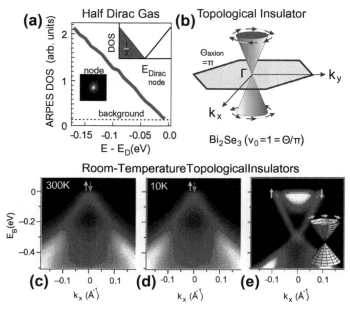

FIGURE 16 Observation of room-temperature (300 K) topological order without applied magnetic field in Bi_2Se_3. (a) Crystal momentum integrated ARPES data near Fermi level exhibit linear fall-off of density of states, which, combined with the spin-resolved nature of the states, suggest that a half Fermi gas is realized on the topological surfaces. (b) Spin-texture map based on spin-ARPES data suggests that the spin chirality changes sign across the Dirac point. (c) The Dirac node remains well defined up a temperature of 300 K suggesting the stability of topological effects up to the room temperature. (d) The Dirac cone measured at a temperature of 10 K. (e) Full Dirac cone. (Adapted from [24]).

surface vortices for quantum computation. The unique physics and applications of the topological superconductor state are distinct from any known material system, and will be an exciting vista for theoretical and experimental exploration if they are achieved for the first time in $Cu_x Bi_2 Se_3$.

Another interesting frontier involves a newly discovered topological phase of matter, namely a topological crystalline insulator where space group symmetries replace the role of time-reversal symmetry in a Z_2 topological insulator [35–37,44,47,94,100,101]. In Ref. [36], we experimentally investigated the possibility of a mirror symmetry protected topological phase transition in the $Pb_{1-x} Sn_x$ Te alloy system with spin-ARPES, which has long been known to contain an even number of band inversions based on band theory. We showed that at a composition below the theoretically predicted band inversion, the system is fully gapped [36], whereas in the inverted regime, the surface exhibits even number of spin-polarized Dirac cone states revealing mirror-protected topological order distinct from that observed in $Bi_{1-x} Sb_x$ and $Bi_2 Se_3$ revealing consistency with mirror Chern number of -2. Fig. 18 presents a comparison including spin between the $Pb_{0.6} Sn_{0.4}$ Te and a single

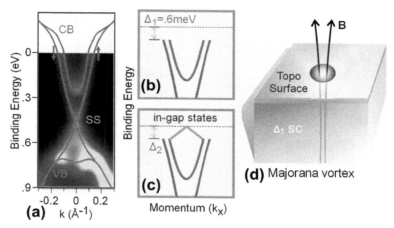

FIGURE 17   A Majorana platform. (a) Topologically protected surface states cross the Fermi level before merging with the bulk valence and conduction bands in a lightly doped topological insulator. (b) If the superconducting wavefunction has even parity, the surface states will be gapped by the proximity effect, and vortices on the crystal surface will host braidable Majorana fermions. (c) If superconducting parity is odd, the material will be a so-called "topological superconductor," and new states will appear below T_c to span the bulk superconducting gap. (d) Majorana fermion surface vortices are found at the end of bulk vortex lines and could be manipulated for quantum computation if superconducting pairing is even. A theoretical (not experimental) scenario is sketched here. (Adapted from [27]).

Dirac cone Z_2 topological insulator (TI) GeBi$_2$Te$_4$ [28]. In clear contrast to GeBi$_2$Te$_4$, which features a single spin-helical Dirac cone enclosing the time-reversal invariant (Kramers') $\bar{\Gamma}$ point [Fig. 18(a)–(d)], none of the surface states of Pb$_{0.6}$Sn$_{0.4}$Te is observed to enclose any of the time-reversal invariant momenta, demonstrating their irrelevance to the time-reversal symmetry related protection. On the other hand, all of the Pb$_{0.6}$Sn$_{0.4}$Te surface states are found to locate along the two independent momentum space mirror line ($\bar{\Gamma} - \bar{X} - \bar{\Gamma}$) directions, revealing their topological protection as a consequence of the crystalline mirror symmetries of the Pb$_{1-x}$Sn$_x$Te system [36,47,94]. The irrelevance to time-reversal symmetry observed in Ref. [36] suggests the potential to realize magnetic yet topologically protected spin surface states on the Pb$_{1-x}$Sn$_x$Te surface, which is fundamentally not possible in the Z_2 topological insulator systems. In general, our discovery of a mirror symmetry protected topological phase first in Bi$_{1-x}$Sb$_x$ Ref. [21], and then in the Pb$_{1-x}$Sn$_x$Te system opens the door for a wide range of exotic new spin physics, such as Lifshitz transition and Fermi surface Fractionalization based on topological surface spin states [101], topological antiferromagnetic insulator [111,112], and topological crystalline superconductor [113,114] via bulk doping or proximity interfacing, as well as new topological crystalline orders protected by other point group symmetries even without spin-orbit interactions [100].

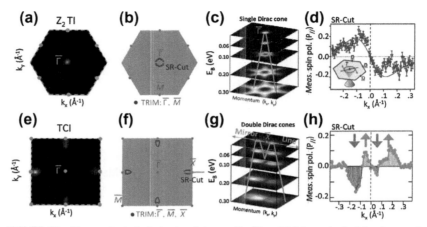

FIGURE 18 The topological distinction between Z_2 (Kane-Mele) topological insulator and topological crystalline insulator phases. (a)–(d) ARPES, spin-resolved ARPES, and calculation results of the surface states of a Z_2 topological insulator GeBi$_2$Te$_4$ [28], an analog to Bi$_2$Se$_3$ [22]. (a) ARPES measured Fermi surface with the chemical potential tuned near the surface Dirac point. (b) First-principles calculated iso-energetic contour of the surface states near the Dirac point. The solid blue line shows the momentum-space cut used for spin-resolved measurements. Right: A stack of ARPES iso-energetic contours near the $\bar{\Gamma}$ point of the surface BZ. (d) Measured spin polarization of Bi$_2$Se$_3$, in which a helical spin texture is revealed. (e)–(h) ARPES and spin-resolved ARPES measurements on the Pb$_{0.6}$Sn$_{0.4}$Te ($x = 0.4$) samples and band calculation results on the end compound SnTe [101]. (e) ARPES measured Fermi surface map of Pb$_{0.6}$Sn$_{0.4}$ Te. (f) First-principles calculated iso-energetic contour of SnTe surface states near the Dirac point. The solid blue line shows the momentum-space cut near the surface BZ edge center $\bar{X}$ point, which is used for spin-resolved measurements shown in panel (h). (g) A stack of ARPES iso-energetic contours near the $\bar{X}$ point of the surface BZ, revealing the double Dirac cone contours near each $\bar{X}$ point on the surface of Pb$_{0.6}$Sn$_{0.4}$ Te. (h) Measured spin polarization of Pb$_{0.6}$Sn$_{0.4}$ Te near the native Fermi energy along the momentum-space cut defined in panel (f), in which two spin-helical Dirac cones are observed near an $\bar{X}$ point. (Adapted from [36]).

ACKNOWLEDGMENTS

The authors acknowledge A. Bansil, R.J. Cava, F.C. Chou, J.H. Dil, A.V. Fedorov, Liang Fu, D. Haldane, S. Jia, C.L. Kane, H. Lin, J. Osterwalder, D. Qian, N. Samarth, and S.G. Wen for discussions and U.S. DOE DE-FG-02-05ER46200, No. AC03-76SF00098, and No. DE-FG02-07ER46352 for support. M.Z.H. acknowledges visiting-scientist support from Lawrence Berkeley National Laboratory and additional support from the A.P. Sloan Foundation, and Princeton University.

REFERENCES

[1] Hasan MZ, Kane CL. Rev Mod Phys 2010;82:3045–67.
[2] Moore JE. Nature Physics 2009;5:378–380.
[3] Hasan MZ, Moore JE. Ann Rev Cond Mat Phys 2011;2:55–78.
[4] Fu L, Kane CL, Mele EJ. Phys Rev Lett 2007;98:106803.
[5] Roy R. Phys Rev B 2009;79:195322.

[6] Moore JE, Balents L. Phys Rev 2007;B 75:121306(R).
[7] Hsieh D et al. Nature 2008;452:970.[Completed and submitted in 2007]. Also see KITP Proceeding at http://online.itp.ucsb.edu/online/motterials07/hasan/ (2007).
[8] Ashcroft NW, Mermin ND. Solid state physics. New York: Holt Rinehart and Winston; 1976.
[9] Klitzing Kv, Dorda G, Pepper M. Phys Rev Lett 1980;45:494.
[10] Tsui DC, Stormer HL, Gossard AC. Phys Rev Lett 1982;48:1559.
[11] Laughlin RB. Phys Rev Lett 1983;50:1395.
[12] Wen X-G. Int J Mod Phys B 1990;4:239.
[13] Thouless DJ et al. Phys Rev Lett 1982;49:405.
[14] Avron JE, Osadchy D, Seiler R. Phys Today 2003;56(8):38.
[15] Kane CL, Mele EJ. Phys Rev Lett 2005;95:146802.
[16] Bernevig BA, Hughes TL, Zhang S-C. Science 2006;314:1757.
[17] König M et al. Science 2007;318:766.
[18] Fu L, Kane CL. Phys Rev B 2007;76:045302.
[19] Fu L, Kane CL, Mele EJ. Phys Rev Lett 2007;98:106803.
[20] Hsieh D et al. Nature 2008;452:970.
[21] Hsieh D et al. Science 2009;323:919.
[22] Xia Y et al. Nature Phys 2009;5:398.
[23] Zhang H et al. Nature Phys 2009;5:438.
[24] Hsieh D et al. Nature 2009;460:1101.
[25] Chen YL et al. Science 2009;325:178.
[26] Hsieh D et al. Phys Rev Lett 2009;103:146401.
[27] Wray LA et al. Nature Phys 2010;6:855.
[28] Xu Su-Yang et al. 2010. Preprint at http://arXiv.org/abs/1007.5111.
[29] Lin H et al. Nature Mat 2011;9:546.
[30] Lin H et al. Phys Rev Lett 2011;105:036404.
[31] Wray LA et al. Nature Phys 2011;7:32.
[32] Xu Su-Yang et al. Science 2011;332:560.
[33] Neupane M et al. Phys Rev B 2012;85:235406.
[34] Xu Su-Yang et al. Nature Phys 2012;8:616.
[35] Dziawa P et al. Nature Mat 2012;11:1023. [This work does not measure the spin texture, which is critical to prove the topological crystalline insulator phase].
[36] Xu Su-Yang et al. Nature Commun 2012;3:1192.
[37] Tanaka Y et al. Nature Phys 2012;8:800.
[38] Zhang Y et al. Nature Phys 2010;6:584.
[39] Chen Y-L et al. Science 2010;329:659.
[40] Sato T et al. Phys Rev Lett 2010;105:136802.
[41] Kuroda K et al. Phys Rev Lett 2010;105:146801.
[42] King PDC et al. Phys Rev Lett 2011;107:096802.
[43] Sato T et al. Nature Phys 2011;7:840.
[44] Okada Y et al. Science 2013;341:1496–1499.
[45] Roushan P et al. Nature 2009;460:1106.
[46] Zhang T et al. Phys Rev Lett 2009;103:266803.
[47] Fang C et al. Phys Rev B 2013;88:125141.
[48] Cheng P et al. Phys Rev Lett 2010;105:076801.
[49] Hanaguri T et al. Phys Rev B 2010;82:081305.
[50] Seo J et al. Nature 2010;466:343.
[51] Beidenkopf H et al. Nature Phys 2011;7:939.
[52] Okada Y et al. Phys Rev Lett 2011;106:206805.
[53] Qu D-X et al. Science 2010;329:821.
[54] Analytis JG et al. Nature Phys 2010;6:960.
[55] Peng HL et al. Nature Mat 2010;9:225.
[56] Steinberg H et al. Nano Lett 2010;10:5032.
[57] Chen J et al. Phys Rev Lett 2010;105:176602.
[58] Ren Z et al. Phys Rev B 2011;82:241306.
[59] He HT et al. Phys Rev Lett 2011;106:166805.
[60] Liu M et al. Phys Rev B 2011;83:165440.

[61] Wang J et al. Phys Rev B 2011;83:245538.
[62] Sacépé B et al. Nature Comm 2011;2:575.
[63] Kim D et al. Nature Phys 2012;8:459.
[64] Checkelsky JG et al. Nature Phys 2012;8:729.
[65] Wang YH et al. Phys Rev Lett 2011;107:207602.
[66] Hsieh D et al. Phys Rev Lett 2011;106:057401.
[67] Park SR et al. Phys Rev Lett 2012;108:046805.
[68] Sobota JA et al. Phys Rev Lett 2012;108:117403.
[69] Wang YH et al. Phys Rev Lett 2012;109:127401.
[70] Schafgans AA et al. Phys Rev B 2012;85:195440.
[71] McIver JW et al. Nature Nanotechnol 2012;7:96.
[72] Valdés Aguilar R et al. Phys Rev Lett 2012;108:087403.
[73] Murakami S. New J Phys 2007;9:356.
[74] Lenoir B et al. Bi-Sb alloys: an update. 15th International Conference on Thermoelectrics 1996:1–13.
[75] Liu Y, Allen E. Phys Rev B 1995;52:1566.
[76] Wolff PA. J Phys Chem Solids 1964;25:1057.
[77] Fukuyama H, Kubo R. J Phys Soc Jpn 1970;28:570.
[78] Buot FA. Phys Rev A 1973;8:1570.
[79] Hebel LC, Smith GE. Phys Lett 1964;10:273.
[80] Zhang Y et al. Nature 2005;438:201.
[81] Zhou S et al. Nature Mat 2007;6:770.
[82] Haldane FDM. Phys Rev Lett 1988;61:2015.
[83] Ast CR, Hochst H. Phys Rev Lett 2001;87:177602.
[84] Hochst H, Gorovikov S. J Electron Spectrosc Relat Phenom 2005;351:144. This work does not measure the surface state along the critical $\bar{\Gamma} - \bar{M}$ direction or detect the bulk Dirac spectrum near L.
[85] Hofmann P. Prog Surf Sci 2006;81:191.
[86] Hirahara T et al. Phys Rev B 2007;76:153305.
[87] Hengsberger M et al. Eur Phys J 2000;17:603.
[88] Ast CR, Hochst H. Phys Rev B 2003;67:113102.
[89] Novoselov KS et al. Science 2007;315:1379.
[90] Teo JCY, Fu L, Kane CL. Phys Rev B 2008;78:045426.
[91] Sakurai JJ. Modern Quantum Mechanics. New York: Addison-Wesley; 1994.
[92] Hufner S. Photoelectron Spectroscopy. Berlin: Springer-verlag; 1995.
[93] Hirahara T et al. Phys Rev B 2007;76:153305.
[94] Wang Y. J Phys Rev B 2013;87:235317.
[95] Sugawara K et al. Phys Rev Lett 2006;96:046411.
[96] Qi X-L, Hughes T, Zhang S-C. Phys Rev B 2008;78:195424.
[97] Hoesch M et al. J Electron Spectrosc Relat Phenom 2002;124:263.
[98] Gay TJ, Dunning FB. Rev Sci Instrum 1992;63:1635.
[99] Meier F et al. Phys Rev B 2008;77:165431.
[100] Fu L. Phys Rev Lett 2011;106:106802.
[101] Hsieh H et al. Nature Commun 2012;3:982.
[102] Fu L, Kane CL. Phys Rev Lett 2008;100:096407.
[103] Leek PJ et al. Science 2007;318:1889.
[104] Kitaev A. Ann Phys (NY) 2003;303:2.
[105] Hor YS et al. Phys Rev B 2010;81:195203.
[106] Ji HW et al. Phys Rev B 2012;85:165313.
[107] Salman Z et al. 2012. Preprint at http://arXiv:1203.4850.
[108] Vobornik I et al. Nano Lett 2011;11:4079.
[109] Eremeev SV et al. 2011. Preprint at http://arXiv:1107.3208.
[110] Xu Su-Yang et al. 2012. Preprint at http://arXiv.org/abs/1206.0278.
[111] Story T et al. Phys Rev Lett 1986;56:777.
[112] Mong RSK et al. Phys Rev B 2010;81:245209.
[113] Sasaki S et al. Phys Rev Lett 2012;109:217004.
[114] Teo JCY, Hughes TL. 2012. Preprint at http://arXiv.org/abs/1208.6303.

Chapter 7

Visualizing Topological Surface States and their Novel Properties using Scanning Tunneling Microscopy and Spectroscopy

Haim Beindenkopf, Pedram Roushan and Ali Yazdani
Joseph Henry Laboratories and Department of Physics, Princeton University, Princeton, NJ 08544, USA

Chapter Outline Head

1 INTRODUCTION

Topological insulators form a class of insulators distinct from all other insulating materials. These materials are distinguished from ordinary insulators by an inverted bulk gap for electronic excitations induced by strong spin-orbit coupling which assures the presence of gapless metallic boundary states, akin to the chiral edge modes in quantum Hall systems, but with helical spin textures [1–3]. Experiments and theoretical efforts have provided strong evidence for both two- and three-dimensional topological insulators, including their

Topological Insulators. http://dx.doi.org/10.1016/B978-0-444-63314-9.00007-X

novel edge and surface states in semiconductor quantum well structures [4–6] and several Bi-based compounds [7–11], respectively. In the case of three-dimensional topological insulators the scanning tunneling microscopy experiments (STM) have played a key role in directly visualizing their two-dimensional topological surface states and in demonstrating some of their novel properties, such as the absence of backscattering from non-magnetic defects and the high transmission of these states through barriers that reflect or absorb conventional surface states [12,13]. They have also provided a microscopic view of how bulk defects may be limiting the mobility of these states because of their modulation of background bulk electronic states [14]. This chapter will review these experimental findings as well as other important contributions spatially resolved measurements with the STM have made to establish the properties of topological surface states. Looking ahead to future developments, measurements with the STM are likely to continue contributing to our understanding of topological surface states. STM could be used to examine the influence of magnetism and superconductivity [15–19] on these states, where interaction with these states is expected to give rise to Majorana fermions or other boundary modes associated with the magnetic-induced gapping of topological surface states.

2 SCANNING TUNNELING MICROSCOPY AND SPECTROSCOPIC MAPPING

Over the past decade, STM has emerged as a powerful tool for studying the electronic structure of novel materials by providing the ability to directly image their electronic states close to the surface. Originally invented as an imaging tool to detect the arrangement of surface atoms, advances in low temperature STM operation have opened the possibility of using tunneling spectroscopy which allows to probe the energy distribution of the electronic density of states with high-energy resolution. STM relies on the ability to position a sharp metallic tip within quantum tunneling distance from the sample surface. The current of tunneling electrons is then proportional to $I(V_B, r) \propto \rho_t(E_F) \int_0^{V_B} \rho_s(E_F + \epsilon, r)d\epsilon$, where ρ_t is the tip's density of states (DOS) assumed to be constant about the Fermi energy E_F, ρ_s is the sample's DOS, and V_B is the applied bias voltage between tip and sample. In the imaging mode, the STM tip is rastered across the surface with piezoelectric elements, while a feedback loop keeps the tunneling current between the tip and the sample constant by adjusting the tip height. The recorded trajectory of the tip (STM topograph) maps contours of constant electronic density and provides an image that contains both structural as well as electronic information about the sample surface. When used as a spectroscopic tool, the STM feedback loop is opened to keep the tip height constant above the sample, while the current versus voltage characteristic of the tunneling between the tip and sample is measured. The tunneling differential conductance (dI/dV) as a function of tip-sample bias is directly proportional to the local density of states (LDOS) of the sample at the location under the tip as a function of energy $dI/dV(V_B, \vec{r}\,) \propto \rho_s(E_F + V_B, \vec{r}\,)$. The unique advantage of the STM is its ability to obtain such information with high-energy resolution (limited only by thermal broadening) over large regions of the sample, in which the conductance maps provide direct information on how the LDOS varies both spatially and as a function of energy. Such LDOS measurements contain critical information about many aspects of the electronic states, and can be used for example to determine how electronic states behave near defects and other scatterers. As we describe below, the Fourier analysis of such data sets, $dI/dV(V_B, \vec{q}\,) = \int dI/dV(V_B, \vec{r}\,)e^{-i\vec{q}\cdot\vec{r}}d\vec{r}$, can provide important information on the properties of electronic states of the sample and can be used to make precise measurements of their scattering mechanisms and interaction with defects. Moreover, the surface sensitivity of STM measurements is a major advantage for examining the properties of surface states such as those of topological insulators.

3 ABSENCE OF BACKSCATTERING

A key predicted characteristic of spin-textured topological surface states is their insensitivity to nonmagnetic scattering, which is expected to protect them from backscattering and localization. This feature distinguishes topological surface states from other ordinary two-dimensional states which become localized in the presence of even weak disorder. A simple way to understand the lack of backscattering is to consider that forward- and backward-moving electrons in a topological surface state have orthogonal spin states. In the absence of a time-reversal symmetry breaking perturbation such as magnetism, there is no matrix element connecting these two states. Alternatively, one can consider the spin rotation of an electron in a strong spin-orbit coupled system as it encounters an impurity, as illustrated in Fig. 1. For a non-magnetic impurity there exist two possible ways for the electron to scatter off the single impurity. Since spin is locked to momentum these two paths acquire a relative π-phase shift which is the topological Berry phase associated with the helical states. Consequently they destructively interfere resulting in zero probability for backscattering of an electron from an ordinary defect.

STM's ability to probe the scattering of surface states, first utilized in imaging the scattering of free electron-like Shockley surface states in noble metals such as Cu(1 1 1) [20, 21], can also be used to probe scattering properties of topological surface states. For ordinary surface states, such as those on Cu(1 1 1), STM conductance maps show standing waves associated with electrons scattering from various impurities on the surface, demonstrated in Fig. 2(a). Such measurements and related studies on noble metal surface states have established the link between the modulation seen in the tunneling conductance at a wavevector $\vec{q}$ and the elastic scattering and interference of quasi-particles

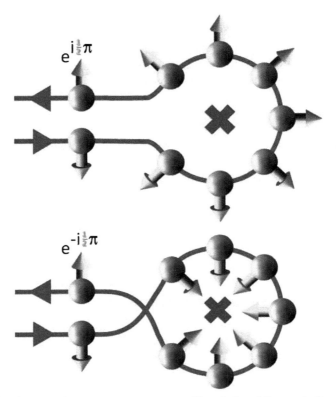

FIGURE 1 Lack of backscattering as a consequence of Berry's phase. Microscopically an electron can backscatter from a non-magnetic impurity in two possible paths. Due to strong spin-orbit coupling the spin is locked to the momentum along these counterrotating paths and thus revolves by either $+\pi$ or $-\pi$ in either of them. For electrons with spin 1/2 a relative phase of π is accumulated which results in destructive interference of the backscattering paths and the elimination of the backscattering channel.

between different momentum states ($\vec{k_1}$ and $\vec{k_2}$, where $\vec{q} = \vec{k_1} - \vec{k_2}$) at the same energy. Elastic scattering between equal energy states gives rise to quasi-particle interference (QPI) patterns in the LDOS, which are being imaged with the STM. The Fourier analysis of such patterns shows a circular pattern in q-space, which results from connecting the initial and final states for a scattering electron on a simple circular Fermi surface as shown in Fig. 2(b) for Cu(1 1 1) Shockley surface states. This circular pattern in Fourier space is a direct visualization of the scattering process for these electrons and is dominated by the backscattering channel from momentum $\vec{k}$ to $-\vec{k}$.

The first experiments to visualize topological surface states and their scattering properties were carried out on $Bi_{1-x}Sb_x$, which was the first

(a)   **(b)**

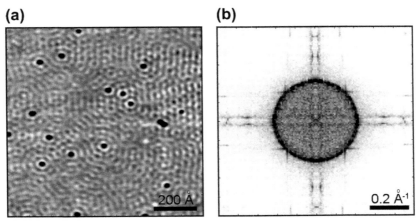

FIGURE 2 QPI on Cu(1 1 1) surface. (a) Topographic image on the surface of Cu(1 1 1) shows real-space interference patterns of the Shockley surface states. (b) Their DFT shows a 2 k_F ring due to the dominant backscattering channel.

three-dimensional topological insulator to be discovered using angle-resolved photoemission spectroscopy (ARPES) [7]. Previous theoretical work had anticipated that doping of Sb results in spin-orbit induced inversion of Bi's band structure and tuning of the chemical potential to realize a relatively small band gap topological insulator. [9] ARPES experiments were used to show that $Bi_{1-x}Sb_x$ surface states have an odd number of Fermi level crossings between time-reversal equivalent points—a predicted feature of topological surface states. Further spin-sensitive ARPES experiments were used to establish that these surface states have a helical spin structure and an associated Berry's phase [8], which makes them distinct from ordinary surface states with strong spin-orbit coupling [22]. It is the influence of this Berry's phase on the scattering that we aim to examine with the STM.

Figure 3 shows an example STM topograph of an in situ cleaved $Bi_{0.92}Sb_{0.08}$ sample measured at 4 K that is dominated by long wavelength (~20 Å) modulations in the LDOS. High-resolution images on smaller scale (inset Fig. 3) show the Bi atomic corrugation and the alloying disorder in this compound caused by Sb substitution. Also shown in this figure are spectroscopic measurements that demonstrate a general suppression of the density of states near the Fermi level. ARPES measurements [7,8] and band structure calculations [11] suggest that within ± 30 meV of the Fermi level, where there is a bulk gap, tunneling would be dominated by the surface states. While tunneling spectroscopy measurements do not distinguish between bulk and surface states, energy-resolved spectroscopic maps shown in Figs. 4(a)–(c) display modulations that are the result of scattering of the surface electronic states. As expected for the scattering and interference of surface states, the observed patterns are not commensurate with the underlying atomic structure.

(a)

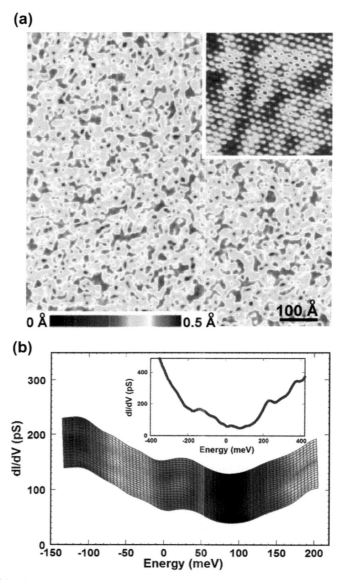

(b)

FIGURE 3 Topography and dI/dV spectroscopy of the $Bi_{0.92}Sb_{0.08}(1\,1\,1)$ surface. (Reproduced with permission from Ref. [12]) STM topograph (+50 meV, 100 pA) of the $Bi_{0.92}\ Sb_{0.08}$ (1 1 1) surface over an area 800 Å by 800 Å. The inset shows an area 80 Å by 80 Å that displays the underlying atomic lattice (+200 mV, 15 pA). b, Spatial variation of the differential conductance (dI/dV) measurements across a line of length 250 Å. Inset, a typical differential conductance measurement over larger energy ranges.

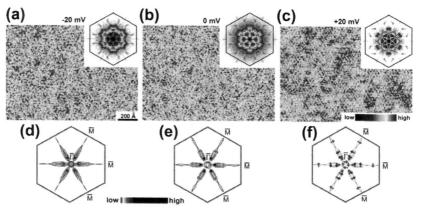

FIGURE 4 dI/dV maps, QPI patterns, and ARPES measurements on the $Bi_{0.92}Sb_{0.08}(1\,1\,1)$ surface. (Reproduced with permission from Ref. [12]) (a)–(c) Spatially resolved conductance maps of the $Bi_{0.92}Sb_{0.08}$ (1 1 1) surface obtained at -20 mV (b), and $+20$ mV (c). (Panels show area 1000 Å by 1300 Å.) Insets, Fourier transforms of the dI/dV maps. The hexagons have the same size as the first Brillouin zone. The Fourier transforms have been symmetrized in consideration of the three-fold rotation symmetry of the (1 1 1) surface. (d) and (e) ARPES intensity map of the surface state at -20 mV (d) and at the Fermi level (e). (f) The spin textures from ARPES measurements are shown with arrows, and high-symmetry points are marked (Γ and three M).

While STM cannot provide direct information on the chemical composition of the scattering defects, the random distribution of substituted Sb atoms is a likely candidate.

Discrete Fourier transform (DFT) of energy-resolved STM conductance maps can be used to reveal the wavelengths of the modulations in the LDOS and to obtain detailed information on the nature of scattering processes for the surface state electrons [21]. Similar to studies described above on noble metal surfaces, the QPI patterns in the conductance maps are due to the scattering processes between different momentum states ($\vec{k_1}$ and $\vec{k_2}$, where $\vec{q} = \vec{k_1} - \vec{k_2}$) at the same energy. For $Bi_{0.92}Sb_{0.08}$, the DFT of the conductance maps shown as insets in Fig. 4(a)–(c) displays a rich pattern of QPI, which has the sixfold rotational symmetry of the underlying lattice, and evolves as a function of energy. These patterns display the allowed wavevectors $\vec{q}$ and the relative intensities for the various scattering processes experienced by the surface state electrons.

Within a simple model of QPI, which ignores the topological helical spin texture of such states, the interference wavevectors connect regions of high density of states on contours of constant energy (or the Fermi surface at the chemical potential). In this model, the QPI patterns should correspond to a joint density of states (JDOS) for the surface state electrons that can be independently determined from ARPES measurements [23,24]. Figure 4(d) and (e) shows contours of constant energy (CCE) in the first Brillouin zone (FBZ), as measured

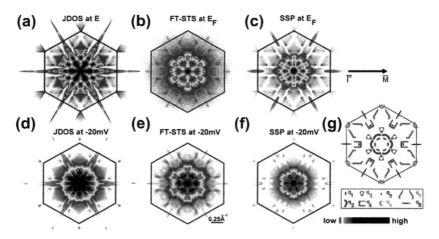

FIGURE 5 Construction of JDOS and SSP from ARPES data and their comparison with FT-STS. (Reproduced with permission from Ref. [12]) (a) The JDOS and SSP calculated at the Fermi energy (E_F), from ARPES data presented in Fig. 4(e). (b) The FT-STS at E_F. (c) The SSP calculated at E_F. (d) The JDOS calculated at −20 mV, from ARPES data presented in Fig. 4(d). (e) The FT-STS at −20 mV. (f) The SSP calculated at −20 mV. (g) Schematization of the features associated with scattering wavevectors q_1–q_8 in the FT-STS data.

with ARPES at two energies on $Bi_{0.92}Sb_{0.08}$ crystals (following procedures described in Ref [7]). The CCE consists of an electron pocket centered on the $\overline{\Gamma}$ point, hole pockets half way to the $\overline{M}$ point, and two electron pockets that occur very close to the $\overline{M}$ point [7,8]. From these measurements, we can determine the JDOS as a function of the momentum difference between initial and final scattering states, $\vec{q}$, using $JDOS(\vec{q}) = \int I(\vec{k})I(\vec{k}+\vec{q})d\vec{k}$, where $I(\vec{k})$ is the ARPES intensity that is proportional to the surface density of states at a specific two-dimensional momentum $\vec{k}$. Figure 5 shows the results of computing the JDOS from ARPES data in panels 5(a) and (d) for two different energies. Contrasting these figures to the corresponding QPI data in Fig. 4(b) and (c), we find a significant suppression of the scattering intensity along the directions equivalent to $\overline{\Gamma} - \overline{M}$ in the FBZ. Backscattering between various electron and hole pockets around the $\overline{\Gamma}$ point should give rise to a continuous range of scattering wavevectors along the $\overline{\Gamma} - \overline{M}$ direction, a behavior not observed in the data (see also the expanded view of the JDOS and QPI in Fig. 6(a)). This discrepancy suggests the potential importance of the surface states' spin texture and the possibility that spin rules are limiting the backscattering for these helical electronic states.

To include spin effects, we use the results of spin-resolved ARPES studies and assign a helical spin texture to the electron and hole pockets as shown in Fig. 4(f). ARPES studies have resolved the spin structure only for the central electron pocket and the hole pockets near the $\overline{\Gamma}$ point in the FBZ; however, we

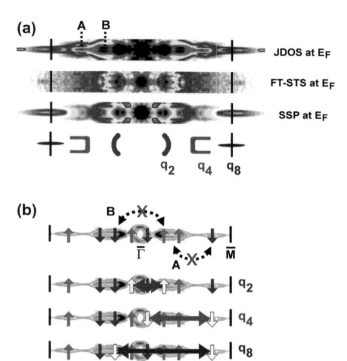

FIGURE 6 Comparison of the various parts of the QPI patterns along the—direction at the Fermi energy. (Reproduced with permission from Ref. [12]) (a) Close-up view of the QPI pattern from JDOS, FT-STS, and SSP at E_F, along the $\overline{\Gamma} - \overline{M}$ direction. The last row shows the schematic representation of $q_2, q_4,$ and q_8, which correspond to scatterings shown in (b). Two high-intensity points which are only seen in JDOS are labeled as A and B. (b) The Fermi surface along the $\overline{\Gamma} - \overline{M}$ direction, with spin orientation of the quasi-particles shown with arrows. The horizontal color-coded arrows show the sources of the scatterings seen in the STM data. Note that all highlighted spins have the same orientation. The bottom row depicts the scatterings which involve opposite spins and are present in JDOS, but absent in FT-STS and SSP.

assign a similar helical structure to states near $\overline{M}$. This assignment is consistent with the presence of a π Berry's phase that distinguishes the spin topology of the $Bi_{0.92}Sb_{0.08}$ surface states from that of surface state bands that are simply split by spin-orbit coupling. In the latter case, the spin-polarized surface bands come in pairs, while for topological surface states there should be an odd number of spin-polarized states between two time-reversal equivalent points in the band structure [9,11,25,26].

To understand scattering and interference for these topological surface states, we determine the spin-dependent scattering probability, $SSP(\vec{q}) = \int I(\vec{k})T(\vec{q},\vec{k})I(\vec{k}+\vec{q})d^2\vec{k}$, which is computed in a similar fashion to the

JDOS using the ARPES-measured density of states, $I(\vec{k})$, but also includes a spin-dependent scattering matrix element $T(\vec{q},\vec{k})$. This matrix element describes the scattering probability as a function of momentum transfer and the spins of states that are connected by the scattering process. Shown in Fig. 5(c) and (f) are the calculated $SSP(\vec{q})$ from ARPES data at two different energies using a matrix element of the form $T(\vec{q},\vec{k}) = \left| \langle \vec{S}(\vec{k}) | \vec{S}(\vec{k} + \vec{q}) \rangle \right|^2$. This simple form of spin-selective scattering reduces scattering between states with non-aligned spins and completely suppresses scattering between states with opposite spin orientations. Comparison of the SSP patterns to the QPI measurements in Fig. 5 shows that including spin effects leads to remarkably good agreement between the scattering wavevectors measured by STM and those expected from the shape of the surface CCE as measured by ARPES. (Wavevectors corresponding to features in QPI and SSP are identified in Fig. 5(g)). A quantitative comparison between the QPI from STM data and the JDOS and SSP derived from ARPES data can be made by computing the cross-correlation between the various patterns. Focusing on the high-symmetry direction, which is shown in Fig. 6(a), we find that QPI data (excluding the central q = 0 section, which is dominated by disorder) is 95% correlated with the SSP in the same region. Similar cross-correlation is found to be 83% between the QPI and JDOS. Therefore, the proposed form of the spin-dependent scattering matrix element is the critical component for understanding the suppression of scattering along the high-symmetry directions in the data.

The proposed scattering matrix elements $T(\vec{q},\vec{k})$ and associated spin-scattering rules are further confirmed by a more detailed analysis of the QPI patterns. An example of such an analysis is shown in Fig. 6, in which we associate features for scattering along the high-symmetry direction in QPI and SSP with specific scattering wavevectors $\vec{q}$ that connect regions of the CCE. The observed wavevectors in QPI and SSP obey spin rules imposed by $T(\vec{q},\vec{k})$, as illustrated schematically in Fig. 6(b). We also depict in Fig. 6(b) an example of scattering processes that, while allowed by the band structure and observed in JDOS, violate the spin-scattering rules and are accordingly absent from the QPI data in Fig. 6(a). A comprehensive analysis of all the features in the QPI data (see supplementary section of Ref. [12]) demonstrates that allowed scattering wavevectors $\vec{q}_1$ to $\vec{q}_8$ (Fig. 5(g)) exclude those that connect states with opposite spins. Remarkably, all the features of the complex QPI patterns and their energy dependence (see Ref. [12] for the energy dispersion) can be understood in detail by the allowed scattering wavevectors based on the band structure of the topological surface states and the spin-scattering rule. This agreement provides a precise demonstration that scattering of electrons over thousands of Angstroms, which underlies the QPI maps, strictly obeys the spin-scattering rules and associated suppression of backscattering.

STM experiments on other topological insulators have also been used to probe the absence of backscattering for topological surface states. In particular, a number of studies have focused on Bi_2Te_3 and Bi_2Se_3 [14,27–29], where topological surface states have simpler characteristics as compared to $Bi_{1-x}Sb_x$. In these materials, the topological surface states consist of a single band with an anisotropic Dirac-like energy dispersion [30–33]. The Fermi surface for these topological surface states starts as circular near the Dirac point and evolves continuously into a warped snow-flake-like shape at higher energies, due to the anisotropic dispersion of the bands along different directions [34]. Figure 7 shows QPI patterns for Ca-doped Bi_2Te_3 and for Mn-doped Bi_2Te_3 and Bi_2Se_3 measured in a similar fashion as those on $Bi_{1-x}Sb_x$ described above. Remarkably, detailed Fourier analysis of the QPI patterns at energies above the Dirac point on the different samples shows patterns that are independent of whether the scattering dopant is magnetic or not (Fig. 7(f)–(j)) or even if magnetic order is established [35]. In fact, the strong resemblance of the QPI patterns and their DFT across dopants (Ca and Mn) and materials (Bi_2Te_3 and Bi_2Se_3) suggests that they can be understood based on the shape of the Fermi surface and the spin texture associated with it [30–34], as sketched in Fig. 8(a). The warped shape of the surface band structure in Bi_2Te_3 and Bi_2Se_3 dominates scattering at high energies relative to the Dirac point. Fermi

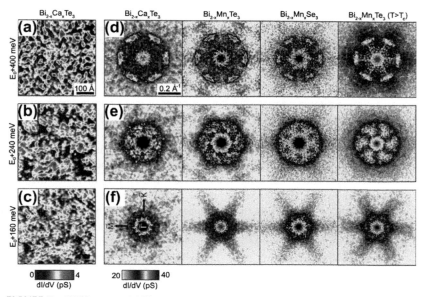

FIGURE 7 dI/dV maps and QPI patterns on magnetically and non-magnetically doped. Bi_2Se_3 and Bi_2Te_3 surfaces. (Reproduced with permission from Ref. [14]) (a)–(c) Real-space QPI patterns on the surface of Ca-doped Bi_2Te_3 at different energies. (d)–(f) DFTs of the QPI patterns at three bias voltages. From left to right: Ca- and Mn-doped Bi_2Te_3, Mn-doped Bi_2Se_3 at 4 K, and Mn-doped Bi_2Te_3 at 20 K (above $T_c = 13$ K). All compounds show similar patterns in q-space, consisting of six strong peaks along the $\bar{\Gamma} - \bar{M}$ directions at high energies and circular patterns at lower ones.

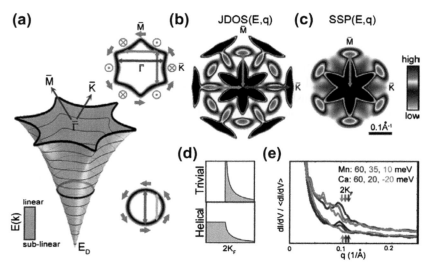

FIGURE 8 Origin of QPI patterns. (Reproduced with permission from Ref. [14]) (a) Schematic drawing of the surface band structure in Bi_2Te_3 and Bi_2Se_3. Top right: warped CEC at high energies and possible quasi-nested scattering modes and the associated spin texture. Bottom right: circular CEC close to the Dirac energy and helical spin texture. (b) Calculated JDOS for the warped CEC. (c) Calculated SSP accounting for the full spin texture. (d) Scattering profile based on transfer-matrix calculation with (bottom) and without (top) helical spin texture. (e) Radial profile of measured QPI pattern at low energies displaying a dispersing $2\,k_F$ peak both in magnetically and non-magnetically doped Bi_2Te_3.

surface warping is consistent with a number of nearly nested parts of the CCE (connecting points with high density of state in k-space) both along the $\overline{\Gamma} - \overline{K}$ and $\overline{\Gamma} - \overline{M}$ directions as captured by the calculated JDOS, shown in Fig. 8(b). The QPI pattern measured at high energies (Fig. 7(d)) has scattering peaks only along the $\overline{\Gamma} - \overline{M}$ direction. As in the case of experiments on $Bi_{1-x}Sb_x$, the suppressed scattering channels can be accounted for by incorporating the spin texture in the calculated SSP. However, to fully suppress the scattering modes along the $\overline{\Gamma} - \overline{K}$ direction the influence of the warped band structure must be considered. Hexagonal warping induces an out-of-plane spin component that oscillates along the $\overline{\Gamma} - \overline{M}$ directions and turns to point fully in plane along $\overline{\Gamma} - \overline{K}$, as sketched in Fig. 8(a). The calculated SSP shown in Fig. 8(c) agrees then well with the measured QPI pattern at high energies.

At low energies closer to the Dirac point, in the region where the dispersion is conic, scattering results in a simple circular pattern in the DFT of the real-space QPI (Fig. 7(f)). In this energy range there are no quasi-nesting conditions to support strong scattering modes. Detailed calculation within the transfer-matrix formalism [36,37] shows that instead of the $2\,k_F$ ring generated by backscattering of trivial surface states, such as those on Cu(1 1 1) shown in Fig. 2, the helical spin texture gives rise to a cusp-like feature at $q = 2\,k_F$ (see calculated profiles in Fig. 8(d)). The latter results from the oblique scattering from the sides of the circular CECs which has a finite spin overlap in the absence of the forbidden backscattering channel. Experimentally we find a clear 2 k_F feature that disperses inwards as the energy is lowered toward the Dirac point both in magnetically and non-magnetically doped samples (Fig. 8(e)). A conclusive observation of these features requires improved signal-to-noise ratio when compared to the QPI associated with the warped Fermi surface at higher energies, and their absence in some previous experiments has been incorrectly associated with an absence of backscattering. Still, we cannot draw the distinction between a 2 k_F peak and a cusp since those features are obscured by a broad background. It is interesting to note, that a 2 k_F scattering peak is also present in the QPI measurements in graphene [38], where the same transfer-matrix calculation holds as well.

Overall, the detailed measurements and analysis of the scattering patterns on various topological insulators (also including Sb, see below) provide strong evidence that spin-selective scattering strongly suppresses backscattering for topological surface states. Although scattering in other directions is allowed and undoubtedly limits the transport of topological surface states when defects are present (see below also for properties of charge defects), we can claim that STM experiments establish the absence of backscattering (especially those on $Bi_{1-x}Sb_x$ and on other topological insulators with wrapped Fermi surfaces) with high precision.

4 FLUCTUATION OF THE DIRAC POINT IN DOPED TOPOLOGICAL INSULATORS

Many of the proposed future research directions using topological insulators require tuning the chemical potential of topological surface states to the Dirac point. For example, interfacing topological insulators tuned to the Dirac point with superconductors is a key ingredient in approaches to realizing Majorana fermions with these materials [16–19]. Similarly, interaction with magnetism is expected to induce a gap at the Dirac energy and turn the massless topological Dirac fermions into massive excitations and result in a quantum spin Hall state with novel transport signatures [15]. A common means of tuning the Fermi energy to the Dirac point and inducing strong interactions is doping the bulk of the topological insulators or dosing their surfaces with various elements. However, examining the properties of topological surface states near the Dirac point shows that they are sensitive to the influence of bulk defects, especially dopants which may be needed to tune the properties of these materials like other semiconductors. These experiments also show similarities to results obtained on graphene [38,39], where Dirac electrons are influenced by the presence of nearby charged defects in the substrates supporting the graphene sheets [40]. Understanding the role of such defect-induced changes, as in the case

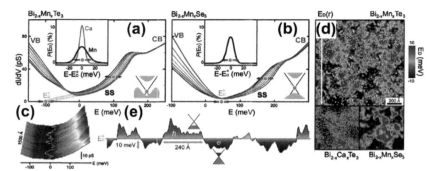

FIGURE 9 Bulk origin of the charge inhomogeneity detected on the surface. (Reproduced with permission from Ref. [14]) (a) and (b) The LDOS measured at various locations on the surface of Mn-doped Bi_2Te_3 and Bi_2Se_3, respectively, showing a rigid shift of the bulk and surface bands (band structures shown schematically). Insets show the Gaussian distribution of the Dirac energy. (c) High-energy resolution LDOS, taken along a line cut that crosses several charge puddles (white line) on Mn-doped Bi_2Se_3. (d) Spatial distribution of Dirac energy extracted from local shift in LDOS. Its structure shows no correlation with the locations of dopants (triangles). (e) Line cut along yellow line in (d) showing electron and hole "puddles" that would form once the chemical potential is tuned to the average Dirac point, E_D^0, in the sample.

of graphene, is important to understanding how the mobility of topological surface states can be improved.

The influence of bulk defects can be seen in STM spectra measured over a range of energies at different locations of the surfaces of doped topological insulators. Figure 9 shows examples of such spectra on the surfaces of doped Bi_2Te_3 and Bi_2Se_3 crystals. From these spectra it is evident that both the bulk and surface state bands are rigidly shifted in energy between different locations on the surface. These changes are due to bulk disorder, such as poorly screened charged dopants further into the bulk, since they are not necessarily correlated with the locations of the surface dopants resolved in STM topographs. Clearly, the Dirac energy in the surface state electronic structure varies spatially together with the bulk-induced nanoscale fluctuations of the electronic band structure, as illustrated by the line in Fig. 9(c). The extent of these variations can be mapped by extracting the energy shifts of the spectra obtained at different locations on the sample as shown in Fig. 9(d). Our measurements show that the Dirac energy follows a Gaussian distribution with a width of between 20 and 40 meV depending on the specific dopant and its actual concentration. A cross-section of the bulk-induced energy landscape, shown in Fig. 9(e), highlights that by tuning the sample's chemical potential close to the Dirac energy, the local electronic structure alternates between electron-like and hole-like doped regions that would have opposing helical spin texture.

The influence of bulk disorder on the topological surface states can be further quantified by the analysis of the QPI data at higher energies. For this analysis we divide the real-space maps into sub-regions, as shown in

Fig. 10(a), according to intervals of the energy shift of the Dirac point in Fig. 9(d). Obtaining the DFT from QPI measurements on two distinct regions (defined by upper and lower halves of the distribution of Dirac energies) we find clear shifts of the QPI's q vectors. An example of such a momentum shift ($\Delta q \sim 0.01$ Å^{-1}) is shown in Fig. 10(b). Clearly, the surface Dirac electrons alter their wavelength to adjust to the underlying bulk disorder potential, and are not immune to such perturbations. While such fluctuations in momentum are a relatively weak perturbation on the Dirac electrons at high energy, near the Dirac point they are comparable to the average value of the momentum. Figure 10(c) indeed shows that the Dirac electrons exhibit such a shift in momentum even close to the Dirac energy where their dispersion is perfectly linear. Combining these results with the Fermi velocity obtained from the QPI dispersion ($v_F = 1.3$ eV Å, in agreement with ARPES [33]) confirms that the extent of shift in momentum is consistent with the energy shift measured for the Dirac point ($\Delta E = v_F \Delta q \sim 16$ meV). Therefore, at energies close to the Dirac point in the presence of such fluctuations the sample is effectively made up of p-n junctions, and as a consequence, the momentum carried by these electrons becomes ill defined. The signature of this phenomenon can be seen in the absence of QPI patterns approaching the Dirac point when the topological surface states' wavelength is comparable to or larger than the length scale imposed by the bulk-induced disorder. Such fluctuations due to charge dopants result in the ill-defined nature of the momentum structure of topological surface

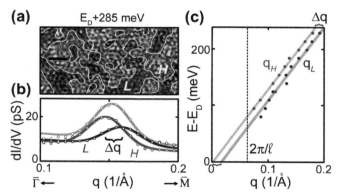

FIGURE 10 Spatial fluctuations of momentum. (Reproduced with permission from Ref. [14]) (a) The conductance map on $Bi_{1.95}Mn_{0.05}Te_3$ divided into sub-regions with high and low LDOS. (b) Profile of a q-space QPI peak along $\bar{\Gamma} - \bar{M}$ (gray), and those peaks when Fourier transformed separately from the different sub-regions in (a) showing a relative shift in momentum $\Delta q = 0.01$ Å^{-1} (lines are Gaussian fits to the data points). (c) Similar analysis in the region of conical energy dispersion of $Bi_{1.95}Mn_{0.05}Se_3$. QPI peaks from both sub-regions show linear dispersion shifted by an energy-independent $\Delta q \sim 0.01$ Å^{-1} with respect to each other. The dashed line marks $q = 2\pi/l$, below which the electronic wavelength is greater than the typical dimension of a puddle, l, causing the QPI pattern to vanish.

states near the Dirac point—an effect which is potentially more significant than that associated with the presence of magnetic dopants in our samples.

The physical picture emerging from these experiments and their more detailed analysis [14] is that despite their protection against backscattering and localization, topological surface states are susceptible to charge disorder caused by bulk defects. This conclusion is consistent with recent transport studies on topological insulators claiming that doped samples, besides having low mobilities, show magneto-fingerprints associated with mesoscopic fluctuations on the length scale of a few tens of nanometers found in macroscopic single crystals [41]. Transport experiments using electrostatic gating on thin films give evidence of a minimum conductivity [42] that has been attributed to a defect-induced residual charge density [43]. Both studies are consistent with our observation of bulk-induced fluctuations that introduce a length scale associated with the screening of defects and preclude tuning the chemical potential to the Dirac energy.

Overall, similar to efforts in graphene, the reduction of bulk charge defects appears to be required to fully realize the potential of topological insulators to carry out fundamental studies associated with properties of this system near the Dirac point and for any potential applications that rely on high carrier mobility. In fact, more recent experimental measurements on the newly realized topological insulator Bi_2Te_2Se [44,45] suggest that reducing charge defects can indeed result in less fluctuations of the topological surface states and their Dirac point. Although this new compound still contains defects that are associated with substitutional disorder (Se-to-Te), the number of charge defects is less than that seen in similarly grown samples of Bi_2Te_3.

5 ENHANCED TRANSMISSION THROUGH DEFECTS

The experiments discussed so far provide strong evidence that topological surface states do not undergo backscattering and hence cannot be localized by ordinary disorder. However, they do not actually probe the transmission of topological surface states nor do they provide any evidence that these states are actually extended in the presence of strong disorder. The extended nature of topological surface states would make these states wrap an entire sample despite surface roughness that would interrupt ordinary two-dimensional electronic states. To test this unique feature of topological surface states, we have used the STM to examine Sb's surface states and to measure their reflection by and transmission through atomic steps. Although Sb is not a topological insulator, instead being a semi-metal, it still has the inverted band structure that makes its surface states topological in nature as verified by previous ARPES studies [8,46].

The STM topographic image in Fig. 11 shows the surface of a cleaved Sb crystal with nanometer-sized atomic terraces that are separated by single atom

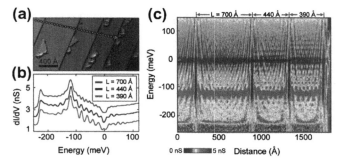

FIGURE 11 The topological surface states of Sb(1 1 1) on atomic terraces. (Reproduced with permission from Ref. [13]) (a) STM topographic image ($V_{bias} = 1$ V, $I = 8$ pA) of a 2,500 Å-by-1,250 Å area showing terraces of various widths separated by 3.7 Å-high single atomic steps. (b) The dI/dV measurement averaged over each terrace shows quantized peaks. The spectra are offset vertically for clarity. (c) Spatially and energetically resolved dI/dV measurements taken along the dotted arrow in (a) demonstrate the interference in space and the quantization in energy.

height steps along the [1 1̄ 0] crystallographic direction (Fig. 11(a)). We have examined the spatial and energy dependence of the LDOS within the terraces of Sb's surface by performing differential conductance (dI/dV) measurements along a line perpendicular to the step edges. In Fig. 11(b) the spatially averaged spectroscopic measurements for terraces of different widths show peaks that signal the occurrence of quantized resonances. These spectroscopic features start at around -225 meV, which is the bottom of the Sb's topological surface state bands as measured by ARPES studies [8,46]. Resolving the spatial variation of the LDOS within each terrace (Fig. 11(c)), we find clear evidence for quantized resonances, with nearly equal spacing in energy. The quantized resonances and the standing wave patterns in the LDOS are caused by scattering of Sb's topological surface states from the atomic step edges and can be analyzed to obtain reflection properties of these steps.

The standing wave patterns and the nearly linear quantized resonances in LDOS can be understood based on ARPES band structure for Sb's surface states and the spin-selection rules, as were discussed earlier in this chapter for other topological surface states. Figure 12 shows the Fourier transform of the dI/dV measurement on the 390 Å-wide terrace (Fig. 12(a)) revealing two different quantized wavevectors, q_1 and q_2, for the standing pattern of the LDOS. Unlike conventional confined states that involve superposition of states with equal and opposite momentum k [47–50], the helical spin texture of Sb eliminates the possibility of superposition of states with orthogonal spins. Consequently, the wavevectors (q) observed correspond to superposition between states with different values of k but similar spin states. From the schematic of ARPES measurements of Sb's constant energy contours (Fig. 2(c)) and their spin texture (Fig. 12(d)), we identify q_1 with scattering between adjacent hole pockets, which accordingly vanishes once the hole pockets overlap

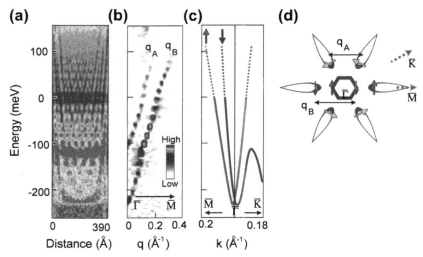

FIGURE 12   Allowed scattering wavevectors and their quantization. (Reproduced with permission from Ref. [13]) (a) The dI/dV measurement on a 390 Å-wide terrace. (b) Energy-resolved Fourier transform of the spatial modulation of the data in (a) reveals the quantization of scattering wavevectors q_A and q_B. (c) The dispersions of q_A and q_B match the dispersion of the surface bands as measured by ARPES along the high-symmetry directions (solid lines) and extend it above the Fermi level (dotted lines). (d), Contours of constant energy of the antimony surface state (ARPES measurements from Ref. [8]). The colored arrows represent spin texture of the surface state. The scattering wavevectors q_A and q_B indicate allowed scattering processes.

each other (around -110 meV in Fig. 12(c)). Similarly, q_2 corresponds to the superposition between the electron and the hole pockets with oppositely oriented momentum and parallel spin as shown in Fig. 12(d) [46]. The scattering from the step edges and geometric constraint set by the terrace width L results in quantization of these allowed superpositions of momentum states, such that $q_n = 2\pi n/L$ ($n > 0$ is an integer). Despite their quantization, the energy dependence of q_1 and q_2 relates to the dispersion measured by ARPES bands since $q_n(E_n) = k_i(E_n) - k_f(E_n)$. The nearly linear energy dependences of q_1 and q_2 have slopes (≈ 1.2 V·Å) that are in excellent agreement with ARPES measurements [8,46].

The energy widths of the quantized resonance in the LDOS contain information on the scattering properties for Sb's topological surface states. In general, energy level broadening, Γ, is inversely proportional to the lifetime of a state τ, $\Gamma \sim \hbar/\tau$. Examining the broadening of the resonances as a function of energy for different terraces (Fig. 13), we find $\Gamma(E) = \Gamma_L + \gamma(E - E_F)^2$, where E_F is the Fermi level (with $\hbar/\gamma = 0.37 \pm 0.01$ fs·eV^2, for terraces where level broadening is smaller than level spacing). Such a functional form illustrates the role of electron-electron scattering in the decoherence of topological surface states, similar to other Fermi liquids [52]. However, the residual finite resonance

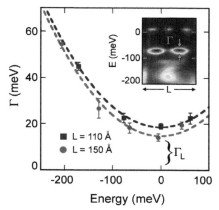

FIGURE 13 Lifetime and leakage of quantized quasiparticles. (Reproduced with permission from Ref. [13]) Each resonant peak is fitted to a Lorentzian function to yield the full-width at half-maximum of the quantized energy peaks, Γ (inset). The plot shows the energy dependence of Γ for two different terraces. The dashed lines are parabolic fits, and Γ_L is the peak broadening at the Fermi energy.

width at E_F, Γ_L characterizes the degree to which step edges can reflect Sb's topological surface states in the same spin channel.

In order to find the reflection probability of topological surface states from the atomic steps, R_0, we use a simple model of energy broadening of the resonant states based on a double barrier potential. We sum over all electron plane waves that get reflected from the step edges, with appropriate momenta for left and right movers, k_L and k_R, respectively. The measured interference pattern is proportional to the absolute value squared of that summation, where the full-width at half-maximum of the resonant states is given by $\Gamma_L = 2C(1 - |R_0|^2)/LR_0$, with $C = (C_L^{-1} + C_R^{-1})^{-1}$ and $E = C_{L,R} k_{L,R}$. This model yields a reflection probability of $42 \pm 4\%$. This reflectivity is much lower than those reported for free-electron-like surface states on Cu, Ag, or Au from STM and other measurements; however, it can be a result of scattering of the surface states into the bulk states rather than their transmission. Indeed, previous studies show that non-topological noble metal surface states are absorbed in the bulk states with 30–50% probability without transmission when scattered by step edges [51–55].

Finite transmission through the atomic step edges should give rise to coupling between electronic states on adjacent terraces. To search for evidence of such effects, we examined a configuration of atomic steps consisting of a narrow (110 Å) and a wide terrace (2500 Å), which resembles a nanoscale Fabry-Perot resonator, as shown in Fig. 14. The narrow terrace shows signature of quantized resonances according to the spin scattering rules described above (Fig. 14(b)). On the wide terrace, which is effectively semi-infinite, the LDOS

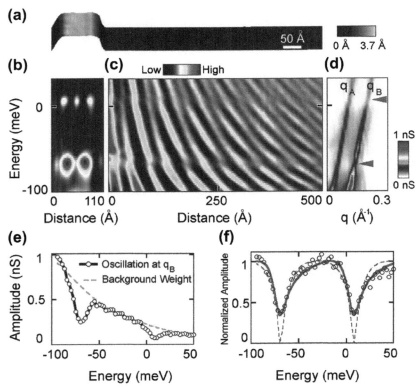

FIGURE 14 Resonant tunneling between adjacent terraces. (Reproduced with permission from Ref. [13]) (a) STM topographic image of a narrow terrace ($L = 110$ Å) and part of an adjacent wide terrace ($L = 2500$ Å). (b) On the narrow terrace, the dI/dV measurement shows the quantization of the energy levels. (c) The dI/dV measurement on the wide terrace shows sudden phase shifts and suppressions of the dI/dV intensities at around $+5$ meV and -70 meV due to resonant tunneling. The averaged background conductance at each energy has been subtracted. (d) The Fourier transform of the spatial modulation of the data in (c). The gray markers indicate suppression of modulation intensities. (e) Spectral weight of the scattering wavevector $\mathbf{q_B}$ as extracted from (d). The dashed line corresponds to the background spectral weight in the absence of resonance tunneling. (f) Spectral weight (circles) normalized by the background. The best fit (red line) yields $\sim 42\%$ reflection, $\sim 35\%$ transmission, and $\sim 23\%$ bulk absorption from the scattering process at the boundary. The blue line, which is the fit based on a model without bulk absorption, is displayed for comparison. (For interpretation of the references to color in this figure legend, the reader is referred to the web version of this book.)

oscillation forms an almost continuous pattern (Fig. 14(c)) with diverging wavelength toward the bottom of the surface state bands. The standing wave patterns on the wide terrace also obey the helical spin texture and are made up of the two wavevectors $\mathbf{q_1}$ and $\mathbf{q_2}$ described above, as illustrated in the Fourier transform shown in Fig. 14(d). Any finite transmission through the step edges results in resonant tunneling of surface state electrons from the wide terrace through the electron states of the narrow terrace.

Evidence for this resonant tunneling of the topological surface states can be seen both in the structure of energy-resolved modulations of the LDOS and its Fourier transform in Fig. 14. At energies corresponding to the quantized resonances of the narrow terrace, there are clear suppressions of the modulation in the LDOS on the wide terrace. Figure 14(e) illustrates this phenomenon by plotting the Fourier intensity of the q_2 peak as a function of energy. Once the overall background changes in the LDOS (dotted green line in Fig. 14(e)) are removed, the intensity of the q_2 Fourier component displays characteristic signs of Fabry-Perot resonant tunneling (Fig. 14(f)) in which the surface state electrons pass through the small terrace without reflection. Deviation from perfect transmission is, however, a signature of scattering of the surface states into the Sb's bulk states. We use a similar summation procedure over all transmitted and reflected plane waves (with probability T_0 and R_0 respectively) to find the oscillating intensity profile of mode q_2. In the present instance, however, we also take into account scattering into the bulk with probability α via the continuity equations: $1 + R_0 = T_0, |R_0|^2 + |T_0|^2 = 1 - \alpha$, from which we can deduce that the probability of this process is $\alpha = 23 \pm 7\%$. Similarly, the width of the Fabry-Perot resonance (Fig. 14f) can be used to extract the transmission and reflection probabilities and to show that they occur at $35 \pm 3\%$ and $42 \pm 4\%$, respectively. In contrast, STM data on Fabry-Perot structures for non-topological surface states of Ag (1 1 1) shows no evidence for resonant tunneling and can be explained in detail by considering only reflections and absorption of the surface states at the step edges [53].

The remarkable finding that topological surface states on Sb are as likely to be transmitted through as reflected by an atomic step edge demonstrates an important difference between topological states and typical surface states in other materials. The transmissibility illustrates the extended nature of topological surface states even in the presence of strong surface barriers. The nearly equal probability for reflection and transmission at the step edges in Sb can perhaps be understood based on the topological surface state's band structure, which includes both forward-and backward-moving momenta with the same spin orientation (Fig. 2).

6 CONCLUSION AND OUTLOOK

In this chapter, we reviewed several STM studies that have established the novel scattering and transmission properties of topological surface states. These measurements confirm the topological nature of these states by comparing our findings to many previous studies of ordinary surface states on noble metal surface states. We have not provided a comprehensive review of all the STM work on topological insulators, such as those on MBE grown sample or the studies carried out in a magnetic field. The studies in a magnetic field have provided for example important evidence for the Dirac dispersion of topological surface states on several Bi-based materials, providing further confirmation of ARPES results but with higher-energy resolution. Clearly STM's high energy and spatial resolution will continue to play a key role in future research on topological surface states. In particular examining the local influence of magnetism and superconductivity in topological surface states appears to be promising future directions.

ACKNOWLEDGMENTS

It is a great pleasure for us to acknowledge other members of our group that have contributed to the results discussed here. In particular, we would like to thank Jungpil Seo and Ilya Drozdov for their contributions to STM measurements on topological insulators. Collaboration with Robert Cava's group at Princeton has been critical to the success of these experiments, where close feedback between measurements and material preparation has made challenging experiments possible. Finally, we acknowledge funding from NSF-DMR1104612 and NSF-MRSEC programs through the Princeton Center for Complex Materials (DMR-0819860), the W.M. Keck foundation, Eric and Linda Schmidt Transformative fund at Princeton, ONR, ARO, and DARPA-SPAWAR Grant N6601-11-1-4110.

REFERENCES

[1] Hasan MZ, Kane CL. Topological insulators. Rev Mod Phys 2010;82:3045.
[2] Moore JE. The birth of topological insulators. Nature 2010;464:194.
[3] Qi XL, Zhang SC. The quantum spin Hall effect and topological insulators. Phys Today 2010;63:33.
[4] Bernevig BA, Zhang SC. Quantum spin Hall effect. Phys Rev Lett 2006;96:106802.
[5] Bernevig BA, Hughes TL, Zhang SC. Quantum spin Hall effect and topological phase Transition in HgTe quantum wells. Science 2006;314:1757.
[6] König M et al. Quantum spin Hall insulator state in HgTe quantum wells. Science 2007;318:766.
[7] Hsieh D et al. A topological Dirac insulator in a quantum spin Hall phase. Nature 2008;452:970.
[8] Hsieh D et al. Observation of unconventional quantum spin textures in topological insulators. Science 2009;323:919.
[9] Teo JCY, Fu L, Kane CL. Surface states and topological invariants in three-dimensional topological insulators: application to $Bi_{1-x}Sb_x$. Phys Rev B 2008;78:045426.
[10] Zhang H et al. Topological insulators in Bi_2Se_3, Bi_2Te_3 and Sb_2Te_3 with a single Dirac cone on the surface. Nat Phys 2009;5:438.
[11] Zhang HJ, Liu CX, Qi XL, Deng XY, Dai X, Zhang SC, et al. Electronic structures and surface states of topological insulator $Bi_{1-x}Sb_x$. Phys Rev B 2009;80:085307.
[12] Roushan P, Seo J, Parker CV, Hor YS, Hsieh D, Qian D, et al. Topological surface states protected from backscattering by chiral spin texture. Nature 2009;460:1106.
[13] Seo J, Roushan P, Beidenkopf H, Hor YS, Cava RJ, Yazdani A. Transmission of topological surface states through surface barriers. Nature 2010;466:343.
[14] Beidenkopf H, Roushan P, Seo J, Gorman L, Drozdov I, Hor YS, et al. Spatial fluctuations of helical Dirac fermions on the surface of topological insulators. Nat Phys 2011;7:939.
[15] Li R, Wang J, Qi XL, Zhang SC. Dynamical axion field in topological magnetic insulators. Nat Phys 2010;6:284.

[16] Linder J, Tanaka Y, Yokoyama T, Sudbø A, Nagaosa N. Interplay between superconductivity and ferromagnetism on a topological insulator. Phys Rev B 2010;81:184525.

[17] Cheng M, Lutchyn RM, Galitski V, Sarma SD. Tunneling of anyonic Majorana excitations in topological superconductors. Phys Rev B 2010;82:094504.

[18] Fu L, Kane CL. Superconducting proximity effect and Majorana fermions at the surface of a topological insulator. Phys Rev Lett 2008;100:096407.

[19] Tanaka Y, Yokoyama T, Nagaosa N. Manipulation of the Majorana fermion, Andreev reflection, and Josephson current on topological insulators. Phys Rev Lett 2009;103:107002.

[20] Crommie MF, Lutz CP, Eigler DM. Imaging standing waves in a two-dimensional electron gas. Nature 1993;363:524.

[21] Petersen L, Sprunger PT, Hofmann P, Lægsgaard E, Briner BG, Doering M, et al. Direct imaging of the two-dimensional Fermi contour: Fourier-transform STM. Phys Rev B 1998;57:R6858.

[22] LaShell S, McDougall BA, Jensen E. Spin splitting of an Au(1 1 1) surface state band observed with angle resolved photoelectron spectroscopy. Phys Rev Lett 1996;77:3419.

[23] Hoffman JE et al. Imaging quasiparticle interference in $Bi_2Sr_2CaCu_2O_{8+}$. Science 2002;297:1148.

[24] Markiewicz RS. Bridging k and q space in the cuprates: comparing angle-resolved photoemission and STM results. Phys Rev B 2004;69:214517.

[25] Kane CL, Mele EJ. Z_2 topological order and the quantum spin Hall effect. Phys Rev Lett 2005;95:146802.

[26] Moore JE, Balents L. Topological invariants of time-reversal-invariant band structures. Phys Rev B 2007;75:121306(R).

[27] Zhang T, Cheng P, Chen X, Jia JF, Ma X, He K, et al. Experimental demonstration of the topological surface states protected by the time-reversal symmetry. Phys Rev Lett 2009;103:266803.

[28] Okada Y, Dhital C, Zhou W, Huemiller ED, Lin H, Basak S, et al. Direct observation of Broken time-reversal symmetry on the surface of a magnetically doped topological insulator. Phys Rev Lett 2011;106:206805.

[29] Alpichshev Z, Analytis JG, Chu JH, Fisher IR, Chen YL, Shen ZX, et al. STM imaging of electronic waves on the surface of Bi_2Te_3: topologically protected surface states and hexagonal warping effects. Phys Rev Lett 2010;104:016401.

[30] Zhang H, Liu CX, Qi XL, Dai X, Fang Z, Zhang SC. Topological insulators in Bi_2Se_3, Bi_2Te_3 and Sb_2Te_3 with a single Dirac cone on the surface. Nat Phys 2009;5:438.

[31] Chen YL, Analytis JG, Chu JH, Liu ZK, Mo SK, Qi XL, et al. Experimental realization of a three-dimensional topological insulator, Bi_2Te_3. Science 2009;325:178.

[32] Hsieh D, Xia Y, Qian D, Wray L, Dil JH, Meier F, et al. A tunable topological insulator in the spin helical Dirac transport regime. Nature 2009;460:1101.

[33] Xia Y, Qian D, Hsieh D, Wray L, Pal A, Lin H, et al. Observation of a large-gap topological-insulator class with a single Dirac cone on the surface. Nat Phys 2009;5:398.

[34] Fu L. Hexagonal warping effects in the surface states of topological insulator Bi_2Te_3. Phys Rev Lett 2009;103:266801.

[35] Hor YS, Roushan P, Beidenkopf H, Seo J, Qu D, Checkelsky JG, et al. Development of ferromagnetism in the doped topological insulator $Bi_{2-x}Mn_xTe_3$. Phys Rev B 2010;81:195203.

[36] Biswas RR, Balatsky AV. Impurity-induced states on the surface of three-dimensional topological insulators. Phys Rev B 2010;81:233405.

[37] Lee WC, Wu C, Arovas DP, Zhang SC. Quasiparticle interference on the surface of the topological insulator Bi_2Te_3. Phys Rev B 2009;80:245439.

[38] Zhang Y, Brar VW, Girit C, Zettl A, Crommie MF. Origin of spatial charge inhomogeneity in graphene. Nat Phys 2009;5:722.

[39] Martin J, Akerman N, Ulbricht G, Lohmann T, Smet JH, von Klitzing K, et al. Observation of electron–hole puddles in graphene using a scanning single-electron transistor. Nat Phys 2008;4:144.

[40] Adam S, Hwang EH, Galitski VM, Das Sarma S. A self-consistent theory for graphene transport. Proc Natl Acad Sci USA 2007;104:18392.

[41] Checkelsky JG et al. Quantum interference in macroscopic crystals of nonmetallic Bi_2Se_3. Phys Rev Lett 2009;103:246601.

[42] Kim D, Cho S, Butch NP, Syers P, Kirshenbaum K, Adam S, et al. Surface conduction of topological Dirac electrons in bulk insulating Bi_2Se_3. Nat Phys 2012;8:460.

[43] Adam S, Hwang EH, Sarma SD. Two-dimensional transport and screening in topological insulator surface states. Phys Rev B 2012;85:235413.

[44] Ren Z, Taskin AA, Sasaki S, Segawa K, Ando Y. Large bulk resistivity and surface quantum oscillations in the topological insulator Bi_2Te_2Se. Phys Rev B 2010;82:241306(R).

[45] Xiong J, Luo Y, Khoo Y, Jia S, Cava RJ, Ong NP. High-field Shubnikov–de Haas oscillations in the topological insulator Bi_2Te_2Se. Phys Rev B 2012;86:045314.

[46] Gomes KK, Ko W, Mar W, Chen Y, Shen ZX, Manoharan HC. Quantum imaging of topologically unpaired spin-polarized Dirac fermions. Preprint at <http://arxiv.org/abs/0909.0921v2> (2009).

[47] Pivetta M, Silly F, Patthey F, Pelz JP, Schneider WD. Reading the ripples of confined surface-state electrons: profiles of constant integrated local density of states. Phys Rev B 2003;67:193402.

[48] Li J, Schneider WD, Berndt R, Crampin S. Electron confinement to nanoscale Ag islands on Ag(111): a quantitative study. Phys Rev Lett 1998;80:3332.

[49] Heller EJ, Crommie MF, Lutz CP, Eigler DM. Scattering and absorption of surface electron waves in quantum corrals. Nature 1994;369:464.

[50] Fiete GA, Heller EJ. Theory of quantum corrals and quantum mirages. Rev Mod Phys 2003;75:933.

[51] Bürgi L, Jeandupeux O, Brune H, Kern K. Probing hot-electron dynamics at surfaces with a cold scanning tunneling microscope. Phys Rev Lett 1999;82:4516.

[52] Bürgi L, Jeandupeux O, Hirstein A, Brune H, Kern K. Confinement of surface state electrons in Fabry-Pérot resonators. Phys Rev Lett 1998;81:5370.

[53] Crommie MF, Lutz CP, Eigler DM. Imaging standing waves in a two-dimensional electron gas. Nature 1993;363:524.

[54] Avouris P, Lyo IW. Observation of quantum size effects at room temperature on metal surfaces with STM. Science 1993;264:942.

[55] Mugarza AA et al. Lateral quantum wells at vicinal Au(111) studied with angle-resolved photoemission. Phys Rev B 2002;66:245419.

Chapter 8

Transport Experiments on Three-Dimensional Topological Insulators

Jeroen B. Oostinga[*] and Alberto F. Morpurgo[†]

[*]*Physikalisches Institut (EP3), University of Würzburg, Am Hubland, D-97074 Würzburg, Germany*
[†]*DPMC & GAP, University of Geneva, 24 quai Ernest-Ansermet, CH1211 Geneva, Switzerland*

Chapter Outline Head

Topological Insulators. http://dx.doi.org/10.1016/B978-0-444-63314-9.00008-1

1 INTRODUCTION

The classification of the electronic properties of crystalline solids as metals or insulators—depending on whether the Fermi level falls in a partially filled band or in the gap between two bands—is one of the great early successes of quantum mechanics, and represents a cornerstone of modern condensed matter physics. For many decades, based on this principle, band-structure calculations of a virtually infinite variety of materials have been performed in the attempt to understand or predict their electronic properties. The recent discovery that this classification scheme is incomplete has therefore come as a true surprise [1,2]. We now know that, in the presence of strong spin-orbit interaction and time-reversal symmetry, topological insulators represent a third class of systems [3,4], characterized by a gapped spectrum in their bulk coexisting with gapless states at their edges—for 2D electronic systems—or at their surfaces—for the 3D topological insulators that we will discuss in this chapter.

The surfaces of 3D strong topological insulators (TIs) are predicted to host an odd number of families of Dirac fermions, that are robust against time-reversal symmetry preserving perturbations [3,4]. In the presence of time-reversal symmetry, these surface Dirac fermions cannot be gapped out completely: different families of Dirac fermions can hybridize in pairs and develop a gap, but—since their number is odd—at least one of the families must remain ungapped. Theory also predicts that, for a single Dirac fermion family, the amplitude of probability for backscattering vanishes, and that Anderson localization cannot occur. When the Fermi level inside the bulk is located in a bandgap, therefore, the low-energy transport properties of topological insulating materials are entirely dominated by the surface states, which remain metallic. Under these conditions, the investigation of Dirac fermions at the surface of 3D topological insulators by means of transport experiments seems straightforward, since at temperatures much lower than the bandgap, the only states contributing to transport are those on the surface.

The experimental situation, unfortunately, is not as clear-cut as the underlying theoretical concepts may suggest. The bulk bandgaps of the 3D topological insulators that are being investigated experimentally are a few hundred milli-electronvolts at the most, and typically less. These materials, therefore, exhibit bulk electrical properties that are characteristic of small-gap semiconductors, including a strong sensitivity to defects, such as in-gap states or dopants unintentionally present. Defects result in parasitic conduction channels, either by shifting the Fermi level out of the bandgap, or by inducing the formation of a highly conducting impurity band, which hinder the observation of surface transport. Additional difficulties originate from the poor chemical stability of the material surfaces and from the moderately low carrier mobility values in most materials available (it is worth pointing out that the topological considerations that warrant the existence of Dirac fermions at the surface of 3D topological insulators do not imply a large value of carrier mobility, as it is sometimes suggested in the literature).

There is little doubt that angle-resolved photoemission spectroscopy (ARPES), reviewed in an earlier Chapter, is currently the best probe of the Dirac fermion states at the surface of 3D topological insulators [3]. ARPES has been used to demonstrate the presence of these states and to measure their dispersion relation, in all of the most relevant materials that have been considered so far. It is also clear, however, that to explore the low-energy properties of 3D topological insulators, the controlled investigation of transport in nanofabricated device structures is essential. This is a main motivation for the considerable effort that is devoted to the investigation of the transport properties of these materials. While no textbook-quality experimental manifestation of Dirac surface fermions in the transport properties of 3D TIs has been reported yet (as compared, for instance, to what has been done in graphene [5]), much evidence has been collected that can only be explained simply in terms of the presence of Dirac surface fermions. It is our purpose, after having introduced the key materials that have been investigated so far, to discuss different types of experiments that have been performed until now, emphasizing the implications—and the limitations—of the reported results.

Since this is a very rapidly developing field of research, it is not possible at this stage to give a complete overview. This chapter should be viewed as an analysis of the current status in this area, after that a first generation of experiments has been performed. While we have tried to include important early contributions, some relevant work may have escaped our attention, for which we apologize in advance. We also have tried to maintain a critical attitude, certainly not with the intention to downplay the contribution of research groups that have devoted hard work to the development of this field, but in the hope to alert the reader to experimental and practical issues that remain to be solved, thereby hoping to provide a stimulus to find solutions.

2 BULK TRANSPORT THROUGH BISMUTH-BASED COMPOUNDS

The search for 3D topological insulators naturally focuses on classes of materials with strong spin-orbit interaction, among which bismuth-based compounds are currently the most common. These compounds have been studied extensively for more than half a century because of their interesting electronic and thermoelectric properties. Experiments exploring the electronic properties of bismuth already started in the 1930s, focusing on the study of the de Haas-van Alphen effect. In the 1950s, research shifted to the binary compounds $Bi_{1-x}Sb_x$, Bi_2Se_3, and Bi_2Te_3 [6–14], with the last one attracting most interest for its potential in room-temperature thermoelectric applications. In the attempt to achieve optimal thermoelectric response, considerable effort has been devoted to the synthesis and investigation of materials of the type $(Bi_{1-x}Sb_x)_2(Te_{1-y}Se_y)_3$, hereafter referred to as BSTS compounds. BSTS crystals can be obtained relatively easily by melting a stoichiometric mixture with the preferred Bi/Sb$(= \frac{1-x}{x})$ and Te/Se$(= \frac{1-y}{y})$ ratios, followed by a cooling procedure (e.g., using a modified Bridgman method or a floating zone technique). Sufficiently good contacts can be readily realized on these small-gap semiconductors, which has enabled electrical and thermal transport experiments on millimeter-sized bulk crystals to be performed already at an early stage. These experiments have provided systematic information about the electrical resistivity (ρ), type of majority charge carriers (n-type/p-type), charge densities (n_e, n_h), and charge carrier mobility (μ_e, μ_h) in this class of materials [6–14].

2.1 Bi$_{1-x}$Sb$_x$

The first material for which a strong 3D topological insulating state has been predicted theoretically is Bi$_{1-x}$Sb$_x$, for x in the range between 0.07 and 0.22 [15]. Together with subsequent theoretical studies predicting that also Bi$_2$Se$_3$, Bi$_2$Te$_3$, and Sb$_2$Te$_3$ are 3D topological insulators [16], this work has strongly revived interest in BSTS compounds. Whereas the focus of all previous investigations was on the bulk electronic properties of these materials, exploring their topological insulating nature has required the study of the surface electronic properties. The unambiguous observation of the surface states expected from the theoretical predictions has been made by means of angle-resolved photoemission spectroscopy (ARPES). The first experimental success has been reported in 2008 on BiSb alloys [17], with ARPES measurements confirming that Bi$_{1-x}$Sb$_x$ crystals (with x in the range 0.07–0.22) exhibit a bandgap (E_g) in their bulk bandstructure coexisting with five—an odd number, as expected on the basis of topological considerations—families of surface Dirac fermions.

Experiments have been performed to investigate the transport properties as a function of temperature T and magnetic field B, for different material compositions. Measurements show that the resistance R of a Bi$_{1-x}$Sb$_x$ crystal with $x \approx 0.06$ has a metallic temperature dependence ($dR/dT > 0$). A crystal with $x \approx 0.09$, on the contrary, exhibits an insulating temperature dependence ($dR/dT < 0$) [18]. While this observation may suggest that a bandgap is opened in the bulk states, the resistance is found to stop increasing when T is lowered below approximately 50 K, and remains finite at lower temperatures. For this same stoichiometry, the magnetic field dependence of the resistance exhibits clear Shubnikov-de Haas (SdH) oscillations [19]. A careful analysis of their angular dependence—together with measurements of the de Haas-van Alphen effect—indicates that the SdH oscillations originate from different sets of states, some of which have a 3D character, while others have a 2D nature. The coexistence of bulk conductance due to a non-vanishing density of bulk (3D) carriers at the Fermi level ($n_b \sim 10^{19}$ cm^{-3}) and of 2D states makes it difficult to investigate the precise origin of the latter. The complicated structure of the surface states observed by ARPES in BiSb materials makes a clean interpretation of transport experiments extremely difficult.

2.2 Bi$_2$Se$_3$ and Bi$_2$Te$_3$

The focus of transport experiments moved to Bi$_2$Se$_3$ and Bi$_2$Te$_3$ after these compounds were predicted to be 3D topological insulators, because of their much simpler properties as compared to Bi$_{1-x}$Sb$_x$ [16]. ARPES measurements rapidly confirmed the key aspects of the theoretical predictions, showing that both Bi$_2$Se$_3$ [20] and Bi$_2$Te$_3$ [21] crystals exhibit topological surface states in the bulk bandgap at the high-symmetry Γ point. The size of the bandgap for Bi$_2$Se$_3$ and Bi$_2$Te$_3$ was found to be $E_g \approx 300$ meV and $E_g \approx 165$ meV, respectively, in good agreement with bandstructure calculations (Fig. 1). These bandgap values are larger than in Bi$_{0.9}$Sb$_{0.1}$, and even than the thermal energy at room temperature, which makes these materials particularly promising for the investigation of transport. Additionally, and in contrast to the case of Bi$_{0.9}$Sb$_{0.1}$ where five families of surface states coexist, the states at the surface of both Bi$_2$Se$_3$ and Bi$_2$Te$_3$ are effectively described by a single Dirac cone. Bi$_2$Se$_3$ appears to be better suited to investigate the surface Dirac fermions by transport experiments, because the Dirac point (i.e., the energy separating Dirac electrons from Dirac holes) resides inside the bulk bandgap of the material, approximately 200 meV below the bottom of the bulk conduction band. In Bi$_2$Te$_3$, on the contrary, the Dirac point is located at the top of the bulk valence band, which makes it more difficult to separate the contribution to transport of the hole part of the Dirac surface states, from that due to the bulk valence band (Fig. 1).

As-grown Bi$_2$Se$_3$ crystals are usually found to be electron doped, as indicated by Hall effect measurements, with bulk carrier densities between $n_b \sim 10^{17}$ cm^{-3} and $n_b \sim 10^{19}$ cm^{-3} [6–9,22–27]. In this case, SdH conductance oscillations are often observed, which can be attributed—from their angular

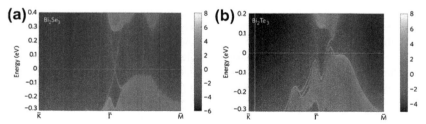

FIGURE 1   Calculated energy and momentum dependence of the local density of states for Bi$_2$Se$_3$ (a) and Bi$_2$Te$_3$ (b) on the [1 1 1] surface. The warmer colors represent higher local densities of states. The red and blue regions indicate bulk energy bands and bulk energy gaps, respectively. The surface states can be clearly seen around the Γ point as red lines dispersing in the bulk gap. The dashed white lines denote the Fermi level ($E \equiv 0$ eV) in undoped Bi$_2$Se$_3$ and Bi$_2$Te$_3$ of perfect crystallinity. (For interpretation of the references to color in this figure legend, the reader is referred to the web version of this book.) Adapted from Ref. [16].

dependence—to electrons residing in 3D bulk states. This indicates that the Fermi level is positioned in the bulk conduction band. An identical conclusion is normally drawn from ARPES measurements [20,23], which, however, is not entirely free of ambiguity, since ARPES measures the sample surface and not the bulk. Band-bending effects in small-gap semiconductors, that locally shift the Fermi level at the surface only, can be relevant and can make it difficult to draw conclusions about bulk properties from what is observed at the surface (for instance, it is often found that the position of the Fermi level observed in ARPES shifts with time, because of modifications of the exposed surface during the measurements [28,29]). Depending on the crystal growth conditions, such as the initial stoichiometry and temperature treatments, and possibly also on the part of the ingot from which a specific sample is taken, crystals exhibiting electron densities as low as $n_b \approx 10^{16}$ cm^{-3} can also be found [22] (Fig. 2). In this case, SdH resistance oscillations are normally not observed.

Finding a rather broad range of carrier densities depending on the details of the crystal growth conditions is not surprising, since in small-gap semiconductors, the density of charge carriers is sensitively dependent on small amounts of defects and dopants. Identifying the specific defects responsible for the presence of charge carriers, however, is not at all straightforward. It is often claimed that electrons in Bi_2Se_3 are introduced in the conduction band by selenium vacancies acting as dopants. Although the presence of selenium vacancies is likely, owing to the high volatility of selenium, the experimental evidence linking them to the doping level of bulk Bi_2Se_3 crystals does not seem to be compelling. For instance, scanning tunneling microscopy (STM) and spectroscopy (STS) experiments at the surface of Bi_2Se_3 crystals performed by Alpichshev and collaborators [30] illustrate some of the complications. These STM measurements enable the identification of defects corresponding to selenium vacancies, which manifest themselves in a characteristic triangular shape, and it is found that their density is in fair agreement with the electron density obtained from Hall effect measurements. Such an observation seems to corroborate the role of selenium vacancies as dopants. However, STM measurements also indicate that the electronic states associated with the selenium vacancies are well below the bottom of the conduction band ($\sim$150–200 meV), which makes it unclear why doped electrons should remain in the conduction band at low temperature, as it is observed in transport experiments. Additionally, STM measurements also detect a very large density of states in the bulk bandgap of Bi_2Se_3 crystals. If it is assumed that this density of states is present in the bulk and not only at the surface, these in-gap states should act as a "sink" for the doped charge carriers, and play a key role in determining the position of the Fermi level in the bulk. That is: whether the Fermi level is at the bottom of the conduction band, or in an impurity band inside the gap should not depend only on the density of dopants, supposedly the selenium vacancies, but also on the density of in-gap states. Indeed, although with less frequency, Bi_2Se_3 crystals are found, for which the Fermi level is located inside the bandgap of the material [31].

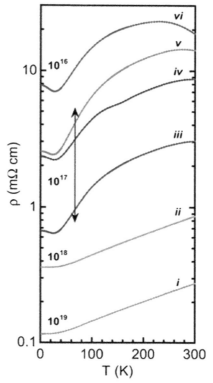

FIGURE 2 Comparison of the temperature-dependent electrical resistivity $\rho(T)$ of Bi_2Se_3 crystals with different carrier densities n, as estimated from Shubnikov-de Haas oscillations: (i) $1 \cdot 10^{19}$ cm^{-3}, (ii) $5.3 \cdot 10^{18}$ cm^{-3}, (iii) $4.9 \cdot 10^{17}$ cm^{-3}, (iv) $3.7 \cdot 10^{17}$ cm^{-3}, (v) $3.3 \cdot 10^{17}$ cm^{-3}, and (vi) $\sim 10^{16}$ cm^{-3}. The crystals with the highest density exhibit a clear metallic temperature dependence, while $\rho(T)$ for crystals of $n < 10^{18}$ cm^{-3} show anomalous behavior. Adapted from Ref. [22].

While Bi_2Se_3 naturally tends to grow as a n-type material, Bi_2Te_3 can be obtained as either n-type or p-type by varying the stoichiometric ratio of the Bi/Te employed in the crystal growth [10–12,27,32]. The Fermi level can be controllably positioned in the conduction (or valence) band by slightly decreasing (or increasing) the Bi/Te ratio. The n-type behavior is often attributed to Te -vacancies which can act as electron donors, while p-type behavior may originate from anti-site defects—Bi positioned at Te sites and vice versa— which can act as electron acceptors (or hole donors). The bulk density can be varied from $n_b \sim 10^{19}$–10^{20} cm^{-3} down to values as low as $n_b \sim 10^{17}$ cm^{-3}. Presumably, the Fermi level is close to the band edge at the lowest densities.

In Bi_2Se_3 and Bi_2Te_3, impurity states in the bandgap are expected to form an impurity band already at rather low densities, because of the large Bohr radius associated with the impurity states [13,23,33] (a Bohr radius of ~ 5 nm originates from the large dielectric constant, $\epsilon_r \approx 100$, and from the small effective mass, $m_{\text{eff}} \approx 0.1 \cdot m_0$). In the presence of an impurity band, a residual bulk conductivity is present even when the Fermi level is positioned in the bandgap, and, as compared to more common semiconductors (e.g., silicon), the impurity band conductivity should be expected to be much larger in Bi_2Se_3 and Bi_2Te_3. Indeed, the large Bohr radius increases the overlap between nearby impurity sites, the small bandgap causes a decay of the bound states on a longer length scale, and—possibly most important—the large dielectric constant strongly suppresses Coulomb interactions, a dominant mechanism responsible for suppressing transport in an impurity band (e.g., through the opening of a so-called Coulomb gap [34]).

Irrespective of the microscopic mechanism responsible for transport, the largest resistivity values that have been observed in as-grown Bi_2Se_3 and Bi_2Te_3 crystals at $T \approx 4$ K are $\rho \sim 1$–10 m$\Omega \cdot$ cm for typical charge densities of $n_b \sim 10^{17}$ cm^{-3} [22,24,35,36]. Such values are rather low for a material that should supposedly have an insulating bulk. The resistivity of these crystals usually shows a metallic temperature dependence ($d\rho/dT > 0$; Fig. 2), compatible with either a finite population of a bulk band [22,23,25,35] (in which case SdH resistance oscillations with an angular dependence expected for 3D bulk band states have been frequently reported) or with the presence of a highly conducting impurity band [31] (with no "bulk" SdH oscillations observed). The presence of such a non-insulating bulk poses severe problems to the investigation of surface transport, and two main strategies have been followed at the level of the material synthesis. They consist of reducing the bulk conductivity by either adding compensating dopants to the crystal or by suitably tuning the crystal composition.

2.3 Controlling the Bulk Conductivity

For Bi_2Se_3 crystals in which the Fermi level E_F is normally positioned in the bulk conduction band, E_F can be shifted all the way into the valence band by adding acceptors to Bi_2Se_3. Suitable atoms, which for small concentrations substitute Bi atoms during the crystal growth, are Ca [26,37,38], Cd [24], and Sb [36]. In a sequence of experiments with increasingly larger Ca concentration, the position of the Fermi level in the valence and the conduction band could be determined using Hall effect and thermopower measurements to identify the nature of charge carriers. E_F could also be tuned in the bandgap by optimizing the concentration of acceptors to compensate for the density of electrons initially present, resulting in a residual charge density as low as $n_b \sim 10^{16}$ cm^{-3} and a resistivity up to $\rho \sim 100$ m$\Omega \cdot$ cm [24,27,36,38]. Contrary to the uncompensated material, which normally shows a metallic temperature dependence of the resistivity, compensated crystals exhibit an insulating temperature dependence of the resistivity [24,27,36,38] (Fig. 3). Interestingly, SdH resistance oscillations due to bulk band states were observed both for crystals having the

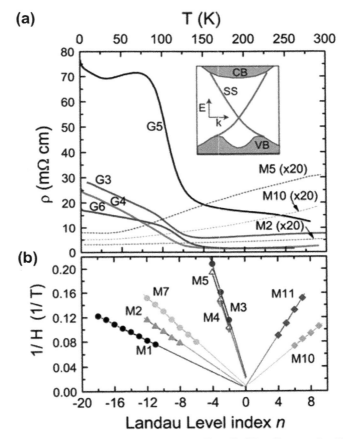

FIGURE 3 (a) Temperature-dependent resistivity $\rho(T)$ of $Ca_x Bi_{2-x} Se_3$ crystals with different Ca concentration. By increasing the concentration x of Ca-dopants in $Ca_x Bi_{2-x} Se_3$, the Fermi level E_F is tuned from the conduction band (samples M1, M2, M3, M4, M5, M7) into the bulk bandgap (samples G3, G4, G5, G6), and then into the valence band (samples M10, M11). $\rho(T)$ displays insulating-like behavior when E_F is tuned in the gap and metallic behavior when E_F is in conduction or valence band (20x magnification). The inset is a sketch of the surface states crossing the bulk gap from the valence band to the conduction band. (b) The inverse magnetic field values of the Shubnikov-de Haas resistance minima plotted versus the Landau level indices of eight metallic samples (M1 − M11). In these plots, negative (positive) index n represents electron (hole) carriers, showing that E_F is tuned to be inside the conduction (valence) band. Adapted from Ref. [38].

Fermi level in the conduction band, and for those having the Fermi level in the valence band. However, all attempts to measure SdH resistance oscillations when the Fermi level was tuned to be inside the bandgap—i.e., oscillations due to the surface states—were not successful. It was concluded that calcium atoms strongly affect the mobility of the surface Dirac fermions, preventing SdH oscillations to be visible at the magnetic field reached in the experiments.

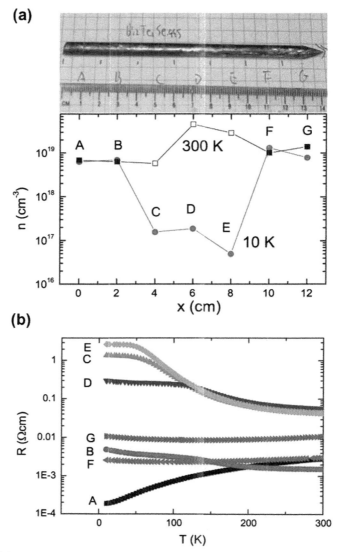

FIGURE 4 (a) Carrier density n at 10 K (circles) and 300 K (squares) of $Bi_2Te_2Se_{0.995}$ crystals taken from different parts of a same ingot, showing large density variations. (b) Temperature-dependent resistivity $\rho(T)$ of the $Bi_2Te_2Se_{0.995}$ crystals, showing a large sample-to-sample variation for crystals taken from different parts of a same ingot. Adapted from Ref. [41].

In Bi_2Te_3, the Fermi level can also be tuned in the bandgap. When p-type samples are annealed (at 410 °C) in an environment of Te vapor, the Te content in the crystals is expected to increase and to reduce anti-site defects. The resistivity of the obtained crystals exhibits an insulating temperature dependence

[27,32,35]. Also in this case, however, no unambiguous observation of surface state transport has been reported.

A different approach that has been followed to control the position of the Fermi level and of the bulk conductivity is to grow BSTS crystals of different compositions and stoichiometry [39]. This can be achieved by varying the Bi/Sb $\left(\frac{1-x}{x}\right)$ and the Te/Se ratios $\left(\frac{1-y}{y}\right)$ of the mixture of elemental constituents used in the crystal growth. Most combinations of x- and y-values yield rather high conductivity (metallic-like) materials with typical electron densities in the range of 10^{18}–10^{19} cm^{-3}. However, for specific compositions, relatively high—for this class of materials—bulk resistivities have been found. The highest resistivity values are obtained for Bi_2Te_2Se [40–43] and $Bi_{1.5}Sb_{0.5}Te_{1.7}Se_{1.3}$ crystals [39,44] with values reaching 1–10 $\Omega \cdot$ cm at $T \approx 4$ K, and bulk carrier densities estimated to be approximately 10^{16}–10^{17} cm^{-3}. These findings, together with the insulating temperature dependence observed for $T > 50$–100 K, indicate that the Fermi level is likely in the bulk bandgap (Fig. 4). These materials, therefore, provide favorable conditions for the experimental detection of surface transport (see also next section). Nevertheless, a remaining difficulty originates from the high sensitivity of the bulk resistivity to small deviations of the stoichiometry from the ideal value. As a result, crystals extracted from a same ingot can exhibit resistivity values that differ by two to four orders of magnitude (Fig. 4), and often only a small fraction of the crystals obtained from a same batch has the desired properties.

3 IDENTIFICATION OF SURFACE TRANSPORT USING BULK SINGLE CRYSTALS

As explained in the previous section, the bulk conductivity of Bi-based topological insulating materials is non-negligible at low temperature, which usually prevents the straightforward observation of surface transport in these systems. One way to observe surface transport is to reduce the material thickness, e.g., by crystal exfoliation or by growing thin nanoribbons or films, to minimize the contribution due to bulk transport to the measured conductance. Such a strategy will be discussed in the next section, while in this section we first discuss work investigating the presence of surface transport in "thick" bulk crystals.

3.1 Different Types of Charge Carriers

A method frequently used to determine the presence of different types of charge carriers is the analysis of the magnetic field dependence of the Hall resistance [45]. When only one type of carrier dominates transport, the transverse Hall resistance R_{xy} usually depends linearly on magnetic field. The corresponding Hall coefficient (i.e., $R_H = R_{xy}/B$) is positive for holes and negative for electrons. When two or more carrier types having different density and mobility contribute to transport, the magnetic field dependence of R_{xy} is often more complex, as it exhibits non-linear behavior and R_H may even change sign (Fig. 5). An analysis of this behavior in terms of multiple types of charge carriers can allow the densities and mobilities of the different carrier types to be extracted.

This type of analysis has been performed for different Bi-based topological insulators. Hall measurements at $T \approx 4$ K on low-resistivity Bi_2Se_3 crystals—which presumably have the Fermi level in a bulk band—show linear $R_{xy} - B$ characteristics [24,25], as expected when transport is dominated by one type of carrier. On the other hand, high-resistivity crystals, for which the Fermi level is likely to be located in the bulk bandgap, exhibit non-linear $R_{xy} - B$ characteristics at $T \approx 4$ K [24,36], from which the presence of multiple types of carriers is inferred. Non-linear Hall characteristics have also been observed in insulator-like crystals of Bi_2Te_3 [35], Bi_2Te_2Se [40–43], and $Bi_{1.5}Sb_{0.5}Te_{1.7}Se_{1.3}$ [39, 44] (Fig. 5). In most cases, the Hall effect data measured on these "insulating" samples are accounted for by assuming the presence of charge carriers in a low-μ 3D band coexisting with carriers in a high-μ 2D band, with meaningful density and mobility values for the two types of charge carriers (typically, $\mu \approx 20$–200 cm^2/Vs for carriers in the 3D band and $\mu \approx 1000$–3000 cm^2/Vs for carriers in the 2D band [24,39,40,44]). Although useful, this type of analysis inputs the assumption that the two families of carriers have 2D and 3D nature, and does not provide direct information as to their specific origin.

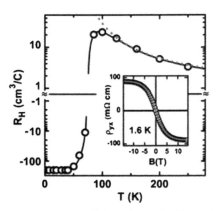

FIGURE 5 Temperature dependence of the Hall coefficient $R_H(T)$ for a highly resistive Bi_2Te_2Se crystal, showing a sign change in the range 50–100 K, indicative of a change in the dominant type of charge carriers from holes to electrons. The inset shows the magnetic field dependence of the low-T Hall resistivity $\rho_{yx}(B)$, exhibiting non-linear behavior fitted by a two-band model (red line), based on the assumption that 2D and 3D electronic states contribute in parallel to transport. (For interpretation of the references to color in this figure legend, the reader is referred to the web version of this book.) Adapted from Ref. [40].

Typically, in high-resistivity samples, ρ increases upon lowering temperature down to $T \sim 50$–100 K, and approximately saturates at lower T (Figs. 3 and 4) [22,24,35,36,38–40,42–44]. Such a temperature dependence is normally interpreted in terms of 3D bulk carriers in the valence or conduction band, which are thermally activated to, or from, in-gap impurity states [39,40]. Their contribution to transport is relevant at high temperatures and is suppressed at sufficiently low T. The residual low-T conductivity due to 3D carriers inferred from the analysis of the Hall effect is attributed to a bulk impurity band with rather delocalized carriers at all energies inside the bandgap, a scenario that supposedly accounts for the small mobility extracted from the Hall measurements. However, it is not obvious how the contribution of the impurity band to the Hall effect should be viewed microscopically, since—owing to the small bandgap of these materials—the electronic states in this band are likely to originate from the disorder-induced hybridization/superposition of states in the valence and conduction bands (and hence it is difficult to determine a priori even the sign of their contribution to the Hall resistance). As to the origin of the 2D carriers that are identified from the analysis of the Hall effect, there are two obvious hypotheses. The first is that they are due to the formation of a charge accumulation/inversion layer caused by band-bending near the material surface (such 2D surface states are regularly observed in ARPES measurements in these systems [28]). The second is that they are the gapless Dirac surface fermions of topological origin. Clearly, the analysis of the Hall effect cannot discriminate between these two possibilities.

3.2 Angle-Dependent Shubnikov-de Haas Resistance Oscillations

A more direct technique—as compared to the analysis of the Hall effect—to identify the presence of electron states of different dimensionality is the study of angle-dependent SdH resistance oscillations. In a 3D conductor, SdH resistance oscillations exhibit a periodicity that varies continuously with varying the direction of the applied magnetic field [45]. For a 2D conductor, on the contrary, the periodicity of the SdH oscillations only depends on the perpendicular component of the magnetic field (unless the electron spin plays a crucial role), and oscillations will be absent when an in-plane magnetic field is applied [46]. The experimental investigation of the SdH periodicity as a function of the angle between the magnetic field and the sample is therefore a conclusive method to distinguish between 2D and 3D electron systems.

Shubnikov-de Haas resistance oscillations are observable if the scattering time is longer than the time it takes for an electron to complete a cyclotron orbit, a condition that can be expressed in terms of the carrier mobility and the applied magnetic field as $\mu \cdot B \gtrsim 1$ [45]. Consequently, SdH oscillations originating from high-μ carriers are observable at much lower magnetic fields than those due to low-μ carriers. In crystals of several different Bi-based

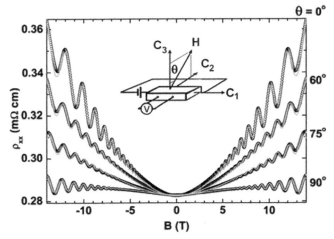

FIGURE 6 Shubnikov-de Haas oscillations of the longitudinal resistivity ρ_{xx} measured at different angles θ in the C2–C3 plane relative to a Bi_2Se_3 crystal (at $T \approx 1.5$ K). The oscillations are attributed to 3D bulk states, because they do not disappear in an in-plane magnetic field ($\theta = 90°$). Adapted from Ref. [25].

topological insulators—e.g., $Bi_{1-x}Sb_x$ [18, 19], Bi_2Se_3 [22–25, 36, 38], Bi_2Te_3 [35], Bi_2Te_2Se [40, 42, 43], and $Bi_{1.5}Sb_{0.5}Te_{1.7}Se_{1.3}$ [44]—SdH oscillations become visible in the magnetoresistance at already moderate values of the applied magnetic field of $B \gtrsim 5$–10 T, which results in an estimated carrier mobility of $\mu \sim 1000$–2000 cm^2/Vs. First experiments on low-resistivity materials mainly detected SdH oscillations that could be attributed—on the basis of their angular dependence on the applied field—to bulk 3D carriers [22, 23, 25] (Fig. 6). In some cases, coexisting SdH resistance oscillations with 2D and 3D character were also observed [18, 19].

Subsequently, SdH resistance oscillations dependent only on the perpendicular component of the applied magnetic field—i.e., with purely 2D character—have been observed [24, 35, 36, 40, 42–44] (Fig. 7). The best quality data have been obtained after the development of high-resistivity Bi_2Te_2Se and $Bi_{1.5}Sb_{0.5}Te_{1.7}Se_{1.3}$ crystals [40, 42–44]. In the majority of cases, the 2D SdH resistance oscillations are superimposed on a magnetoresistance background whose microscopic origin is not yet understood. Removing this background by subtracting a smooth curve, or by taking the derivative of the magnetoresistance with respect to B, is necessary to isolate the oscillations. In the higher resistivity crystals, the experimental results are consistent—at least qualitatively—with the conclusion drawn from the Hall measurements, from which the coexistence of low-μ 3D bulk states and high-μ 2D surface states is inferred. Interestingly, for the crystals in which SdH resistance oscillations with 2D character are observed, no clear indication is found of the presence of two distinct families of 2D charge carriers, as it could be expected from the presence of two distinct crystal

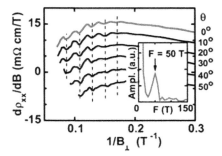

FIGURE 7 Derivative of the longitudinal resistivity with respect to the magnetic field $(d\rho_{xx}/dB)$ plotted versus the inverse perpendicular component of $B (1/B_\perp)$ for different angles between field direction and the plane of a $Bi_{1.5}Sb_{0.5}Te_{1.7}Se_{1.3}$ crystal (at $T \approx 1.5$ K). All curves are shifted for clarity. The angle-dependent measurements show that the positions of the Shubnikov-de Haas minima only depend on $B_\perp$, as is characteristic of 2D states. The inset shows the Fourier transform of the data at $\theta = 0°$. Adapted from Ref. [44].

surfaces perpendicular to the applied magnetic field. Possibly, the two surfaces have nearly the same density of charge carriers, so that a single oscillation frequency is observed. Alternatively, if the bulk resistivity of the measured samples is sufficiently large, the experiments may be mainly sensitive to the transport contribution due to the carriers at the surface in direct contact with the electrodes.

3.3 Landau Level Indexing

Having established that 2D charge carriers are present at the surface of Bi-based topological insulators, is not sufficient to conclude that these are the expected Dirac fermions. Indeed, as we mentioned earlier, 2D surface accumulation layers can be present due to band-bending effects, causing spatial confinement of bulk states at the surface. It is in principle possible to discriminate between this type of 2D electrons, and the Dirac surface fermions of topological origin, by looking at the structure of the Landau levels responsible for the SdH resistance oscillations.

It is well known from graphene [47,48] that the wave functions of Dirac fermions acquire a Berry phase of π upon completing a cyclotron orbit in the presence of a magnetic field. The Berry phase modifies the quantization condition for these orbits, resulting in a zero-energy Landau level (i.e., a Landau level located at the "Dirac point") shared by Dirac electron and hole states. Because of this peculiar zero-energy Landau level ($N \equiv 0$), the amount of carrier density that is needed to completely fill the Nth Landau level is shifted as compared to the case of conventional fermions. Specifically, for a density n of Dirac fermions, the Fermi level is positioned between the Nth and $(N + 1)$th Landau levels when $B = \frac{n}{|N|+1/2}h/e$ (for common electrons the "1/2" term at the denominator is absent). In a four-terminal measurement, it should then be expected that the minima in the longitudinal resistance in the SdH oscillations occur when $B = \frac{n}{|N|+1/2}h/e$.

By plotting the measured SdH resistance oscillations as a function of $1/B$, an integer Landau level index N can be assigned to each one of the resistance minima and, in the ideal case, the relation between N and $1/B$ is linear. For Dirac fermions, the extrapolated value of N for $1/B \to 0$ is a half-integer, whereas it is an integer for conventional fermions. This difference enables Dirac fermions to be discriminated from conventional carriers, as shown for graphene [47,48]. This strategy has also been applied to SdH resistance oscillations measured in bulk crystals of Bi_2Se_3 [36], Bi_2Te_3 [35], Bi_2Te_2Se [40,42,43], and $Bi_{1.5}Sb_{0.5}Te_{1.7}Se_{1.3}$ [44], and—in certain cases—it has been used to attribute the observed SdH oscillations to the presence of Dirac fermion Landau levels. In practice, however, the experimental situation on 3D topological insulators is complex, which makes the conclusions obtained from this type of analysis not as clear-cut as it would be desirable.

A first important point is that the Landau level indexing is often done after removing from the measured signal a large, typically non-linear, background magnetoresistance. Indeed, the surface contribution to the total longitudinal conductance that has been measured in thick crystals is less than $\sim 1\%$ for Bi_2Se_3 [22,25,36,38,40] and Bi_2Te_3 [35] and $\sim 6\%$ for Bi_2Te_2Se [40,42,43]. As the background is very frequently much larger than the oscillation amplitude, its subtraction causes the position (in magnetic field) of the resistance minima to shift, which can largely influence the indexing analysis. Better results have been obtained for $Bi_{1.5}Sb_{0.5}Te_{1.7}Se_{1.3}$ crystals [44] thinned down to a thickness of 8 μm leading to a surface contribution as large as $\sim 70\%$ of the total conductance. Here, however, the best quality resistance oscillations are usually observed in the transverse resistivity rather than in the longitudinal one, a fact that remains to be understood, and whose implication in relation to the indexing—which should normally be done on the longitudinal magnetoresistance—has not yet been considered. Additionally, unless a large number of minima are observed, the extrapolation of N for $1/B \to 0$ is often affected by a non-negligible error, which makes it difficult to establish conclusively whether the extrapolated value is integer or half-integer. Note that, in the case of graphene, the data analysis takes considerable advantage of the possibility to gate-tune the density of charge carriers, enabling the consistency of the Landau level indexing to be checked for many different values of n (which is important, as it does happen—also in graphene—that accidental deviations of the extrapolated N from the expected half-integer value occur).

Possibly not unrelated to the previous points, several groups have also found that the analysis of the SdH oscillations yields non-linear $N - 1/B$ characteristics over the full range of magnetic field investigated [49] (Fig. 8). Different physical mechanisms have been proposed as a possible explanation. For instance, the dispersion relation of the surface fermions that deviates

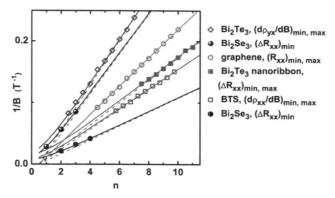

FIGURE 8 Landau level indexing of Shubnikov-de Haas oscillations observed in various topological insulators and graphene. Symbols correspond to measured data. Dashed lines are calculations for ideal Dirac fermions without Zeeman coupling. Solid lines are calculations taking into account the non-linearity of the dispersion relation of the surface states and the Zeeman coupling to an external magnetic field. Adapted from Ref. [49].

pronouncedly from linearity, and the Zeeman effect, which can become important at high magnetic field [36,49]. A third important effect to be considered is the presence of bulk states inside the gap (i.e., the impurity band) that are uniformly distributed in the crystals, and are therefore also present at the surface [30]. The coexistence of Dirac fermions with a finite density of states of different origin is important, because the position in energy of the Landau levels changes with varying magnetic field. Such a situation— which is not encountered in graphene—results in charge transfer between the Dirac states and the impurity band at/near the surface upon varying magnetic field, effectively causing the density n of carriers in the Dirac band to depend on B.

For all these reasons, claims that the 2D electrons at the surface of bulk Bi-based topological insulators are Dirac fermions, based on the extrapolation of the "N versus $1/B$" Landau level indexing, often do not come across as entirely convincing. The case of the highly resistive Bi_2Te_2Se and $Bi_{1.5}Sb_{0.5}Te_{1.7}Se_{1.3}$ crystals measured in a sufficiently large magnetic field range may be a first exception, since for the best samples, surface transport contributes a large part of the measured conductance and a large number of oscillation periods can be observed experimentally [40,42–44].

4 ELECTRONIC TRANSPORT THROUGH NANOFABRICATED DEVICES

The possibility to equip 3D topological insulators with gate electrodes that give control over the density of charge at the surface allows more systematic transport measurements to be made, that are particularly useful to detect different manifestations of the presence of topological Dirac fermions. Bulk single crystals, however, are not particularly well suited for the development of gated devices, because the deposition of metal contacts, a gate insulating layer, and a metal gate electrode on top of an irregularly shaped crystal is not straightforward. More importantly, to investigate the topological Dirac fermions, it is desirable to decrease the thickness of the material as much as possible, to increase the relative contribution of surface conduction. This is essential in gated devices, where a large parasitic bulk conductance would decrease the experimental effectiveness of the gate action.

Gated devices based on 3D topological insulators have been realized in different ways. Possibly the simplest strategy is to exfoliate bulk crystals to obtain thin crystalline flakes that can be transferred onto suitable substrates, and subsequently patterned using conventional nanofabrication techniques (Fig. 9). Next to its simplicity, this technique enables experiments to be performed on the same material from which larger crystals are obtained. Alternatively, thin layers of 3D topological insulators can be obtained by using epitaxial thin film growth techniques, or bottom-up synthesis strategies to grow nanoribbons or nanowires.

4.1 Top-Down Approach: Mechanical Exfoliation of Macroscopic Crystals

Bi_2Se_3 and Bi_2Te_3 crystals exhibit a layered structure consisting of so-called quintuple layers (QLs) stacked on top of each other, and held together by weak Van der Waals interactions [16]. Within each quintuple layer (exhibiting Se-Bi-Se-Bi-Se or Te-Bi-Te-Bi-Te stacking), the atoms are bonded covalently. Owing to the weakness of the Van der Waals interaction, Bi_2Se_3 and Bi_2Te_3 crystals can be easily exfoliated using an adhesive tape, adapting the same strategy used to extract graphene from graphite [50]. However, Bi_2Se_3 and Bi_2Te_3 are much more brittle than graphite, so that crystals more easily crack during exfoliation, resulting in flakes with much smaller linear dimensions—at most 1–10 μm— than what can be now routinely achieved for graphene. The exfoliated thin crystals are usually transferred onto a substrate (Fig. 9), for instance a highly doped conducting Si wafer (to be used as gate electrode) coated with SiO_2, and processed to attach electrical contacts, and possibly a second gate electrode on the top surface [31,51–54]. In most cases, the thickness of the layers used was 10 nm or larger, but transport through gated layers as thin as a few nanometers (corresponding to a few QLs) has also been investigated [51].

A word of caution has to be spent about device reproducibility. As it should be clear by now, crystals of Bi-based topological insulators tend to exhibit a rather broad range of properties, even when they are grown under nominally identical conditions. Device fabrication results in the exposure of the material surface to different solvents and polymers, and to moderately elevated temperatures (~150–180 °C), all factors that may additionally affect

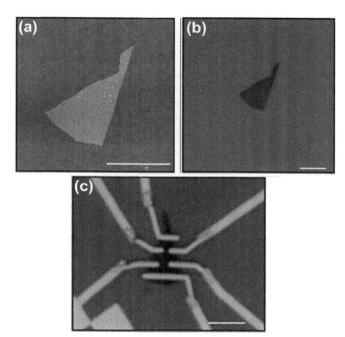

FIGURE 9 Atomic force micrograph (a) and optical micrographs (b, c) of a $\sim$3.5 nm thick exfoliated Bi_2Se_3 flake on a SiO_2/Si substrate. Panel (c) shows a six-terminal device with Pd electrodes, fabricated by e-beam lithography on top of the same flake shown in panels (a, b). All scale bars correspond to 4 μm. Adapted from Ref. [51].

reproducibility. As a consequence, in many cases, the number of stable devices exhibiting interesting phenomena is only a small fraction of the total number of fabricated devices. For instance, in many research groups it has been found that the surface of Bi_2Se_3 crystals often degrades during the device fabrication [55], resulting in the formation of an insulating layer that needs to be removed to achieve electrical contacts, and appears to lower the device quality (e.g., the carrier mobility). It has also been found, however, that in several cases this phenomenon does not occur, and devices with very stable electrical characteristics (over a period of months), exhibiting clear signatures of surface states, were realized. Just like for bulk crystals, it remains to be understood why devices realized with nominally identical crystals and following nominally identical fabrication processes can behave so differently.

4.2 Epitaxial Films, Nanoribbons, and Nanowires

Molecular beam epitaxy (MBE) is a widely used method to grow high-quality, crystalline semiconductor layers and heterostructures, which is also suitable to grow Bi-based thin films. Thin films have been grown successfully on different substrates, such as SiC [56], SrTiO$_3$ [57], Si [58–61], Al$_2$O$_3$ [62], and sapphire [63]. Layer-by-layer growth makes it possible to obtain ultrathin layers of Bi$_2$Se$_3$ or Bi$_2$Te$_3$ from single to multiple QLs in a controllable way. Alternatively, layers can be epitaxially grown by pulsed laser deposition [64], and laser ablation [65]. Crystalline films on a substrate have proven useful to probe different kind of properties of surface Dirac fermions. For instance, MBE-grown films of controlled thickness down to a single QL have enabled the ARPES investigation of a gap opening due to hybridization of Dirac fermions on opposite surfaces of a 3D topological insulator [56]. MBE-grown thin films have also been used to perform STM measurements in the presence of a magnetic field, which have led to the spectroscopic observation of the Dirac fermion Landau levels [66]. These high-quality experiments, however, probe mainly local properties (e.g., local density of states). They prove that Dirac fermions are present at the surface of the films, but they cannot guarantee that the film quality is sufficient for probing Dirac fermions in transport experiments.

Next to MBE deposition, alternative techniques have been developed to grow quasi-one-dimensional structures of Bi$_2$Se$_3$ and Bi$_2$Te$_3$. For instance, nanowires have been grown successfully by either metal-catalyzed vapor-liquid-solid (VLS) growth method [55,67–74] or chemical synthesis [75]. To catalyze VLS growth, a substrate is usually covered with a distribution of metal nanoparticles (e.g., a Si substrate with Au nanoparticles of $\sim$20 nm diameter [67]). When the substrate is placed in a hot furnace in the vicinity of the source material, wire-like structures of Bi$_2$Se$_3$ or Bi$_2$Te$_3$ grow at the nanoparticle sites. The preferred growth direction—which can be in-plane or perpendicular to the quintuple layers—as well as the final dimensions of the nanowires, depend sensitively on the environmental conditions in the furnace (e.g., temperature and gas flow). As-grown nanowires exhibit typical cross-sectional dimensions in the range $\sim$10–100 nm, while their length can reach values up to $\sim$10–100 μm. After VLS growth, the nanowires are usually mechanically transferred to a substrate to enable the fabrication of mesoscopic devices for electronic transport measurements [67].

The VLS-growth technique has also been used successfully to grow Bi$_2$Se$_3$ and Bi$_2$Te$_3$ nanoribbons [55,67–74]. These stripe-like structures grow in the in-plane direction, yielding a typical ribbon width of several up to tens of microns (while the thickness is in the range $\sim$10–100 nm). Alternatively, it has been demonstrated that Bi$_2$Se$_3$ and Bi$_2$Te$_3$ nanoribbons can be grown by a catalyst-free vapor-solid (VS) growth method [55,74,76,77]. An advantage of this latter method is that the nanoribbons can be grown directly on a substrate suitable for device fabrication (e.g., a SiO$_2$/Si wafer [76]).

4.3 Ambipolar Transport in Gated Devices

One of the characteristic aspects of massless Dirac fermions is the possibility to shift continuously the Fermi level E_F from the electron to the hole band, across the Dirac point, without ever crossing a gap in the density of states. Depending on the position of E_F, the low-temperature transport is then dominated by either Dirac electrons or Dirac holes. As E_F is swept to cross over from one type of charge carrier to the other, a (finite) resistance maximum occurs, which is a direct manifestation of ambipolar transport of Dirac fermions. For Dirac fermions in graphene, this behavior is observed routinely when measuring the resistivity as a function of gate voltage, which provides a simple way to determine at which gate voltage charge neutrality occurs. The peak resistivity corresponds to a minimum conductivity of $\sigma_{min} \approx 4e^2/h$, with the factor 4 counting a contribution of e^2/h for each of the spin- and valley-degenerate families of Dirac fermions [47,48]. These same measurements also enable the estimation of the carrier mobility in graphene, by simply measuring the change in conductivity due to a known change in carrier density induced by the gate voltage.

A main conceptual difference between graphene and Bi-based 3D topological insulators is that, for the latter, bulk states at the same energy as the surface states are usually present. In many of the Bi-based 3D topological insulators, this is because the Dirac point at the surface occurs at an energy that falls inside one of the bulk bands. This is the case of Bi_2Te_3 [21] and—in a broad stoichiometric range—of $(Bi_xSb_{1-x})_2Te_3$ [77]. Under these conditions, it is not possible—even in principle—to isolate surface transport from bulk transport in the vicinity of the Dirac point by simply sweeping E_F. For Bi_2Se_3, on the contrary, the Dirac point is in the middle of the bandgap, not overlapping—in a perfect crystal—with other bulk states. This is a property that makes Bi_2Se_3 particularly appealing for the realization of gated devices.

Electrical measurements on gated exfoliated flakes [31,52–54] or epitaxially grown layers [57,60,71,78,79] of Bi_2Se_3 indeed show ambipolar transport characteristics (Fig. 10). The small amount of surface charge that needs to be accumulated to cross over from electron to hole transport provides an indication that the observed behavior originates from the ambipolar character of gapless surface states. Note that when using electrolyte gates—which enable charge densities in excess of 10^{14} cm^{-2} to be accumulated at the material surface—it is also possible to observe ambipolar transport due to E_F crossing over from the valence to the conduction bulk bands. In that case, very large conductivity changes—two orders of magnitude—can be observed [61], much larger than what is normally achieved with common solid state gate dielectrics, which only allow a much smaller modulation of surface charge.

Even for Bi_2Se_3, however, the situation is not as ideal as for graphene. Although the Dirac point is "exposed"—i.e., it is in the bulk bandgap (Fig. 1)—bulk states are still present at the same energy due to disorder (i.e., the states forming the impurity band). The density of these states is rather large, as directly measured in STM experiments [30]. Upon applying a gate voltage, not only the Dirac surface states are filled, but also the impurity band within the Thomas-Fermi screening length ($\sim$3–4 nm for Bi_2Se_3 [31]) from the surface. Since the density of states of the Dirac fermions vanishes at the Dirac point, at low

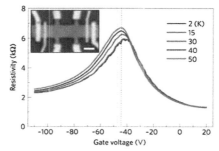

FIGURE 10   Gate voltage dependence of the resistivity $\rho(V_G)$ of an exfoliated Bi_2Se_3 flake ($\sim$10 nm thick) at different temperatures, showing ambipolar behavior. The maximum resistivity is close to h/e^2, as expected for ideal Dirac fermions. Note that the bulk conduction is highly suppressed by using a gate electrolyte. The inset shows an optical micrograph of the device (the scale bar corresponds to 2 μm). Adapted from Ref. [54].

energy most of the carriers accumulated by the gate end up occupying states in the impurity band near the surface. This results in a large broadening of all gate voltage dependent properties [31].

The impurity band also provides a transport channel insensitive (or only weakly sensitive) to the gate voltage, which conducts in parallel to the surface. This bulk transport channel can dominate even for thin crystals ($\sim$10 nm thick), in which case the relative change in resistance induced by the gate voltage is rather small and the resistance per square at the peak is much smaller than the expected value $\sim h/e^2$. A notable exception is represented by recent experiments performed by the Maryland group [54]. In these experiments, the bulk transport channel could be strongly (nearly entirely) suppressed by using a gate electrolyte, to shift the Fermi level inside the bandgap of crystals several nanometers thick. As a result, a peak resistivity of $\rho_{max} \sim h/e^2$ was observed (Fig. 10), as expected for ideal Dirac fermions. The same group had also previously observed a gate-induced highly resistive state in very thin Bi_2Se_3 crystals—probably only three quintuple layers thick—which was attributed to the opening of a large gap at the Dirac point ($\gtrsim 250$ meV) due to hybridization of topological Dirac fermions present on opposite surfaces [51]. Although the estimated gap value seems rather large for hybridization of states on two surfaces separated by three quintuple layers, and the resistance measured upon accumulation of charge carriers seems higher than what could be expected in the proposed scenario, these experiments indicate that in the crystals used by the Maryland group, the density of states in the gap is significantly smaller than in the crystals used by others.

4.4 Gate-Tunable Shubnikov-de Haas Resistance Oscillations

The ambipolar transport characteristics observed in gated devices show that surface conduction occurs through electron- and hole-like states, and that a smooth crossover occurs as a function of gate voltage, upon accumulating only a modest amount of charge carriers ($\sim 10^{12}$ cm^{-2}). This is indeed what is expected for the massless Dirac fermions that are predicted to live at the surface of a 3D topological insulator. However, the observed ambipolar transport may also be argued to originate from the properties of the impurity band, that has more pronounced electron or hole character depending on whether the Fermi level is gate-shifted closer to the conduction or the valence band. It is important to exclude this possibility experimentally and to find more direct indications that the properties of the surface carriers is consistent with those expected for Dirac fermions.

Gate-controlled SdH resistance oscillations that have been observed in bottom-gated Bi$_2$Se$_3$ exfoliated crystals do provide these indications [31]. Magnetotransport measurements as a function of gate voltage V_G were performed by Sacépé et al. in a two-terminal configuration, in devices with superconducting contacts that also exhibited Josephson supercurrent at sufficiently low temperature (see following section). A large magnetoresistance is observed, exhibiting regular features that disperse symmetrically (for B up to approximately 10 T) in opposite directions when V_G is swept at opposite sides of the Dirac point (Fig. 11). These features form the so-called fan diagram of Landau levels that is routinely observed in graphene [47,48]. Although the features are visible directly in the measured resistance, for a quantitative analysis a background was removed by looking at the quantity $-\frac{d^2 R(V_G, B)}{dB^2}$ (analyzing data by removing a smooth background gave fully consistent results). The oscillations of this quantity are found to be periodic in $1/B$ as expected for the SdH effect. Interestingly, when looking at the Fourier spectrum of the oscillations as a function of gate voltage, two different families of SdH oscillations are detected. Both families have two-dimensional character, as oscillations only appear in a perpendicular magnetic field, and not when the field is applied in-plane. For one of the families, the frequency depends on V_G, while the frequency of the other is gate voltage independent (at $V_G = 0$ V, the two frequencies approximately coincide, i.e., the density of charge carriers responsible for the SdH oscillations is the same at $V_G = 0$ V).

These observations are naturally explained if the measured SdH oscillations originate from Dirac fermions present on the two opposite surfaces of the Bi$_2$Se$_3$ crystal. As the crystal thickness is larger than the Thomas-Fermi screening length ($\sim$3–4 nm [31]), only the carrier density at the surface of the crystal closer to the gate electrode is modulated by the applied gate voltage, which is why one of the two families of SdH oscillations has a V_G-dependent frequency and the other not. The observation that the different SdH oscillations have approximately the same frequency at $V_G = 0$ V is also naturally explained, since in the absence of a gate voltage the two surfaces are approximately equivalent (not identical, as one of them is in contact with SiO$_2$ and the other not) and host therefore the same density of surface charge carriers.

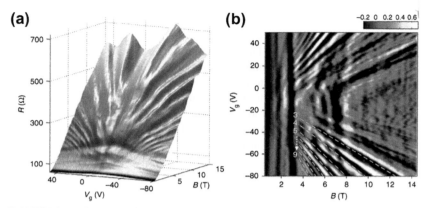

FIGURE 11 (a) Resistance R of an exfoliated Bi_2Se_3 flake ($\sim$10 nmthick) as a function of gate voltage V_G and magnetic field B applied perpendicular to the flake ($T \approx 4$ K). $R(B)$ at fixed V_G increases approximately linearly, exhibiting clear oscillations that are enhanced by deriving the data with respect to V_G or B. (b) Color plot of $-d^2R/dB^2$ (in arbitrary units) as a function of B and V_G. The features dispersing with V_G originate from the formation of Landau levels of Dirac fermions on the surface of the Bi_2Se_3 crystal closer to the gate electrode. Features with a V_G-independent position are associated with Landau levels on the surface far away from the gate electrode. The white dashed lines (and the numbers indexing the Landau levels N) indicate the condition for the Fermi level to fall between two Landau levels for holes on the surface closer to the gate electrode. They correspond to $B = \frac{n}{|N|+1/2}(h/e)$, as expected for Dirac fermions (where n is the density of Dirac holes). Adapted from Ref. [31].

The measured data could be accounted for in terms of a simple physical picture in which the Dirac fermions coexist with a large density of in-gap states associated with an impurity band (whose density of states could be estimated and was found to be in fair agreement with values estimated from STM experiments [30]). From this analysis, the carrier mobility was found to be approximately 2500 cm^2/Vs for the lower surface in contact with SiO_2 and 5000 cm^2/Vs for the upper one. These values are in agreement with the observation that the two families of SdH oscillations start to be visible at about 3–4 T and 1 T, respectively. The indexing of Landau levels was also performed for many different values of the gate voltage, with E_F both in the conduction and valence band of the surface Dirac fermions. Having data for many different gate voltages allowed the consistency with the half-integer quantum Hall effect to be checked (Fig. 11). Unfortunately, however, since the measurements were performed in a two-terminal configuration, the indexing analysis required assumptions to be made to reach this conclusion. Nevertheless, the qualitative aspects of the experimental results are unambiguous, and exclude the possibility that the ambipolar character of transport originates from states in the impurity band, since the occurrence of SdH oscillations requires carriers to have a well-defined dispersion relation and a sufficiently long mean free path.

4.5 Topological Surface States in Epitaxially Grown HgTe Layers

Whereas most experimental investigations of 3D topological insulators have so far relied on the use of Bi-based materials, it has been predicted that a different system realization of a 3D topological insulator with a single Dirac cone at each surface is provided by HgTe layers with in-plane strain [15,80]. Recently, high-quality HgTe layers have been grown epitaxially on a CdTe substrate [81]. The lattice mismatch with the CdTe substrate gives rise to in-plane strain in the HgTe layer, which causes the opening of a bandgap in the bulk states [15,80] of about 20 meV [81], an order of magnitude smaller than in the best BSTS compounds. For transport experiments, a small gap represents a disadvantage, as it limits the energy range in which surface transport due to Dirac fermions can be observed without being affected by bulk states (for instance, the Dirac fermion Landau levels hybridize with bulk states at already moderately low magnetic fields). Nevertheless, since the bulk of epitaxially grown, strained HgTe films is virtually impurity-free, the bulk contribution to their conductance vanishes at low temperatures ($kT \ll 20$ meV). This makes strained HgTe films particularly appealing in comparison to BSTS compounds, whose residual bulk conductance is normally large and difficult to suppress experimentally.

Low-temperature transport experiments have been carried out on Hall bars etched from a 70 nm strained HgTe layer [81]. The quantum Hall effect is observed at high magnetic fields, with plateaus in the Hall resistance coinciding with minima of the longitudinal resistance, as expected in the ideal case for transport through 2D states. At moderate fields ($B \approx$ 3–7 T), Hall plateaus develop which correspond to an odd Landau level filling index: $\nu = 9,7,5$. This indicates that transport is mediated by Dirac states, which are predicted to reside at the two opposite surfaces of the film. However, odd as well as even plateaus (i.e., $\nu = 4,3,2$) appear at higher fields ($B \approx$ 7–14 T). This observation is attributed to a difference of the density of states at both surfaces, originating from the difference in environment (vacuum at top versus CdTe at bottom surface). The mobility of the surface carriers is estimated from low-field Hall measurements yielding $\mu \sim 3.4 \cdot 10^4$ cm^2/Vs [81], which is an order of magnitude higher than typically found in BSTS compounds. The high mobility of the Dirac-like surface carriers and the vanishing bulk conductance make epitaxially grown, strained HgTe layers a promising material system to study transport through topological surface states. Work is ongoing to investigate transport through strained HgTe using devices equipped with gate electrodes, to control the surface density of carriers.

5 MISCELLANEOUS TRANSPORT PROPERTIES

The identification of Dirac fermions at the surface of 3D topological insulators in transport experiments is not essential to establish the presence of these states, for which purpose ARPES experiments clearly do a better job. It is, however, crucial to explore the low-energy electronic properties of 3D topological insulators using electronic devices providing excellent energy resolution and a high degree of experimental flexibility. In this section, we present a brief overview of selected recent experiments that illustrate research in this direction.

5.1 Aharonov-Bohm Effect in Bi_2Se_3 and Bi_2Te_3 nanowires

In confined geometries, Dirac fermions at the surface of a 3D topological insulator can give rise to unusual phenomena. An ideal nanowire made out of a topological insulator, for instance, can be considered as an insulating core surrounded by a thin conducting shell. A magnetic field applied along the nanowire axis should then result in Aharonov-Bohm (AB) conductance oscillations if the phase coherence length of the electrons at the surface is larger than the circumference of the nanowire—similar to what happens in carbon nanotubes [82]. These AB conductance oscillations are periodic in magnetic field with a periodicity of $\Delta B = \frac{\Phi_0}{A}$, where A is the cross-section of the nanowire and $\Phi_0 = h/e$ the flux quantum [83].

To investigate the appearance of the AB-effect in the surface states of a topological insulator, magnetotransport through Bi_2Se_3 [69] and Bi_2Te_3 [75] nanowires has been measured. Interestingly, the measurements show periodic resistance oscillations with a periodicity that is consistent with the cross-section of the nanowire ($\Delta B = \frac{\Phi_0}{A}$; Fig. 12). This observation provides a clear indication of the presence of surface states at the nanowire surface. Next to the periodic oscillations, aperiodic ones are also observed in these data, which are attributed to universal conductance fluctuations [84] due to the residual charge carrier density in the bulk of the nanowires.

5.2 Josephson Junctions

When a superconductor (S) is contacted to a normal conductor, superconductivity can be induced in the latter by proximity effect. This is also true when the normal metal is the Dirac 2D electron system at the surface of a 3D topological insulator [85]. In that case, proximity effect can have a particularly interesting nature, owing to the fixed helicity of Dirac electrons (whose spin is locked to the momentum). As a result, induced superconductivity may lead to a so-called spinless p-wave order, with Majorana fermions predicted to appear at S-TI interfaces in certain junction configurations [86,87].

Experiments on S-TI and S-TI-S junctions indeed show the occurrence of proximity-induced superconductivity in Bi_2Se_3 [31,65,72,88,89], Bi_2Te_3 [90], and Bi_2Te_2Se [65], and a Josephson supercurrent has been observed in Bi_2Se_3 and Bi_2Te_3 Josephson junctions [31,89,90]. Since normal transport is still mainly occurring through bulk states, an important issue is to determine whether and how the surface states are involved in the proximity-induced superconductivity. An important part of this question could be answered by

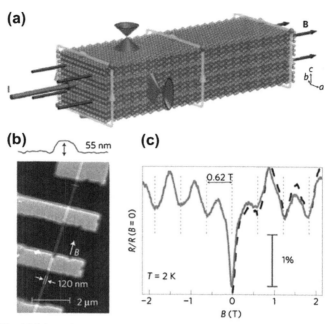

FIGURE 12 (a) Schematic drawing of 2D topological surface states of a layered Bi_2Se_3 nanowire under a magnetic field along the wire length. The red and black arrows correspond to the electric current and magnetic field lines, respectively. The two cones on the top and side surfaces illustrate the Dirac surface states propagating on all surfaces with linear dispersion. The green loops illustrate phase coherent paths through which the surface electrons interfere. (b) Scanning electronic micrograph of a Bi_2Se_3 nanowire, 120 nm in width, contacted by four Ti/Au electrodes. The thickness of the nanowire is measured by an atomic force microscope (see the line cut in the inset) to be 55 nm. (c) Normalized magnetoresistance of the nanowire, showing a clear modulation of the resistance with a period of 0.62 T, corresponding to one flux quantum (h/e) threaded into the cross-section of the nanowire. The solid red (dashed black) trace is an up-sweep (down-sweep) with a scan rate of 3 mT/s (10 mT/s). (For interpretation of the references to color in this figure legend, the reader is referred to the web version of this book.) Adapted from Ref. [69].

recent experiments on Bi_2Se_3 Josephson junctions equipped with a bottom gate, which enabled tuning both the normal and the superconducting transport in the junction [31]. Gate-dependent magnetoresistance measurements on these junctions with the superconducting electrodes in the normal state (i.e., at a temperature above the critical temperature of the superconductor) show the manifestation of ambipolar transport through topological surface states. Electron (hole) transport occurs when the Fermi level is tuned above (below) the Dirac point, with a resistance maximum at the Dirac point. In the presence of an applied magnetic field, electron and hole Landau levels are also observed, forming a fan diagram characteristic of Dirac fermions (see previous section; Fig. 11). When the temperature is lowered well below the critical temperature of the superconducting electrodes (and no magnetic field is applied), a supercurrent

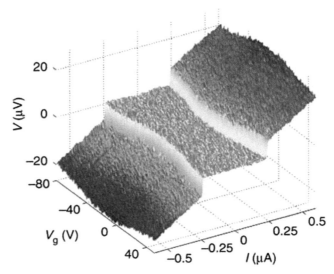

FIGURE 13 Voltage V across a Josephson junction as a function of the applied current I, measured at different gate voltages V_G (at $T \approx 30$ mK and $B = 0$). The junction consists of two closely spaced ($\sim$400 nm) superconducting (Ti/Al) electrodes attached to an exfoliated Bi_2Se_3 flake ($\sim$10 nm thick). When the current is smaller than a critical value, a Josephson supercurrent flows through the junction due to the proximity effect. The critical current is gate-dependent, and exhibits a minimum at $V_G \approx -10$ V, corresponding to the gate voltage at which the maximum of the normal state resistance occurs (i.e., the Dirac point of states on the bottom surface; see Fig. 11). This shows that at least part of the supercurrent is through the Dirac states on the bottom surface (i.e., the gated surface). Adapted from Ref. [31].

appears (Fig. 13). The key observation is that the critical current of the junction can be gate-tuned, and exhibits ambipolar behavior conforming to the expectations based on the normal transport properties. The critical current I_c is minimum at the same gate voltage at which the normal state resistance is maximum, and at which the apex of the Landau level fan diagram occurs (i.e., the Dirac point). I_c increases when the gate voltage is tuned away from the Dirac point either on the hole or on the electron side. The conclusion can therefore be drawn that at least part of the supercurrent is carried by surface Dirac fermions: since the observed modulation of the critical current in the explored gate voltage range is approximately 15%, at least this fraction of the supercurrent is carried by surface states (it is likely that the impurity band states, that dominate the normal transport, also contribute to part of the supercurrent).

Also for Josephson junctions realized on $\sim$200 nm thick exfoliated Bi_2Te_3 crystals—exhibiting SdH oscillations in the normal transport regime—the Josephson effect has been clearly observed using Nb electrodes [90]. The Fermi level could not be controlled in these crystals, because the devices were not equipped with a gate. Nevertheless, the length, temperature, and magnetic field dependence of the critical current was investigated. The trends in the data were found to be consistent with a theoretical analysis, in which ballistic transport between the two superconductors was assumed. Based on the characterization of the normal state transport properties, it was argued that ballistic transport could only be mediated by surface Dirac electrons—which were found to have high mobility—and not by carriers occupying bulk states—whose mobility was estimated to be much lower. On this basis, it was claimed that the supercurrent is entirely carried by surface states, a very interesting conclusion that would certainly deserve confirmation through more direct experimental observations.

5.3 Doped Topological Insulators

Following the discovery of topological insulators, increasing effort has been devoted to the exploration of non-trivial topological aspects in the electronic structure of a broad variety of systems. Among these, ferromagnetic and superconducting materials have attracted considerable attention. In the simplest case, the occurrence of ferromagnetism in a 3D topological insulator leads to the opening of a gap at the Dirac point in the spectrum of the Dirac surface fermions, through breaking of time-reversal symmetry caused by the coupling of the exchange field and the electron spin. Such a phenomenon is of great interest, as it would provide a way to confine the surface Dirac electrons spatially. Indeed, theoretical suggestions for devices in which Majorana fermions at the surface of 3D topological insulators can be generated and confined in one-dimensional structures have heavily relied on the possibility to open a gap through the Zeeman effect [85–87]. The search for superconducting phases with unconventional topological properties has also been pursued intensively. Here as well, one of the main driving forces is the possibility to generate and control Majorana fermions [3,4].

It has been discovered that phase transitions to a superconducting or a ferromagnetic state can be realized by doping, or intercalating, Bi_2Se_3 and Bi_2Te_3 crystals with specific elements. For instance, Cr-doped Bi_2Se_3 shows a magnetic phase without long-range order [91], and Mn-doped Bi_2Te_3 exhibits a ferromagnetic phase transition with long-range order [92]. Measurements of the magnetization of $Bi_{2-x}Mn_xTe_3$ (with doping levels of $x \approx 0.04$ and 0.09) have shown that the ferromagnetic phase sets in at critical temperatures in the range $T_c \approx 9$–12 K. So far, magnetically doped Bi_2Se_3 and Bi_2Te_3 crystals have not been studied in detail, and their magnetic properties are currently being investigated. However, ARPES experiments have already led to the observation that a gap in the Dirac surface states of magnetically doped Bi_2Se_3 seems indeed to open [93] (~7 meV for Mn-dopants and ~50 meV for Fe-dopants).

Transport experiments have shown that a superconducting phase transition can be achieved by intercalating Pd in Bi_2Te_3 [27] or Cu in Bi_2Se_3 [27,94–98]. The superconducting phase has been observed in $Cu_xBi_2Se_3$ with doping levels in the range $x \approx 0.09$–0.64. The critical temperature varies from $T_c \approx 3.8$ K at $x \approx 0.12$ down to $T_c \approx 2.2$ K at $x \approx 0.64$. Two ways to introduce Cu-doping are possible in Bi_2Se_3: as an electron donor when it intercalates between Bi_2Se_3 layers or as an electron acceptor when it substitutes Bi in a Bi_2Se_3 layer; Cu-intercalated Bi_2Se_3 is the only one that exhibits a superconducting phase transition. The electron density of $Cu_xBi_2Se_3$ is $n_e \sim 10^{20}$ cm^{-3} and, very surprisingly, was found to be almost independent of the intercalated Cu concentration (in the range $x \approx 0.09$–0.64). Magnetization and transport experiments have shown that $Cu_xBi_2Se_3$ is a Type II superconductor. The first critical magnetic field has a very low value of $B_{c1} \approx 0.4$ mT, whereas the second critical magnetic field is much larger and very strongly anisotropic: $B_{c2} \approx 1.7$ T (or 3–4.6 T) for fields perpendicular (or parallel) to the $Cu_xBi_2Se_3$ layers. The anisotropy as well as the symmetry of the superconducting order parameter in $Cu_xBi_2Se_3$ are being investigated. Experimental evidence for unconventional behavior has been reported through the observation of zero-energy surface states that have been detected in spectroscopic transport experiments [96].

6 CONCLUSIONS AND OUTLOOK

The intense research effort that has been devoted during the past few years to experimental research on topological insulators has led to very significant progress. Following theoretical predictions, a number of material systems have been investigated and proven to be 3D topological insulators by means of angle-resolved photoemission spectroscopy. Transport studies have been mainly focused on those compounds possessing a larger bandgap, an exposed Dirac point, and smaller bulk conductivity.

The goal so far has been to discriminate the contribution to transport due to the surface Dirac fermions from other parallel transport channels, mainly due to the bulk. Most research has focused on a variety of Bi-based materials, but alternative systems—such as epitaxially grown, strained HgTe layers—are emerging. No textbook-quality demonstration of the transport properties of Dirac electrons at the surface of a 3D topological insulator has been reported so far. However, at the current stage it would not be easy—if possible at all—to find a realistic scenario capable of explaining all the observations made in transport experiments, without invoking the presence of Dirac surface states. The first successful experiments tuning transport mediated by Dirac surface fermions by means of gate electrodes have also been performed. What needs to be done in the future is to build on these results to realize nanofabricated devices providing better experimental control.

It seems clear that to proceed in this direction, significant progress is needed in our material understanding and control. It is important to drastically improve the material reproducibility, suppress the disorder-induced density of states present inside the bulk bandgap of the materials used, bring the Fermi level inside the gap, and increase the carrier mobility. In individual crystals of several of the Bi-based compounds that have been investigated, one or more of these conditions could be fulfilled. However, not all the conditions could be fulfilled at the same time. More importantly, it is remaining unclear why some crystals exhibit better properties than others, despite having been prepared under nominally identical conditions. The problem is difficult, as it is typically the case when dealing with small-gap semiconductors, where even small concentrations of impurities have very large effects on the electronic properties. Hopefully, material science studies of these compounds will enable a more systematic progress, by finding and eliminating the dominant defects that introduce charge carriers and electronic states in the bulk of crystals, by using alternative techniques to grow better quality samples, or by identifying new 3D topological insulating materials for which crystals of better quality can be grown. Certainly, the underlying physical concepts provide a strong drive to undertake this research, since the investigation of quantum transport in nanofabricated devices based on 3D topological insulators is a rich, virtually unexplored area of research, hinging on some fundamental concepts—the topological aspects of the electronic properties of materials—that are only now starting to be broadly appreciated in many different contexts of condensed matter physics.

REFERENCES

[1] Kane CL, Mele EJ. Z_2 topological order and the quantum spin Hall effect. Phys Rev Lett 2005;95:146802.
[2] Fu L, Kane CL, Mele EJ. Topological insulators in three dimensions. Phys Rev Lett 2007;98:106803.
[3] Hasan MZ, Kane CL. Colloquium: topological insulators. Rev Mod Phys 2010;82:3045.
[4] Qi XL, Zhang SZ. Topological insulators and superconductors. Rev Mod Phys 2011;83:1057.
[5] Castro Neto AH, Guinea F, Peres NMR, Novoselov KS, Geim AK. The electronic properties of graphene. Rev Mod Phys 2009;81:109.
[6] Köhler H, Landwehr G. Constant energy surfaces of n-type bismuth selenide from the Shubnikov-de Haas effect. Phys State Solid B 1971;45:K109.
[7] Köhler H. Conduction band parameters of Bi_2Se_3 from Shubnikov-de Haas investigations. Phys State Solid B 1973;58:91.
[8] Hyde GR, Beale HA, Spain IL, Woollam JA. Electronic properties of Bi_2Se_3 crystals. J Phys Chem Solids 1974;35:1719.
[9] Köhler H, Fisher H. Investigation of the conduction band Fermi surface in Bi_2Se_3 at high electron concentrations. Phys State Solid B 1975;69:349.
[10] Satterthwaite CB, Ure RW. Electrical and thermal properties of Bi_2Te_3. Phys Rev 1957;108:1164.
[11] Pawlewicz WT, Rayne JA, Ure RW. Resistivity of Bi_2Te_3 from 1.3 K to 300 K. Phys Lett 1974;48A:391.
[12] Köhler H. Non-parabolicity of the highest valence band of Bi_2Te_3 from Shubnikov-de Haas effect. Phys State Solid B 1976;74:591.
[13] Lenoir B, Cassart M, Michenaud JP, Scherrer H, Scherrer S. Transport properties of Bi-rich Bi-Sb alloys. J Phys Chem Solids 1996;57:89.
[14] Kulbachinskii VA, Miura N, Nakagawa H, Arimoto H, Ikaida T, Lostak P, et al. Conduction-band structure of $Bi_{2-x}Sb_xSe_3$ mixed crystals by Shubnikov-de Haas and cyclotron resonance measurements in high magnetic fields. Phys Rev B 1999;59:15733.
[15] Fu L, Kane CL. Topological insulators with inversion symmetry. Phys Rev B 2007;76:045302.

[16] Zhang H, Liu CX, Qi XL, Dai X, Fang Z, Zhang SC. Topological insulators in Bi_2Se_3, Bi_2Te_3
      and Sb_2Te_3 with a single Dirac cone on the surface. Nat Phys 2009;5:438.
[17] Hsieh D, Qian D, Wray L, Xia Y, Hor YS, Cava RJ, et al. A topological Dirac insulator in a
      quantum spin Hall phase. Nature 2008;452:970.
[18] Taskin AA, Ando Y. Quantum oscillations in a topological insulator $Bi_{1-x}Sb_x$. Phys Rev B
      2009;80:085303.
[19] Taskin AA, Segawa K, Ando Y. Oscillatory angular dependence of the magnetoresistance in
      a topological insulator $Bi_{1-x}Sb_x$. Phys Rev B 2010;82:121302.
[20] Xia Y, Qian D, Hsieh D, Wray L, Pal A, Lin H, et al. Observation of a large-gap topological-
      insulator class with a single Dirac cone on the surface. Nat Phys 2009;5:398.
[21] Chen YL, Analytis JG, Liu JH, Mo SK, Qi XL, Zhang HJ, et al. Experimental realization of
      a three-dimensional topological insulator, Bi_2Te_3. Science 2009;325:178.
[22] Butch NP, Kirshenbaum K, Syers P, Sushkov AB, Jenkins GS, Drew HD, et al. Strong
      surface scattering in ultrahigh-mobility Bi_2Se_3 topological insulator crystals. Phys Rev B
      2010;81:241301.
[23] Analytis JG, Chu JH, Chen Y, Corredor F, McDonald RD, Shen ZX, et al. Bulk Fermi
      surface coexistence with Dirac surface state in Bi_2Se_3: a comparison of photoemission and
      Shubnikov-de Haas measurements. Phys Rev B 2010;81:205407.
[24] Ren Z, Taskin AA, Sasaki S, Segawa K, Ando Y. Observations of two-dimensional quantum
      oscillations and ambipolar transport in the topological insulator Bi_2Se_3 achieved by Cd
      doping. Phys Rev B 2011;84:075316.
[25] Eto K, Ren Z, Taskin AA, Segawa K, Ando Y. Angular-dependent oscillations of the
      magnetoresistance in Bi_2Se_3 due to the three-dimensional bulk Fermi surface. Phys Rev
      B 2010;81:195309.
[26] Hor YS, Richardella A, Roushan P, Xia Y, Checkelsky JG, Yazdani A, et al. p-type
      Bi_2Se_3 for topological insulator and low-temperature thermoelectric applications. Phys Rev
      B 2009;79:195208.
[27] Hor YS, Checkelsky JG, Qu D, Ong NP, Cava RJ. Superconductivity and non-metallicity
      induced by doping the topological insulators Bi_2Se_3 and Bi_2Te_3. J Phys Chem Solids
      2011;72:572.
[28] Zhu ZH, Levy G, Ludbrook B, Veenstra CN, Rosen JA, Comin R, et al. Rashba spin-splitting
      control at the surface of the topological insulator Bi_2Se_3. Phys Rev Lett 2011;107:186405.
[29] Liu ZK, Chen YL, Analytis JG, Mo SK, Lu DH, Moore RG, et al. Robust topological surface
      state against direct surface contamination. Physica E 2012;44:891.
[30] Alpichshev Z, Biswas RR, Balatsky AV, Analytis JG, Chu JH, Fisher IR, et al. STM imaging
      of impurity resonances on Bi_2Se_3. Phys Rev Lett 2012;108:206402.
[31] Sacépé B, Oostinga JB, Li J, Ubaldini A, Couto NJG, Giannini E, et al. Gate-tuned normal and
      superconducting transport at the surface of a topological insulator. Nat Commun 2011;2:575.
      http://dx.doi.org/10.1038/ncomms1586.
[32] Hor YS, Qu D, Ong NP, Cava RJ. Low temperature magnetothermoelectric and
      magnetoresistance in the Te vapor annealed Bi_2Te_3. J Phys Condens Matter 2010;22:375801.
[33] Bejenari I, Kantser V, Balandin AA. Thermoelectric properties of electrically gated bismuth
      telluride nanowires. Phys Rev B 2010;81:075316.
[34] Shklovskii BI, Efros AL. Electronic properties of doped semiconductors. Berlin, Heidelberg:
      Springer; 1984.
[35] Qu DX, Hor YS, Xiong J, Cava RJ, Ong NP. Quantum oscillations and Hall anomaly of
      surface states in the topological insulator Bi_2Te_3. Science 2010;329:821.
[36] Analytis JG, McDonald RD, Riggs SC, Chu JH, Boebinger GS, Fisher IR. Two-dimensional
      surface state in the quantum limit of a topological insulator. Nat Phys 2010;6:960.
[37] Hsieh D, Xia Y, Qian D, Wray L, Dil JH, Meier F, et al. A tunable topological insulator in
      the spin helical Dirac transport regime. Nature 2009;460:1101.
[38] Checkelsky JG, Hor YS, Liu MH, Qu DX, Cava RJ, Ong NP. Quantum interference in
      macroscopic crystals of nonmetallic Bi_2Se_3. Phys Rev Lett 2009;103:246601.
[39] Ren Z, Taskin AA, Sasaki S, Segawa K, Ando Y. Optimizing $Bi_{2-x}Sb_xTe_{3-y}Se_y$ solid
      solutions to approach the intrinsic topological insulator regime. Phys Rev B 2011;84:165311.
[40] Ren Z, Taskin AA, Sasaki S, Segawa K, Ando Y. Large bulk resistivity and surface quantum
      oscillations in the topological insulator Bi_2Te_2Se. Phys Rev B 2010;82:241306.

[41] Jia S, Ji H, Climent-Pascual E, Fuccillo MK, Charles ME, Xiong J, et al. Low-carrier-concentration crystals of the topological insulator Bi_2Te_2Se. Phys Rev B 2011;84:235206.

[42] Xiong J, Petersen AC, Qu D, Hor YS, Cava RJ, Ong NP. Quantum oscillations in a topological insulator Bi_2Te_2Se with large bulk resistivity (6 Ω cm). Physica E 2012;44:917.

[43] Xiong J, Luo Y, Khoo Y, Jia S, Cava RJ, Ong NP. High-field Shubnikov-de Haas oscillations in the topological insulator Bi_2Te_2Se. Phys Rev B 2012;86:045314.

[44] Taskin AA, Ren Z, Sasaki S, Segawa K, Ando Y. Observation of Dirac holes and electrons in a topological insulator. Phys Rev Lett 2011;107:016801.

[45] Ashcroft NW, Mermin ND. Solid State Phys. Saunders College Publishing; 1976.

[46] Beenakker CWJ, van Houten H. Quantum transport in semiconductor nanostructures. Solid State Phys 1991;44:1.

[47] Novoselov KS, Geim AK, Morozov SV, Jiang D, Katsnelson MI, Grigorieva IV. Two-dimensional gas of massless Dirac fermions in graphene. Nature 2005;438:197.

[48] Zhang YB, Tan YW, Stormer HL, Kim P. Experimental observation of the quantum Hall effect and Berry's phase in graphene. Nature 2005;438:201.

[49] Taskin AA, Ando Y. Berry phase of nonideal Dirac fermions in topological insulators. Phys Rev B 2011;84:035301.

[50] Novoselov KS, Geim AK, Morozov SV, Jiang D, Zhang Y, Dubonos SV, et al. Electric field effect in atomically thin carbon films. Science 2004;306:666.

[51] Cho S, Butch NP, Paglione J, Fuhrer MS. Insulating behavior in ultrathin bismuth selenide field effect transistors. Nano Lett 2011;11:1925.

[52] Steinberg H, Gardner DR, Lee YS, Jarillo-Herrero P. Surface state transport and ambipolar electric field effect in Bi_2Se_3 nanodevices. Nano Lett 2010;10:5032.

[53] Checkelsky JG, Hor YS, Cava RJ, Ong NP. Bulk band gap and surface state conduction observed in voltage-tuned crystals of the topological insulator Bi_2Se_3. Phys Rev Lett 2011;106:196801.

[54] Kim D, Cho S, Butch NP, Syers P, Kirshenbaum K, Adam S, et al. Fuhrer MS. Surface conduction of topological Dirac electrons in bulk insulating Bi_2Se_3. Nat Phys 2012;8:459.

[55] Kong D, Cha JJ, Lai K, Peng H, Analytis JG, Meister S, et al. Rapid surface oxidation as a source of surface degradation factor for Bi_2Se_3. ACS Nano 2011;5:4698.

[56] Zhang Y, He K, Chang CZ, Song CL, Wang LL, Chen X, et al. Crossover of the three-dimensional topological insulator Bi_2Se_3 to the two-dimensional limit. Nat Phys 2010;6:584.

[57] Zhang G, Qin H, Chen J, He X, Lu L, Li Y, et al. Growth of topological insulator Bi_2Se_3 thin films on $SrTiO_3$ with large tunability in chemical potential. Adv Func Mater 2011;21:2351.

[58] Li HD, Wang ZY, Kan X, Guo X, He HT, Wang Z, et al. The van der Waals epitaxy of Bi_2Se_3 on the vicinal Si(111) surface: an approach for preparing high-quality thin films of a topological insulator. New J Phys 2010;12:103038.

[59] Li YY, Wang G, Zhu XG, Liu MH, Ye C, Chen X, et al. Intrinsic topological insulator Bi_2Te_3 thin films on Si and their thickness limit. Adv Mater 2010;22:4002.

[60] Steinberg H, Laloë JB, Fatemi V, Moodera JS, Jarillo-Herrero P. Electrically tunable surface-to-bulk coherent coupling in topological insulator thin films. Phys Rev B 2011;84:233101.

[61] Yuan H, Liu H, Shimotani H, Guo H, Chen M, Xue Q, et al. Liquid-gated ambipolar transport in ultrathin films of a topological insulator Bi_2Te_3. Nano Lett 2011;11:2601.

[62] Bansal N, Kim YS, Brahlek M, Edrey E, Oh S. Thickness-independent transport channels in topological insulator Bi_2Se_3 thin films. Preprint available at arxiv:1104.5709v3 (To appear in Phys Rev Lett 2012)

[63] Liu M, Chang CZ, Zhang Z, Zhang Y, Ruan W, He K, et al. Electron interaction-driven insulating ground state in Bi_2Se_3 topological insulators in the two-dimensional limit. Phys Rev B 2011;83:165440.

[64] Onose Y, Yoshimi R, Tsukazaki A, Yuan H, Hidaka T, Iwasa Y, et al. Pulsed laser deposition and ionic liquid gate control of epitaxial Bi_2Se_3 thin films. Appl Phys Exp 2011;4:083001.

[65] Koren G, Kirzhner T, Lahoud E, Chashka KB, Kanigel A. Proximity-induced superconductivity in topological Bi_2Te_2Se and Bi_2Se_3 films: Robust zero-energy bound state possibly due to Majorana fermions. Phys Rev B 2011;84:224521.

[66] Cheng P, Song C, Zhang T, Zhang Y, Wang Y, Jia JF, et al. Landau quantization of topological surface states in Bi_2Se_3. Phys Rev Lett 2010;105:076801.

[67] Kong D, Randel JC, Peng H, Cha JJ, Meister S, Lai K, et al. Topological insulator nanowires and nanoribbons. Nano Lett 2010;10:329.

[68] Hong SS, Kundhikanjana W, Cha JJ, Lai K, Kong D, Meister S, et al. Ultrathin topological insulator Bi_2Se_3 nanoribbons exfoliated by atomic force microscopy. Nano Lett 2010;10:3118.

[69] Peng H, Lai K, Kong D, Meister S, Chen Y, Qi XL, et al. Aharonov-Bohm interference in topological insulator nanoribbons. Nat Mater 2010;9:225.

[70] Cha JJ, Williams JR, Kong D, Meister S, Peng H, Bestwick AJ, et al. Magnetic doping and Kondo effect in Bi_2Se_3 nanoribbons. Nano Lett 2010;10:1076.

[71] Hong SS, Cha JJ, Kong D, Cui Y. Ultra-low carrier concentration and surface-dominant transport in antimony-doped Bi_2Se_3 topological insulator nanoribbons. Nat Commun. 2012;3:757. http://dx.doi.org/10.1038/ncomms1771.

[72] Zhang D, Wang J, DaSilva AM, Lee JS, Gutierrez HR, Chan MHW, et al. Superconducting proximity effect and possible evidence for Pearl vortices in a candidate topological insulator. Phys Rev B 2011;84:165120.

[73] Tang H, Liang D, Qiu RLJ, Gao XPA. Two-dimensional transport-induced linear magneto-resistance in topological insulator Bi_2Se_3 nanoribbons. ACS Nano 2011;5:7510.

[74] Cha JJ, Kong D, Hong SS, Analytis JG, Lai K, Cui Y. Weak antilocalization in $Bi_2(Se_xTe_{1-x})_3$ nanoribbons and nanoplates. Nano Lett 2012;12:1107.

[75] Xiu F, He L, Wang Y, Cheng L, Chang LT, Lang M, et al. Manipulating surface states in topological insulator nanoribbons. Nat Nanotech 2011;6:216.

[76] Kong D, Dang W, Cha JJ, Li H, Meister S, Peng H, et al. Few-layer nanoplates of Bi_2Se_3 and Bi_2Te_3 with highly tunable chemical potential. Nano Lett 2010;10:2245.

[77] Kong D, Chen Y, Cha JJ, Zhang Q, Analytis JG, Lai K, et al. Ambipolar field effect in the ternary topological insulator $(Bi_xSb_{1-x})_2Te_3$ by composition tuning. Nat Nanotech 2011;6:705.

[78] Che J, Qin HJ, Yang F, Liu J, Guan T, Qu FM, et al. Gate-voltage control of chemical potential and weak antilocalization in Bi_2Se_3. Phys Rev Lett 2010;105:176602.

[79] Chen J, He XY, Wu KH, Ji ZQ, Lu L, Shi JR, et al. Tunable surface conductivity in Bi_2Se_3 revealed in diffusive electron transport. Phys Rev B 2011;83:241304.

[80] Dai X, Hughes TL, Qi XL, Fang Z, Zhang SC. Helical edge and surface states in HgTe quantum wells and bulk insulators. Phys Rev B 2008;77:125319.

[81] Brüne C, Liu CX, Novik EG, Hankiewicz EM, Buhmann H, Chen YL, et al. Quantum Hall effect from the topological surface states of strained bulk HgTe. Phys Rev Lett 2011;106:126803.

[82] Saito R, Dresselhaus G, Dresselhaus MS. Physical Properties of Carbon Nanotubes. London: Imperial College Press; 1998.

[83] Washburn S, Webb RA. Aharonov-Bohm effect in normal metal quantum coherence and transport. Adv Phys 1986;35:375.

[84] Lee PA, Stone AD. Universal conductance fluctuations in metals. Phys Rev Lett 1985;55:1622.

[85] Fu L, Kane CL. Superconducting proximity effect and Majorana fermions at the surface of a topological insulator. Phys Rev Lett 2008;100:096407.

[86] Fu L, Kane CL. Probing neutral Majorana fermion edge modes with charge transport. Phys Rev Lett 2009;102:216403.

[87] Akhmerov AR, Nilsson J, Beenakker CWJ. Electrically detected interferometry of Majorana fermions in a topological insulator. Phys Rev Lett 2009;102:216404.

[88] Yang F, Ding Y, Qu F, Shen J, Chen J, Wei Z, et al. Proximity effect at superconducting Sn-Bi_2Se_3 interface. Phys Rev B 2012;85:104508.

[89] Williams JR, Bestwick AJ, Gallagher P, Hong SS, Cui Y, Bleich AS, et al. Unconventional Josephson effect in hybrid superconductor-topological insulator devices. Phys Rev Lett 2012;109:056803.

[90] Veldhorst M, Snelder M, Hoek M, Gang T, Guduru VK, Wang XL, et al. Josephson supercurrent through a topological insulator surface state. Nat Mater 2012;11:417.

[91] Liu M, Zhang J, Chang CZ, Zhang Z, Feng X, Li K, et al. Crossover between weak antilocalization and weak localization in a magnetically doped topological insulator. Phys Rev Lett 2012;108:036805.

[92] Hor YS, Roushan P, Beidenkopf H, Seo J, Qu D, Checkelsky JG, et al. Development of ferromagnetism in the doped topological insulator $Bi_{2-x}Mn_xTe_3$. Phys Rev B 2010;81:195203.
[93] Chen YL, Chu JH, Analytis JG, Liu ZK, Igarashi K, Kuo HH, et al. Massive Dirac fermion on the surface of a magnetically doped topological insulator. Science 2010;329:659.
[94] Wray LA, Xu SY, Xia Y, Hor YS, Qian D, Fedorov AV, et al. Observation of topological order in a superconducting doped topological insulator. Nat Phys 2010;6:855.
[95] Hor YS, Williams AJ, Checkelsky JG, Roushan P, Seo J, Xu Q, et al. Superconductivity in $Cu_xBi_2Se_3$ and its implications for pairing in the undoped topological insulator. Phys Rev Lett 2010;104:057001.
[96] Sasaki S, Kriener M, Segawa K, Yada K, Tanaka Y, Sato M, et al. Topological superconductivity in $Cu_xBi_2Se_3$. Phys Rev Lett 2011;107:217001.
[97] Kriener M, Segawa K, Ren Z, Sasaki S, Ando Y. Bulk superconducting phase with a full energy gap in the doped topological insulator $Cu_xBi_2Se_3$. Phys Rev Lett 2011;106:127004.
[98] Kriener M, Segawa K, Ren Z, Sasaki S, Wada S, Kuwabata S, et al. Electrochemical synthesis and superconducting phase diagram of $Cu_xBi_2Se_3$. Phys Rev B 2011;84:054513.

Part III

The Road Ahead

Chapter 9

New Materials

Ke He[*], Xucun Ma[*], Xi Chen[†] and Qi-Kun Xue[†,*]

[*]Beijing National Laboratory for Condensed Matter Physics, Institute of Physics,
The Chinese Academy of Sciences, Beijing 100190, China
[†]State Key Laboratory of Low-Dimensional Quantum Physics, Department of Physics,
Tsinghua University, Beijing 100084, China

Chapter Outline Head

In spite of significant progress in the prediction and preparation of topological insulator (TI) materials in different forms by various techniques such as self-flux, molecular beam epitaxy (MBE), and metal-organic chemical vapor deposition (MOCVD) [1,2], the major novel quantum phenomena predicted theoretically in TIs have not been experimentally demonstrated yet. The main problem lies in the quality of materials: the samples are not sufficiently uniform (in the case of thin films), the electron mobility is not high enough, and more

Topological Insulators. http://dx.doi.org/10.1016/B978-0-444-63314-9.00009-3
237

critically, the bulk carriers, which often dominate the transport properties, cannot be completely depleted. TI materials are narrow-gap semiconductors with strong spin-orbit coupling (SOC) [3–5], which are naturally vulnerable to the formation of vacancies and antisite defects, and thus are difficult to be made intrinsically insulating. Historically, these materials have long been studied for infrared and thermoelectric applications [6], where the requirement for sample quality is not so high as that for observation of quantum phenomena. So far, the only two successful examples are two-dimensional (2D) TIs, namely, HgTe/CdTe [3,7] and InAs/GaSb/AlSb [8,9] quantum wells (QWs) in which quantum spin Hall (QSH) effect was observed through transport measurements [7,9]. It is simply because very high quality samples could be prepared with the decades-long effort in the MBE growth and micro-fabrication of the two systems, even before they were known as TIs [10,11]. It is clear that the future development of TI field strongly depends on how its material science develops: improving the quality of the existing TI materials, finding new TI materials of unique properties that are relatively easy to prepare, and developing appropriate techniques for electronic structure engineering and micro-fabrication of the materials. In this chapter, we review the recent progress in the TI materials and the experiments on tuning the electronic properties with various methods toward the experimental demonstration of fascinating quantum phenomena. The materials to be discussed in the chapter include three-dimensional (3D) TIs (Section 1), 2D TIs (Section 2), and magnetic TIs (Section 3). The material issues in superconductor/TI hybrid structures and topological superconductors [1,2] are not covered here since they are introduced in detail elsewhere in this book.

1 THREE-DIMENSIONAL TOPOLOGICAL INSULATORS

1.1 Early Discovered Three-Dimensional Topological Insulator Materials

A 3D TI has its bulk band gapped at the Fermi level but exhibits gapless surface states at all of its surfaces [12]. The surface bands in momentum space have the dispersion of a 2D Dirac cone and are spin-polarized except for some high symmetry points, for example $\bar{\Gamma}$ and $\bar{M}$ points of the surface Brillouin zone of a surface with 6-fold symmetry. The first 3D TI material theoretically proposed [4] and experimentally demonstrated [13] is $Bi_{1-x}Sb_x$ alloy with x between ~ 0.07 and ~ 0.22. Angle-resolved photoemission spectroscopy (ARPES) [13] and spin-resolved ARPES [14] measurements have revealed the special band and spin structures of the Dirac surface states. However, this material has a rather small bulk gap (several tenths meV), complex surface band structure and random distribution of Bi and Sb atoms, which significantly reduces the electron mobility. The later found Bi_2Se_3, Bi_2Te_3, and Sb_2Te_3 were proved to be a class of much better TI materials for their ordered lattice structures, relatively large bulk gaps (up to 0.3 eV in Bi_2Se_3), and surface states composed

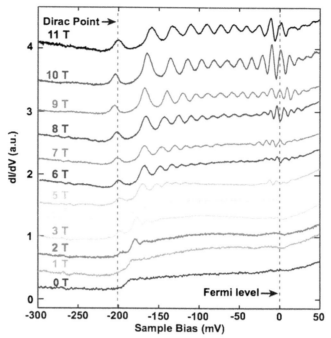

FIGURE 1 Landau levels formed in the Dirac surface band of Bi_2Se_3 measured by STM (from Ref. [21]).

of only single Dirac cone [5,15–18]. These properties, especially the third one, make it much easier to study the quantum effects associated with the Dirac surface states. For example, a peculiar property of the Dirac surface state is the forbidden back-scattering [1,2], which can be revealed by the analysis of the quasiparticle interference (QPI) pattern measured by scanning tunneling microscopy (STM), discussed in Chapter 7. In $Bi_{1-x}Sb_x$ alloy, the complex surface band structure makes the pattern rather complex and difficult to be understood straightforwardly [19]. On the contrary, for Bi_2Te_3 one only needs a simple qualitative analysis of the QPI pattern to reveal the forbidden back-scattering of the surface states [20]. Another interesting phenomenon observed in Bi_2Se_3 family TIs is the Landau quantization of the Dirac surface states in magnetic field. Landau level peaks have clearly been revealed by STM in Bi_2Se_3 [21,22] (see Fig. 1), Sb_2Te_3 [23] and Bi_2Te_3 [24], with non-even spacing between Landau level peaks and magnetic field independent zero-order Landau levels, two hallmark features of the Dirac surface states.

The observation of well-defined Landau levels of the Dirac surface states promises the observation of the quantum Hall (QH) effect associated with the Dirac surface states in Bi_2Se_3 family TIs in transport measurements, one of the most basic quantum phenomena expected in 3D TIs [1]. However, despite intensive efforts by many groups worldwide, surface QH effect has not been realized experimentally. The main problems preventing the observation of the surface QH effect and other predicted quantum phenomena in TIs through transport measurements are listed as follows.

First, measurement of the surface QH effect from one surface will be influenced by the conductance contribution from other ones that are not perpendicular to the applied magnetic field, since gapless surface states exist at every surface of a 3D TI.

Second, the Landau levels observed in ultra high vacuum can be destroyed by adsorption and/or oxidization in ambient condition due to the reduced electron mobility of the surface states.

Third, Bi_2Se_3 family materials are usually heavily doped by defects or band bending and therefore the bulk bands contribute much more conductance than surface states.

Finally, the spatial inhomogeneity of the chemical potential can prevent the observation of QH effect in macroscopic-sized samples.

The last problem can in principle be solved by smaller samples using micro-fabrication techniques, which will be not discussed here. For other three problems, one can find solutions by using thin films or heterostructures of 3D TIs grown by MBE technique. Next, we will review the recent progress on MBE-grown thin films and heterostructures of Bi_2Se_3 family TIs.

1.2 Molecular Beam Epitaxy Grown Thin Films of Topological Insulators

Although 3D TIs, as the name implies, are in principle bulk materials, thin film is the more favored form to reveal their unique properties. Almost all the unique properties of 3D TIs are carried by the Dirac surface states. Thus the large surface/bulk ratio of thin film geometry is beneficial to reveal the contribution of the surface states from the background of bulk bands, especially for non-surface-sensitive techniques such as transport. For the observation of the surface QH effect and other magnetoelectric effects, thin film structure can minimize the conductance contribution from side surface states and bulk bands. Another merit of thin films is that their electronic structures can be easily tuned by various parameters such as surface and interface conditions, film thickness (through finite size effect), and electrical field effect. Furthermore, many quantum effects, e.g., topological exciton condensation [25], large thermoelectric effect [26], and oscillatory topological transition [27], only take place in thin films of 3D TIs. MBE is a standard technique used to grow high quality semiconductor thin films, which in principle can provide samples with the highest mobility and lowest defect density. With MBE, it is also very convenient to fabricate heterostructures between TIs and other materials such as topologically trivial insulators, magnets, and superconductors, which exhibit rich phenomena or functions [1,2].

So far the MBE growth of 3D TIs has been focused on Bi_2Se_3 family materials (Bi_2Te_3, Bi_2Se_3, and Sb_2Te_3). Their simple composition makes the MBE growth relatively easy [5]. The MBE growth of Bi_2Se_3 family materials is carried out by co-evaporation of Bi/Sb and Se/Te sources. Similar to As, Se/Te has much higher vapor pressure than the cation elements (Bi and Sb); therefore, the growth kinetics of Bi_2Se_3 family should be very similar to that well established for GaAs and other III–V compound semiconductors. During growth the substrate is kept at a temperature high enough to efficiently dissociate Se/Te molecules into single atoms and prevent the Se/Te molecules from condensation on the surface, but low enough to avoid re-evaporation of Bi/Sb atoms and breaking of Bi (Sb)–Se (Te) bonds [28–32]. The flux of Se/Te is set much higher than that of Bi (Sb) to minimize Se (Te) vacancy density in the films since $Bi(Sb)_2Se(Te)_3$ is the most Se (Te)-rich phase in phase diagram.

Bi_2Se_3 family TIs are layered materials with the interlayer bonding close to van der Waals type (each layer of the materials actually includes five atomic layers, i.e., quintuple layer (QL)). The layered structure leads to very low surface free energy, which favors the 2D MBE growth mode under standard MBE growth conditions. As a result, 2D growth of the TIs can be easily realized on various substrates, even with large lattice mismatch with the films. The only requirements are that the substrates should have 3- or 6-fold symmetry and be inert to the reactive Se or Te atoms. The substrates supporting 2D growth of Bi_2Se_3 family TIs include Si (1 1 1) (passivated with Bi, H, GaSe, ZnSe, or In_2Se_3 for growth of Bi_2Se_3), graphene-terminated SiC (0 0 0 1), sapphire (0 0 0 1), $SrTiO_3$ (1 1 1), CdS (0 0 0 1), GaN (0 0 0 1), GaAs (1 1 1) B, InP (1 1 1), and mica [28–40]. The rich choice of substrates makes it convenient to apply TI materials in different studies and applications. Figure 2 shows an example of the MBE growth of Bi_2Se_3 film [29]. The sharp streaky RHEED pattern (Fig. 2(a)) indicates the 2D nature of the film. The RHEED intensity oscillation recorded during growth (Fig. 2(b)) characterizes the layer-by-layer growth mode: each intensity maximum represents the stage when the film has uniform thickness. The resulting film shows atomically flat morphology with low defect density as seen from the STM image in Fig. 2(c) and a rather low doping level (the Fermi level locates $\sim$120 meV above the Dirac point) according to the ARPES data displayed in Fig. 2(d). The high quality of the film guarantees the observability of Landau quantization of the Dirac surface states (see Fig. 1) [21,23].

Realization of the growth of 3D TI thin films in layer-by-layer mode enables us to address the question naturally raised for thin films of 3D TIs: what will a 3D TI evolve into when it is too thin to be considered as a 3D system any more? Figure 3 displays thickness dependence of ARPES spectra of Bi_2Se_3 films from 1 QL to 6 QL [30]. We observe clear and sharp quantum well states formed in conduction and valence bands, which indicates the uniform thickness of the films. As the thickness is reduced below 6 QL, a gap opens at the Dirac surface states and increases with the decreasing thickness. At the same time, Rashba-type splitting, i.e., splitting of two spin subbands of an energy band along k

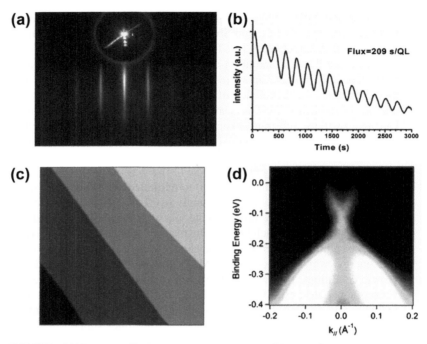

FIGURE 2 MBE growth of Bi_2Se_3. (a) RHEED pattern of a MBE-grown Bi_2Se_3 film. (b) RHEED oscillation measured in MBE growth of the Bi_2Se_3 film, indicating the layer-by-layer growth. (c) STM image of the surface of the Bi_2Se_3 film. (d) ARPES bandmap of the Dirac surface states of the Bi_2Se_3 film (from Ref. [29]).

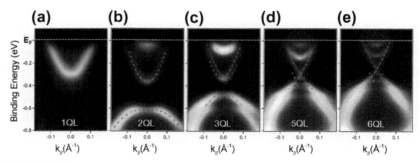

FIGURE 3 Thickness dependent ARPES of MBE-grown Bi_2Se_3 thin films from 1 QL to 6 QL (from Ref. [30]).

axis, is resolved in the gapped surface states. The gap-opening results from the hybridization between the Dirac surface states from the opposite two surfaces as the film thickness is thin enough for the surface state wave functions from the two opposite surfaces to overlap. If the two Dirac surface states have different chemical potentials due to band bending of the film, the hybridization will lead to gapped surface states with Rashba-like splitting [30,41]. Similar quantum well films and gap-opening in the Dirac surface states were also observed in Bi_2Te_3 and Sb_2Te_3, with the minimum thickness of 3 QL and 4 QL for being 3D TI, respectively [23,28,31]. The thin films of 3D TI around 3D-2D crossover region are of special interest for they provide an opportunity for exploring rich topological phase transitions. For example, the film can be tuned between 3D TI, 2D TI, and 2D trivial insulator phases by film thickness [27] and external electrical field [42]; doping magnetic impurities may turn a thin film of 3D TI into a quantum anomalous Hall (QAH) insulator that shows quantum Hall (QH) effect even without external magnetic field [43]. We will discuss these effects later in this chapter.

1.3 Defects in Epitaxial Films of Bi_2Se_3 Family Topological Insulators

Even though 2D growth of Bi_2Se_3 family TIs is not a difficult task, in order to obtain epitaxial films with high quality so as to reveal various quantum phenomena, one needs to reduce all kinds of defects and types of disorder in the films, from vacancies and antisite defects to dislocations and domain walls, all of which significantly reduce the electron mobility and bring extra bulk charge carriers. Below we summarize the defects and disorders commonly seen in MBE-grown thin films of Bi_2Se_3 family TIs and the possible ways to reduce them.

1.3.1 Dislocations

Due to the layered nature of Bi_2Se_3 family TIs, lattice mismatch has a week effect on the 2D growth mode of Bi_2Se_3 family TIs. However, it does deteriorate the sample quality by introducing dislocations in the films, analogous to the MBE-grown GaN films on sapphire substrate with large lattice mismatch [44]. One way to reduce the density of dislocations is by growth at lower substrate temperature, shifting the process away from the thermal equilibrium condition. Figure 4 shows the STM topographic images of Bi_2Se_3 films grown on the sapphire $(0\,0\,0\,1)$ at different substrate temperatures [35]. We can see that the density of islands on the surfaces, most of which result from screw dislocations, obviously decreases with decreasing substrate temperatures. At the temperature 190 °C, there are few screw dislocations remaining, but 3D elongated islands appear on the surface. The islands are often seen in growth of Bi-based materials [45] and are actually Bi islands formed by extra Bi atoms on the surface due to inadequate reaction with Se at low substrate temperature. So growth of Bi_2Se_3 (maybe Bi_2Te_3 and Sb_2Te_3 too) films with fewer dislocations is actually limited by the low reactivity of Se (Te) molecules. To promote the reaction between

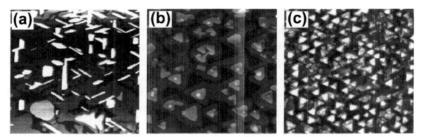

FIGURE 4 STM images of the surface morphology (1 × 1 μm) of 8 QL Bi_2Se_3 film grown on sapphire (0 0 0 1) with the substrate temperatures of 190 °C (a), 220 °C (b), and 250 °C (c), respectively (from Ref. [35]).

Bi (Sb) and Se (Te), one can utilize cracker cells for evaporation of Se and Te, which will efficiently dissociate the Se (Te) molecules into more reactive form, and thus reduce the substrate temperature in growth and increase the effective Se (Te) flux at the same time. Such a condition can not only help to reduce the density of dislocations but also reduce the Se (Te) vacancies in the films. A more direct solution to lattice mismatch problem is by choosing lattice-matched substrates. CdS (0 0 0 1) substrate, commonly used in MBE growth of semiconductors, has a very close lattice constant with Bi_2Se_3, which is expected to significantly reduce the density of screw dislocations. Indeed, epitaxial Bi_2Se_3 films grown on CdS (0 0 0 1) with high electron mobility have been reported [37].

1.3.2 Point Defects

Another class of defects commonly seen in Bi_2Se_3 family TIs is point defects such as vacancies and antisite defects between anions and cations. Combining STM and density functional theory (DFT), Wang et al. [46] and Jiang et al. [47] systematically studied the point defects in MBE-grown Bi_2Te_3 and Sb_2Te_3 films, respectively. It was found that the point defects dominating Bi_2Te_3 films are Bi-on-Te and Te-on-Bi antisite defects, acting as acceptors and donors, respectively. In the MBE growth with higher Te flux and lower substrate temperature, which means extra Te atoms on the surface, Te-on-Bi antisite defects are energetically favored and the film tends to be electron-doped. On the contrary, if the growth is under the condition of lower Te flux and higher substrate temperature, Bi-on-Te antisite defects dominate. In this case, the Bi_2Te_3 films are hole-doped (see Fig. 5). The result provides a way of controlling the densities of different defects and the chemical potential of Bi_2Te_3 by growth parameters instead of introducing other elements. In the Sb_2Te_3 case, three types of defects are found in the films: Sb vacancies, Te-on-Sb antisite defects, and Sb-on-Te ones. The former two favor the Te-rich condition, while the latter one favors the Sb-rich one. Both Sb vacancies and Sb-on-Te antisite defects have lower formation energies than Te-on-Sb antisite defects

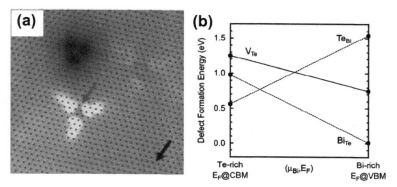

FIGURE 5 Point defects in Bi_2Te_3. (a) STM image of the two kinds of dominating defects in Bi_2Te_3: Te on Bi antisite defect (triangular depression) and Bi on Te antisite defect (clover-shaped protrusion). (b) Formation energies of the various defects and their dependence on chemical potential (from Ref. [46]).

and contribute holes to the film, which makes Sb_2Te_3 films always p-doped, as observed experimentally. However, at certain Te/Sb flux ratio and substrate temperature, all the three types of defects have relatively high formation energy, which leads to rather lower defect density [47]. As for Bi_2Se_3, Se vacancies are dominating, and the samples are always electron-doped [29]. Lower substrate temperature with higher effective Se flux helps to decrease the density of Se vacancies which can be realized by cracker cell evaporators.

1.3.3 Domain Walls

Bi_2Se_3 family bulk crystals have 3-fold symmetry about the axis perpendicular to the cleavage plane. On a substrate with 6-fold symmetry, there will be two crystalline domains opposite to each other in the films. The different crystalline domains manifest themselves as two classes of triangle-shaped islands pointing in the opposite directions, which are observed by STM in MBE-grown films of most Bi_2Se_3 family TIs (see Fig. 6(a)) [34]. One may expect to remove the domain walls from the films by choosing substrates with 3-fold symmetry. However because the interaction between the films and substrates is weak due to the layered nature of Bi_2Se_3 family TIs, in most cases only the topmost atomic layer of the substrate, usually showing 6-fold symmetry, plays an important role in MBE growth. Thus, twin domains still exist, as shown in Bi_2Se_3 films grown on InP (1 1 1) [40] and Bi-passivated Si (1 1 1) [34]. Wang et al. found that the straight and regularly arranged steps resulting from vicinal Si (1 1 1) substrates act as the guiding lines for the initial nucleation stage of growth of Bi_2Se_3, which promotes the formation of a single domain film [34] (see Fig. 6(b)).

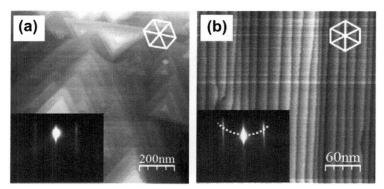

FIGURE 6 STM images and RHEED pattern (inset) of the surface morphology of Bi_2Se_3 films grown on flat (a) and vicinal (b) In_2Se_3 covered Si (1 1 1) substrates (from Ref. [34]).

1.4 Chemical Potential Tuning

In addition to the carrier density control method related directly to the intrinsic defects in Bi_2Se_3 family TIs mentioned above, there are other approaches to tune the Fermi level into the bulk gap, and to the Dirac point of the surface states: by chemical doping or by electric field effect. Soon after the discovery of Bi_2Se_3 and Bi_2Te_3 as topological insulators, several doping elements have been tested to tune their chemical potentials, since the as-grown materials are always heavily n-doped. Three groups of dopants have been used: group II elements (Ca and Mg) [48–50], group IV elements (Pb and Sn) [17,51], as well as the same group elements Bi (Sb) [51–54] or/and Te (Se) [55–58]. The former two groups of dopants exhibit $+2$ valence when substituting for Bi or Sb, reducing the density of n-type carriers (electrons) significantly. The advantage of doping these elements is that only a very small amount of dopants is needed, and the influence on band structure of the films is small. The disadvantages are that Ca and Mg are not stable in ambient conditions while Sn and Pb do not have enough solubility in Bi_2Se_3 to make it into p-type.

Sb doping in Bi_2Te_3 or Bi_2Se_3 can also decrease the density of n-type carriers. The mechanism is through the different formation energies of the defects in different matrix materials, as discussed above. Doping Sb in Bi_2Se_3 can increase the formation energy of Se vacancies, and thus reduce the electron doping level, which, however, cannot change the sample to p-type since the doping degrades the crystalline structure [51]. In Bi_2Te_3, unlimited Sb doping can be realized until Sb_2Te_3 is formed, which can tune the sample from n-type to p-type continuously [53,54]. At the same time, the Dirac surface band structure experiences a continuous change from that of Bi_2Te_3 to that of Sb_2Te_3 (see Fig. 7) [53,54]. The carrier density values measured by Hall effect and ARPES (by measuring the size of Fermi surface of the Dirac surface states) are well consistent with the p-n crossover region, indicating that the carriers are mainly

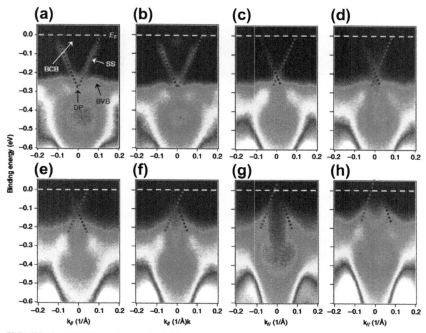

FIGURE 7   ARPES results on the 5 QL $(Bi_{1-x}Sb_x)_2Te_3$ films with $x = 0$ (a), 0.25 (b), 0.62 (c), 0.75 (d), 0.88 (e), 0.94 (f), 0.96 (g), and 1.0 (h), respectively (from Ref. [53]).

of the Dirac surface states. Interestingly, the electron mobility measured by Hall measurement does not decrease by Sb doping as in usual alloys. Rather, it shows a peak near the p-n crossover region [53]. The observation is similar to the case of graphene and consistent with the longer quasiparticle life of the zeroth order Landau level (at the Dirac point) observed by STM [23], confirming the dominance of Dirac surface state in the transport results.

Random distribution of dopants scatters electrons in the surface states and reduces their mobility. It is more favored that the dopants are arranged in ordered way. The ordered doping has been realized in the ternary TI Bi_2Te_2Se. In this material Se atoms substitute the middle Te atomic layer of each QL of Bi_2Te_3 (see the structural model shown in Fig. 8(a)). Bi_2Te_2Se single crystal samples with carrier density down to $\sim 3 \times 10^{16}$ cm^{-2} and the surface mobility up to ~ 2800 cm^2/Vs have been obtained. The samples exhibit a clear SdH oscillation in magnetoresistance (MR) [55,56]. However, the ARPES result of Bi_2Te_2Se indicates that the Dirac point is very close to the valence band and exhibits k_z dependent behavior, which suggests significant contribution from bulk bands (see Fig. 8(b)–(d)). It seriously influences applications in many quantum phenomena [57,58].

Topological Insulators

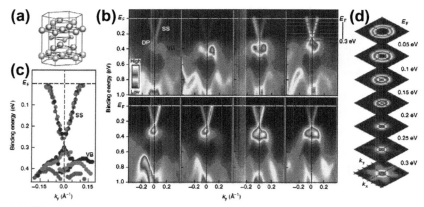

FIGURE 8 (a) Structural model of Bi_2Te_2 Se. Green, purple, and yellow circles represent Se, Bi, and Te atoms, respectively. (b) Photon energy dependent ARPES spectra. (c) Comparison of the ARPES results obtained at different photon energies. (d) Isoenergy circles of the surface states measured by ARPES (from Ref. [57]). (For interpretation of the references to color in this figure legend, the reader is referred to the web version of this book.)

Electrical field effect is another effective method to modify the chemical potential of TIs and can be applied to make field effect transistor (FET) devices. Since the Dirac surface states of a 3D TI film locate at both the top and bottom surfaces, dual-gate structure, in which the chemical potentials of both top and bottom surfaces are tuned independently, is more favored. Top gate was realized by depositing Al_2O_3 or HfO_2 films with atomic layer deposition (ALD) [59–61] or by the liquid gate technique [62,63]. For flake samples of Bi_2Se_3 family TIs, similar to exfoliated graphene samples, silicon oxide covered Si wafers were used as bottom gate [64–69]. However, this approach cannot work for epitaxial TI thin films. By using dielectric $SrTiO_3$ (1 1 1) as the substrate, Chen et al. successfully tuned the carrier density of epitaxial Bi_2Se_3 film by $\sim 3 \times 10^{13}$ cm^{-2} with back-gate voltage (V_g) between ± 210 V for the typical substrate thickness of 0.5 mm [36].

1.5 New Systems of Topological Insulator Materials

Extensive efforts have been made to search for new systems of TI materials, which are summarized as follows.

1.5.1 Ternary Chalcogenides

The structure of Bi_2Se_3 family TIs is composed of elementary atomic layers stacking in certain sequence. Substituting other elements for some of the atomic layers and modifying the stacking sequences can lead to a series of potential 3D TI materials. One example is ternary compounds $TlBiVI_2$ and $TlSbVI_2$, where VI indicates the VI group elements Te, Se, or S [70–73]. In these materials,

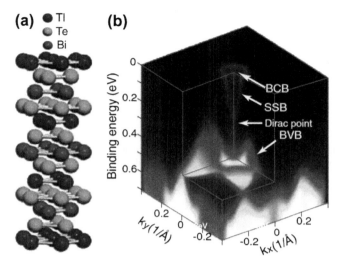

FIGURE 9 Structural model (a) and ARPES measured band dispersion (b) of TlBiTe$_3$ (from Ref. [73]).

elementary layers are arranged in the sequence of -Tl-VI-Bi (Sb)-VI- (see Fig. 9(a) for the structural model), with all the interlayer bonding being of strong covalent type, different from Bi$_2$Se$_3$ family TIs in which the interaction between QLs is of van der Waals type. Therefore, unlike the previously found 3D TIs, these materials are not layered. This character makes it much easier to obtain surfaces of different crystalline orientations with Dirac surface states of different band dispersions and symmetries. The result of first-principle calculations implies that except for TlSbS$_2$, all the materials of this class are TIs [71]. ARPES experiments have revealed the Dirac surface states of TlBiTe$_3$ and TlBiSe$_3$ (see Fig. 9(b)). The latter one exhibits a bulk gap of 0.35 eV, the largest ever reported in TIs. Other possible superlattice structures including REXTe$_3$ [74], AB$_2$X$_4$, A$_2$B$_2$X$_5$, MN$_4$X$_7$, and A$_2$X$_2$X' [58], where RE = rare earth elements; A, B = Pb, Ge, Sb, Bi; M, N = Pb, Bi; and X,X' = Chalcogen family. Many of these have been predicted or confirmed to be a 3D TI [58].

1.5.2 Oxides in Pyrochlore Structure

Guo and Franz predicted that pyrochlore materials with strong SOC and without magnetic order can be 3D TIs [75]. The materials can be found in 5d metal oxides in the form of A$_2$B$_2$O$_7$, with A indicating rare-earth elements (Sm and Eu), and B representing 5d element (Ir). Since 5d atoms contribute both strong SOC and strong electron correlation, richer properties are expected in the materials than usual TI materials. For example, topological Mott insulating phase has been predicted in medium correlation regime of the materials [76].

(a)   **(b)**

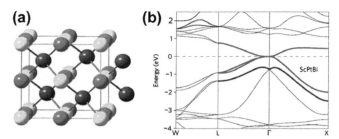

FIGURE 10   (a) Structural model of half Heusler compounds. (b) 3D band structure of half Heusler compound ScPtBi obtained by first-principle calculation.

1.5.3 Heusler Compounds

Heusler compounds are a class of ternary intermetallics, including full Heusler X_2YZ with the $L2_1$ structure, and half Heusler XYZ with $C1_b$ structure (see the structural model shown in Fig. 10(a)), in which X and Y represent transition or rare-earth metals, and Z represents the main-group element, respectively. They have long been known as magnetic materials, but can also be superconductors and heavy fermion systems if rare-earth elements are incorporated. First-principle calculations show that many of the half Heusler compounds are zero gap semiconductors (see the calculated band structure of ScPtBi in Fig. 10(b)), similar to HgTe, which are potential 3D TIs under strain and 2D TIs in quantum well structures [77,78]. Since Heusler compounds have various ordered phases such as ferromagnetic and superconducting phases, it is possible to fabricate TI-based heterostructures that exhibit interesting topological magnetoelectric phenomena and Majorana fermions.

1.5.4 Cerium Filled Skutterudite Compounds

By first-principle calculations, Yan et al. found that cerium filled skutterudite (FS) compounds $CeOs_4As_{12}$ and $CeOs_4Sb_{12}$ are parent compounds of TIs, similar to the Heusler compounds mentioned above [79]. The materials become topological Kondo insulators at low temperature, that is, Kondo insulators with Dirac surface states. Similar to Heusler compounds, they also exhibit various phases, from ferromagnetic, antiferromagnetic, to superconducting. The materials provide a good platform to investigate novel quantum phenomena of TIs.

1.5.5 Chalcopyrite Semiconductors

Another class of potential TIs is I-III-VI$_2$ and II-IV-V$_2$ chalcopyrite semiconductors [80]. They have a similar structure to that of 3D HgTe except that the cubic symmetry is broken by substitution of different cations. The broken symmetry makes them a 3D TI even without strain, different from the cases of HgTe and Heusler compounds. Another merit is that many of them have their lattice constants very close to those of common III–V semiconductors like GaAs. In this case, one can make integrated circuits by combining them with conventional semiconductors.

1.5.6 Topological Crystalline Insulators

The above-mentioned TIs are all protected by time-reversal symmetry. Fu predicted that there exist TIs protected by both time-reversal symmetry and crystalline symmetry, named as topological crystalline insulators [81]. Topological crystalline insulators only show gapless surface states at the surfaces with certain rotation or mirror symmetry. Another feature of a topological crystalline insulator is that the band dispersion of the gapless surface states is not necessarily linear, but can be quadratic. Hsieh et al. predicted that SnTe is a topological crystalline insulator protected by mirror symmetry [82]. The surfaces symmetric about (1 1 0) mirror planes, e.g., (1 1 1), (1 0 0), and (1 1 0), have the topologically protected gapless surface states.

1.5.7 Ag_2Te

Zhang et al. predicted that β-Ag_2 Te with distorted antifluorite structure is a 3D TI with gapless Dirac-type surface states [83]. The Dirac surface state at the surface perpendicular to the c-axis is highly anisotropic, in contrast to most of other TI materials. Its surface band structure leads to a unique nested Fermi surface with chiral spin texture. This feature not only enhances the effect of forbidden back-scattering on transport properties, but also leads to phase transitions such as spin density wave.

2 TWO-DIMENSIONAL TOPOLOGICAL INSULATORS

A 2D TI is a quantum well (QW) film with gapped 2D bulk band (QW states) and a pair of 1D, spin-filtered, and Dirac-type edge states at each edge, in which electrons with different spins propagate in opposite directions [84,85]. The pair of edge states contributes to quantized longitudinal resistance ($h/2e^2$, $\sim$12.9 kΩ for six-terminal measurements) in transport measurements and quantized spin-accumulation at the two edges parallel to the applied current, known as quantum spin Hall (QSH) effect. Graphene was first theoretically predicted to be a 2D TI when the Dirac points of its energy bands are gapped by SOC [84]. However, the SOC of graphene is too weak to open a gap large enough to support detectable QSH effect under realistic conditions. HgTe/CdTe QW is the first experimentally confirmed 2D TI system [3]. The system exhibits characteristic QSH effect: the plateau of longitudinal resistance at $h/2e^2$ [27], non-local transport result supporting the existence of edge states [86], and spin-accumulation at the edges [87]. The results have been described in detail in previous chapters, and thus will not be repeated here.

In spite of the great success in HgTe/CdTe QW 2D TI, for future applications, HgTe/CdTe QW has some shortcomings. First, the MBE growth of high quality HgTe/CdTe QW is very difficult, which can only be done by few groups in the world now. Second, the thermal stability of the material is not good so that unconventional low temperature lithography technique has to be used in the device fabrication. Third, both Hg and Cd are highly toxic and volatile, which makes large-scale productions and applications of HgTe/CdTe based devices difficult. Below we summarize several other proposals for 2D TI materials.

2.1 InAs/GaSb/AlSb Quantum Wells

InAs/GaSb/AlSb is a type-II QW, in which narrow-gap InAs and GaSb layers, adjacent to each other, are confined by barriers of wide-gap AlSb (1.6 eV) which has large band offset with InAs (see the scheme diagram of the band structure of the QW shown in Fig. 11(a)). As a conventional III–V semiconductor material, similar to HgTe/CdTe QW, the system shows very high electron mobility. The band structure and alignment of InAs, GaSb, and AlSb make the QW subbands formed in InAs and GaSb layers electron- and hole-type, respectively [8]. The lowest electron subband in InAs and the highest hole subband in GaSb are very close in energy. When the electron subband is tuned lower than the hole subband, for example, by film thickness, a gap is opened by the hybridization between them (see Fig. 11(b)). The process changes the system into a 2D TI with a pair of QSH edge states in the gap.

InAs/GaSb/AlSb QW has a rather different electronic structure compared to HgTe/CdTe in that its electron and hole subbands are located in different regions of the sample. The inversion symmetry is broken and thus an electric field is built in the sample. One can utilize a perpendicular external electrical field to tune InAs/GaSb/AlSb QW between 2D TI phase and usual 2D insulating phase so that a topological phase transition is realized. This effect can be used

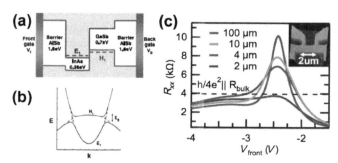

FIGURE 11   (a) Diagram showing band gap and offset of InAs/GaSb/AlSb QW (from Ref. [8]). (b) Schematic diagram of the inverted band structure around InAs/GaSb (from Ref. [8]). (c) QSH effect measured in InAs/GaSb/AlSb QW (from Ref. [9]).

to develop QSH field effect transistor, in which QSH phase and usual insulating phase correspond to on and off states, respectively.

QSH effect was experimentally observed in InAs/GaSb/AlSb QW by Knez, Du, and Sullivan [9]. They used a QW of 12.5 nm InAs/5 nm GaSb, which gives an inverted gap of ~ 4 meV. With the device of four-contact geometry, they observed the longitudinal resistance of $h/4e^2$ (Fig. 11(c)) and the non-local transport behavior, which confirms the existence of the QSH edge states. As a QW structure composed of conventional III–V group semiconductor materials, the 2D TI will be very easy to be integrated into traditional semiconductor technology.

2.2 Bi Thin Films

A much simpler proposal for 2D TI material is Bi thin films. Bi bulk crystal is composed of honey-comb-structured Bi bilayers (BLs) stacking in its (1 1 1) orientation. The material is semi-metallic in bulk and has long been expected to show metal-insulator transition in the thin film regime. Murakami predicted that a Bi (1 1 1) film of only one BL is not only an insulator but also a 2D TI [88]. A later theoretic work claimed that the 2D TI phase can exist in thicker Bi (1 1 1) films [89].

Hirahara et al. have successfully grown single BL Bi (1 1 1) film on the surface of Bi_2Te_3, a 3D TI material [90]. The band structure revealed by ARPES is basically a overlap of the bands of single BL Bi (1 1 1) and the Dirac surface state of Bi_2Te_3 except for the hybridization at the band crossing region (Fig. 12(a) and (b)). Since the band structure of the single BL Bi (1 1 1) film is little changed by the Dirac surface states of Bi_2Te_3, the film should be still a 2D TI as a freestanding Bi film. A very recent work has reported the observation of high density of states near the edges of single BL thick Bi islands with STM, which is attributed to the QSH edge states (Fig. 12(c)–(e)) [91]. However, none of these experimental results are convincing enough to confirm the non-trivial character of single BL Bi (1 1 1). The direct evidence is the QSH effect by transport measurement, which however is rather challenging due to the existence of the Dirac surface states of Bi_2Te_3 and the large band offset between Bi thin film and Bi_2Te_3. Better substrates for Bi thin films are needed to observe the QSH effect in this material.

2.3 Single Atomic Layer of Silicon and Germanium

As mentioned above, the only factor preventing graphene from showing the QSH effect is its very small gap resulting from the weak SOC. In principle, enhancing the SOC of the system can increase the gap size to an experimentally accessible level, which can be done by substituting heavier elements of the same group, silicon or germanium, for carbon. Through first-principle calculations, Liu et al. predicted that single atomic layers of Si and Ge with low-buckled

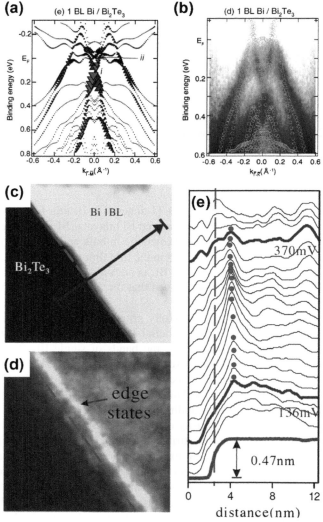

FIGURE 12   (a and b) Calculated (a) and ARPES measured (b) surface band structure of 1 BL Bi grown on Bi_2Te_3 (from Ref. [90]). (c)–(e) Edge states detected by STM at the edge of a 1 BL Bi island on Bi_2Te_3 (from Ref. [91]).

honeycomb structure are 2D TIs, with a gap size of 1.5 and 23.9 meV, respectively. The values are much larger than that of graphene and support measurable QSH effect [92]. The single layer Si, named silicene, has been prepared on the surface of Ag (1 1 1), on which the Dirac-type band structure has been observed by both ARPES and STM [93,94]. However, to observe the QSH effect, insulating substrates must be used.

2.4 Thin Films of Three-Dimensional Topological Insulator

A 3D TI can becomes a 2D TI in the thin film regime. As shown above, when the thickness of a thin film of 3D TI is reduced to such a level that the overlapping between the surface state wave functions of the opposite surfaces is significant, a hybridization gap opens in the Dirac surface state. The topological property of the gapped 2D system, 2D TI or trivial insulator, is determined by how many pairs of QW subbands (each pair includes an electron subband and a hole one) are in inverted regimes (see the scheme diagram in Fig. 13(a)) [27]. Therefore, there are periodic topological phase transitions between 2D TI and trivial insulator phases, accompanied by an oscillation in the gap size, as the thickness decreases from the 3D-2D crossover point (see Fig. 13(b)). In reality, since the thickness of a material can only be changed by a finite value, for example, 1 QL for Bi_2Se_3 family materials, the oscillation in gap size is not necessarily observed. By fitting the band dispersions of the gapped surface states of Bi_2Se_3 thin film with four band model, Zhang et al. found that at least the 2 QL film shows the band dispersion that looks like in inverted regime and thus should be in the 2D TI phase [30]. In Bi_2Se_3 thin films grown on Si (1 1 1) surface, Sakamoto et al. observed an oscillatory behavior of gap size with decreasing thickness, a phenomenon consistent with the predicted oscillatory phase transition [95]. Although the QSH effect has not been observed, the above experimental results suggest that it is possible to realize 2D TIs in thin films of 3D TI materials, which significantly expands the scope of 2D TI materials.

In summary, beside HgTe/CdTe QW, several other materials or structures have been proposed to be 2D TIs. Among them, InAs/GaSb/AlSb QWs have been confirmed to be a 2D TI by showing the QSH effect in transport experiments. In some simpler systems, band structures likely supporting the QSH effect have been observed by ARPES. However it should be pointed out that the observation of the QSH effect requires smaller sample size than the phase coherence length, different from the case of quantum Hall effect which

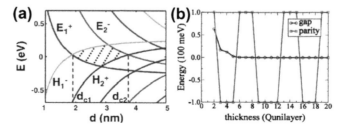

FIGURE 13   (a) Schematic diagram of the evolution of QW states of a thin film of 3D TI with the thickness. The shaded area indicates the 2D TI regime. (b) Evolution of gap size and parity with the thickness. The values of 1.0 and −1.0 of parity indicate the topologically trivial insulator and 2D TI, respectively (from Ref. [27]).

in principle can be observed even in macro-sized samples. Therefore, samples with high electron mobility are crucial to show the QSH effect.

3 MAGNETIC TOPOLOGICAL INSULATORS

Most of the TIs mentioned in the recent literature are the time-reversal-invariant TIs, which, as the name suggests, reject ferromagnetic order in them. Paradoxically a large portion of the unique quantum effects predicted in the time-reversal-invariant TIs actually result from breaking of their time-reversal symmetry (TRS) by introducing ferromagnetism in them [96]. For example, breaking the TRS in a 3D TI opens a gap at the Dirac point of its 2D Dirac surface band, changing it into a quantum Hall (QH) system with half QH conductance [96]. It means that in a thin film of 3D TI with ferromagnetism one expects to observe a special kind of QH effect which is independent of Landau levels, and therefore in principle free from the demanding conditions for conventional QH effect: strong external magnetic field, low temperature, and high mobility of samples. Such an easy-access QH system, also known as quantized anomalous Hall (QAH) insulator, can greatly facilitate the studies and future applications of QH physics [43, 96–103]. Furthermore, other novel phenomena, e.g., topological magnetoelectric effect [96, 104] and image magnetic monopole [105], have been predicted in TRS-broken TIs. The proposals have attracted significant attention not only from condensed matter physics, but also from high-energy physics [106].

There are basically two ways of utilizing ferromagnetism to break the TRS of a TI: attach a ferromagnetic layer on the surface of the TI, or make the TI itself ferromagnetic by doping magnetic impurities. For both approaches, it is crucial that the ferromagnetic material—the ferromagnetic layer in the former case, and the TI itself in the latter one—should be insulating to guarantee the dominance of the QAH edge states, which are located in both the bulk and surface gaps in transport measurements. So far many efforts have been made on the second approach, inspired by the success in diluted magnetic semiconductors (DMSs) based on III–V or II–VI compounds in the past decades [107, 108]. However the ferromagnetic coupling in most of the conventional DMSs is of RKKY-type which requires itinerant carriers of certain density and type as the medium for ferromagnetic coupling [109]. One can never expect a ferromagnetic insulator with the RKKY-type coupling. A totally different mechanism of ferromagnetic coupling which should be independent of carriers is crucial for practical magnetic TIs.

Carrier-independent ferromagnetism is possible in magnetically doped Bi_2Se_3 family TIs, as shown in the theoretic work by Yu et al. [43]. The Bi_2Se_3 family TI materials have inverted band structure in which the original conduction and valence bands are mixed up and re-gapped by the SOC [5]. The resulting bands above and below the gap have quite similar components. The special band structure leads to a rather large van Vleck magnetic susceptibility of valence electrons even when the Fermi level lies in the bulk gap. Hence in these TIs the magnetic moments of the magnetic impurities far from each other can be ferromagnetically coupled by the large susceptibility in the insulating regime [43].

Many experimental attempts have been made in magnetically doped Bi_2Se_3 family TIs, among which Bi_2Se_3 is the first choice with its largest bulk gap and the Dirac point residing in the gap. Using ARPES, Chen et al. observed gap-opening at the Dirac point in Fe and Mn-doped Bi_2Se_3 and attributed it to the broken TRS induced by ferromagnetism [19]. However the result of magnetic and transport measurements could not show clear ferromagnetism in the material. A possible interpretation is that ferromagnetism only forms near the surface through the surface state mediated RKKY mechanism, which does not contribute to the bulk sensitive magnetic and transport measurements [110,111]. A unique property of the magnetically induced gap in the Dirac surface state is that around the edges of the gap, the spin orientation becomes perpendicular to sample surface, with the upper and lower edges having opposite spin directions. In a very recent work, Xu et al., by spin-resolved ARPES, did observe such a "hedgehog" spin texture in MBE-grown Mn-doped Bi_2Se_3 films. Although the result confirms that the gap is magnetically induced, the hysteresis loop of the sample obtained by X-ray magnetic circular dichroism (XMCD) measurement exhibits a small remanent magnetization and coercivity, casting doubt on the formation of the long-range ferromagnetic order in the material [112].

On the other hand, well-established long-range ferromagnetic order has been found in 3d metal doped Bi_2Te_3 [113] and Sb_2Te_3 [114]. In MBE-grown Cr and V doped Sb_2Te_3 films, Chien found ferromagnetism with perpendicular magnetic anisotropy with both magnetic and transport measurements, even before Sb_2Te_3 was recognized as a 3D TI. The samples show perfect hysteresis loops, large remanent magnetization, and significant coercivity, supporting a long-range ferromagnetic order (Fig. 14(a)) [114]. Similar ferromagnetism with square-shaped hysteresis loops and perpendicular magnetic anisotropy was also observed in Mn-doped Bi_2Te_3 bulk crystal samples (Fig. 14(b)) [113]. The perpendicular magnetic anisotropy is very important for the zero field QH effect since only magnetization perpendicular to the surface opens a gap at the Dirac point of the surface band.

After confirming the long-range ferromagnetic order in magnetically doped Sb_2Te_3 and Bi_2Te_3, one needs to know if the ferromagnetism is of carrier-independent van Vleck type. As shown above, since Sb_2Te_3 and Bi_2Te_3 are p- and n-doped, respectively, by alloying Bi_2Te_3 and Sb_2Te_3, one could tune the carriers from n-type to p-type. With this idea, Chang et al. systematically studied the carrier dependent magnetic and transport properties of 5 QL thick Cr-doped $(Bi_xSb_{1-x})_2Te_3$ films with different Bi:Sb ratio (see Fig. 15(a)–(f))

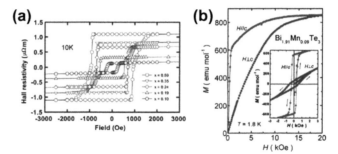

FIGURE 14 (a) Hysteresis loops of Cr-doped Sb_2Te_3 obtained by Hall effect measurement (from Ref. [114]). (b) Magnetization curves and hysteresis loops of Mn-doped Bi_2Te_3 measured by SQUID (from Ref. [113]).

[115]. It was found that in spite of the variation in carrier type from p-type to n-type and in carrier density by one order of magnitude, the films always show good long-range ferromagnetic order with the Curie temperature nearly constant (Fig. 15(g)). At the same time, the anomalous Hall resistance shows an abrupt increase with the decreasing carrier density, qualitatively consistent with the predicted behavior of the QAH effect (Fig. 15(h)). By applying bottom-gate electrical field effect with $SrTiO_3$ (1 1 1) substrate, they observed the similar phenomena of carrier-independent ferromagnetism and carrier-sensitive anomalous Hall effect (Fig. 15(i)). An anomalous Hall resistance of 6.1 kΩ, ~1/4 quantum Hall resistance, was reached, implying the approaching QAH regime (Fig. 15(j) and (k)). Maybe due to the remaining bulk carriers that have not been completely depleted yet, the QAH effect has not been observed. Inhomogeneous chemical potential distribution, band bending, and impurities in the films all can contribute to the remaining bulk carriers. By optimizing the film thickness, growth parameters, and fabricating the films into devices of micrometer or sub-micrometer size with dual-gate structure, it is promising to further reduce the bulk carrier density of the films to reach the QAH regime. Similar result has also been observed in the flake samples of Mn-doped Bi_2Te_3 [116].

In this chapter, we summarized the progress made on the material aspect of TIs in the recent years. Many new TI systems with different properties have been proposed or confirmed. Many methods of improving the quality of the existing TI materials have been introduced. It is to note that for TIs it is the several-nanometer-thick region around the interface that plays the central role in quantum phenomena associated with topological insulators. Therefore, in the future much attention should be paid to the interface regions of TIs with ordinary insulators, magnetic or superconducting materials to obtain well-defined, atomically flat and controllable heterostructures that support interesting quantum phenomena.

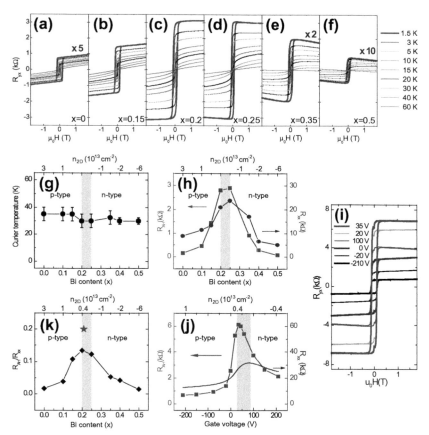

FIGURE 15 (a)–(f) Hysteresis loops of AH resistance of 5 QL $Cr_{0.22}(Bi_x Sb_{1-x})_{1.78}Te_3$ films of different Bi content (x). (g) Dependence of Curie temperature on Bi content (bottom axis) and carrier density (top axis). (h) Dependence of AH resistance (red solid squares) and longitudinal resistance (blue solid circles) at 1.5 K on Bi content (bottom axis) and carrier density (top axis). (i) Hysteresis loops of Hall resistance of a 5 QL $Cr_{0.22}(Bi_{0.2}Sb_{0.8})_{1.78}Te_3$ film grown on $SrTiO_3$ (1 1 1) at different back-gate voltages measured at 250 mK. (j) Dependence of R_{AH} (red solid squares) and R_{xx} (blue solid line) of the 5 QL $Cr_{0.22}(Bi_{0.2}Sb_{0.8})_{1.78}Te_3$ film on $SrTiO_3$ (1 1 1) substrate on back-gate voltages at 250 mK. (k) Dependence of (R_{AH}/R_{xx}) on Bi content (bottom axis) and carrier density (top axis). Black solid diamonds represent the data obtained from (h). The red star represents the value obtained from a Hall measurement at 250 mK, 15 T on the $Cr_{0.22}(Bi_{0.2}Sb_{0.8})_{1.78}Te_3$ film grown on $SrTiO_3(1 1 1)$ with back-gate voltage 35 V. The blue areas in a, b, d, and e indicate the approximate p-n crossover regions. (For interpretation of the references to color in this figure legend, the reader is referred to the web version of this book.)

ACKNOWLEDGMENTS

The authors would thank the collaborations and discussions with Jin-Feng Jia, Yayu Wang, Lili Wang, Shou-Cheng Zhang, Xi Dai, Zhong Fang, Xiao-Liang Qi, Xincheng Xie, Jian Wang, Moses Chan, Shengbai Zhang, Feng Liu, Chao-Xing Liu, Bangfen Zhu, Wenhui Duan, Jian

Wu, Ying Liu, Shunqing Shen, Qian Niu, Nitin Samarth, Kang L. Wang, and Maohai Xie. This work was supported by NSFC (Grant Nos. 11174343 and 91021006), 973 program (Grant Nos. 2009CB929400 and 2012CB921300), and the Knowledge Innovation Program of Chinese Academy of Sciences.

REFERENCES

[1] Hasan MZ, Kane CL. Rev Mod Phys 2010;82:3045.
[2] Qi XL, Zhang SC. Rev Mod Phys 2011;83:1057.
[3] Bernevig BA, Hughes TL, Zhang SC. Science 2006;314:1757.
[4] Fu L, Kane CL. Phys Rev B 2007;76:045302.
[5] Zhang HJ et al. Nat Phys 2009;5:438.
[6] Chu Jbibauthor, Sher A. PPhysics and properties of narrow gap semiconductors. Springer; 2008.
[7] König M et al. Science 2007;318:766.
[8] Liu C et al. Phys Rev Lett 2008;100:236601.
[9] Knez I, Du R-R, Sullivan G. Phys Rev Lett 2011;107:136603.
[10] Becker CR et al. Phys Status Solidi C 2007;4:3382.
[11] Kroemer H. Physica E 2004;20:196.
[12] Fu L, Kane CL, Mele EJ. Phys Rev Lett 2007;98:106803.
[13] Hsieh D et al. Nature 2008;452:970.
[14] Hsieh D et al. Science 2009;323:919.
[15] Xia Y et al. Nat Phys 2009;5:398.
[16] Hsieh D et al. Nature 2009;460:1101.
[17] Chen YL et al. Science 2009;325:178.
[18] Hsieh D et al. Phys Rev Lett 2009;103:146401.
[19] Roushan P et al. Nature 2009;460:1106.
[20] Zhang T et al. Phys Rev Lett 2009;103:266803.
[21] Cheng P et al. Phys Rev Lett 2010;105:076801.
[22] Hanaguri T et al. Phys Rev B 2010;82:081305.
[23] Jiang Y et al. Phys Rev Lett 2012;108:016401.
[24] Okada Y et al. Phys Rev Lett 2012;109:166407.
[25] Seradjeh B, Moore JE, Franz M. Phys Rev Lett 2009;103:066402.
[26] Ghaemi P, Mong RSK, Moore JE. Phys Rev Lett 2010;105:166603.
[27] Liu C-X et al. Phys Rev B 2010;81:041307(R).
[28] Li Y-Y et al. Adv Mater 2010;22:4002.
[29] Song CL et al. Appl Phys Lett 2010;97:143118.
[30] Zhang Y et al. Nat Phys 2010;6:584.
[31] Wang G et al. Nano Res 2010;3:874.
[32] Chen X et al. Adv Mater 2011;23:1162.
[33] Zhang G et al. Appl Phys Lett 2009;95:053114.
[34] Wang ZY et al. J Crys Growth 2011;334:96.
[35] Chang C-Z et al. SPIN 2011;1:21.
[36] Chen J et al. Phys Rev Lett 2010;105:176602.
[37] Kou XF et al. Appl Phys Lett 2011;98:242102.
[38] Richardella A et al. Appl Phys Lett 2011;97:262104.
[39] Peng H et al. Nat Chem 2012;4:281.
[40] Tarakina NV et al. Cry Growth Des 2012;12:1913.
[41] Shan W-Y, Lu H-Z, Shen S-Q. New J Phys 2010;12:043048.
[42] Kim M, Kim CH, Kim H-S. Ihm J PNAS 2012;109:671.
[43] Yu R et al. Science 2010;329:61.
[44] Liliental-Weber Z, Jasinski J, Zakharov DN. Opto-Electron Rev 2004;12:339.
[45] Scott SA, Kral MV, Brown SA. Appl Surf Sci 2006;252:5563.
[46] Wang G et al. Adv Mater 2011;23:2929.
[47] Jiang Y et al. Phys Rev Lett 2012;108:066809.
[48] Hor YS et al. Phys Rev B 2009;79:195208.

[49] Hsieh D et al. Nature 2009;460:1101.
[50] Kuroda K et al. Phys Rev Lett 2010;105:076802.
[51] Zhang Y et al. Appl Phys Lett 2010;97:194102.
[52] Analytis JG et al. Nat Phys 2010;6:960.
[53] Zhang J et al. Nat Commun 2011;2:574.
[54] Kong D et al. Nat Nanotechnol 2011;6:705.
[55] Ren Z et al. Phys Rev B 2010;82:241306.
[56] Xiong J et al. Physica E 2012;44:917.
[57] Arakane T et al. Nat Commun 2011;3:636.
[58] Xu S-Y et al. arXiv:1007.5111.
[59] Steinberg H et al. Nano Lett 2010;10:5032.
[60] Steinberg H et al. Phys Rev B 2011;84:233101.
[61] Liu H, Ye PD. Appl Phys Lett 2011;99:052108.
[62] Yuan H et al. Nano Lett 2011;11:2601.
[63] Kim D et al. Nat Phys 2012;8:460.
[64] Checkelsky JG et al. Phys Rev Lett 2011;106:196801.
[65] Xiu F et al. Nat Nanotechnol 2011;6:216.
[66] Cho S et al. Nano Lett 2011;11:1925.
[67] Hossain MZ et al. Nano Lett 2011;5:2657.
[68] Sacépé B et al. Nat Commun 2011;2:575.
[69] Wang Y et al. Nano Lett 2012;12:1170.
[70] Lin H et al. Phys Rev Lett 2010;105:036404.
[71] Yan B et al. EPL 2010;90:37002.
[72] Sato T et al. Phys Rev Lett 2010;105:136802.
[73] Chen YL et al. Phys Rev Lett 2010;105:266401.
[74] Yan B et al. Phys Rev B 2010;82:161108(R).
[75] Guo H-M, Franz M. Phys Rev Lett 2009;103:206805.
[76] Pesin DA, Balents L. Nat Phys 2010;6:376.
[77] Chadov S et al. Nat Mater 2010;9:541.
[78] Lin H et al. Nat Mater 2010;9:546.
[79] Yan B et al. Phys Rev B 2012;85:165125.
[80] Feng W et al. Phys Rev Lett 2011;106:016402.
[81] Fu L. Phys Rev Lett 2011;106:106802.
[82] Hsieh TH et al. Nat Commun 2012;3:982.
[83] Zhang W et al. Phys Rev Lett 2011;106:156808.
[84] Kane CL, Mele EJ. Phys Rev Lett 2005;95:226801.
[85] Bernevig BA, Zhang S-C. Phys Rev Lett 2006;96:106802.
[86] Roth A et al. Science 2009;325:294.
[87] Brüne C et al. Nat Phys 2010;6:448.
[88] Murakami S. Phys Rev Lett 2006;97:236805.
[89] Liu Z et al. Phys Rev Lett 2011;107:136805.
[90] Hirahara T et al. Phys Rev Lett 2011;107:166801.
[91] Yang F et al. Phys Rev Lett 2012;109:016801.
[92] Liu C-C et al. Phys Rev Lett 2011;107:076802.
[93] Vogt P et al. Phys Rev Lett 2012;108:155501.
[94] Chen L et al. Phys Rev Lett 2012;109:056804.
[95] Sakamoto Y et al. Phys Rev B 2010;81:165432.
[96] Qi XL, Hughes TL, Zhang SC. Phys Rev B 2008;78:195424.
[97] Liu CX, Qi XL, Dai X, et al. Phys Rev Lett 2008;101:146802.
[98] Nomura K, Nagaosa N. Phys Rev Lett 2011;106:166802.
[99] Haldane FDM. Phys Rev Lett 1988;61:2015.
[100] Onoda M, Nagaosa N. Phys Rev Lett 2003;90:206610.
[101] Qiao Z et al. Phys Rev B 2010;82:161414(R).
[102] Tse WK et al. Phys Rev B 2011;83:155447.
[103] Garate I, Franz M. Phys Rev Lett 2010;104:146802.
[104] Qi X-L et al. Nat Phys 2008;4:273.
[105] Qi X-L et al. Science 2009;323:1184.
[106] Li R et al. Nat Phys 2010;6:284.
[107] Ohno H. Science 1998;281:951.

[108] Dietl T. Nat Mater 2010;9:965.
[109] Chen YL et al. Science 2010;329:659.
[110] Liu Q et al. Phys Rev Lett 2009;102:156603.
[111] Rosenberg G, Franz M. Phys Rev B 2012;85:195119.
[112] Xu S-Y et al. Nat Phys 2012;8:616.
[113] Hor YS et al. Phys Rev B 2010;81:195203.
[114] Chien YJ. PhD thesis of the University of Michigan; 2007.
[115] Chang C-Z et al. Adv Mater 2013;25:1065.
[116] Checkelsky JG et al. Nat Phys 2012;8:729.

Chapter 10

Theoretical Design of Materials and Functions of Topological Insulators and Superconductors

Naoto Nagaosa[*],[†]

[*]Department of Applied Physics, University of Tokyo, Tokyo 113-8656, Japan
[†]RIKEN Center for Emergent Matter Science (CEMS), Wako 351-0198, Japan

Chapter Outline Head

1 INTRODUCTION

Symmetry and conservation law are the two most important principles in physics, which are related to each other. In addition to these, the vital role of the topology has been recognized recently in high energy physics, and condensed matter physics. Especially, the topological aspects of electronic states in solids have stimulated the search for new phenomena in new materials as well as in *conventional* materials [1]. This has been initiated by the discovery of quantum Hall effect in 2D electrons in semiconductor MOSFET system at low temperature under a strong magnetic field [2]. The exact quantization of the Hall conductance is directly related to the topological invariant called Chern number characterizing the global topological property of the electronic states

Topological Insulators. http://dx.doi.org/10.1016/B978-0-444-63314-9.00010-X

[3]. The topological invariant is an integer and cannot be changed continuously. This implies certain stability against the external disturbances protected by the energy gap in the bulk states. Another important feature of the topological states is the existence of the edge or surface channels and the *electron fractionalization* there. Generally speaking, the change of the topology is usually associated with the gap closing. This can occur at the surface or edge of the sample of the topologically nontrivial system. Namely, the vacuum has no topological invariant, and hence the "transition" between the topological state and the vacuum is always accompanied by the zero energy states which correspond to the surface or edge states. These surface/edge states are topologically protected by the bulk gap and have several novel physical properties. In the case of quantum Hall system [2], the chiral edge channels appear at the edge of the sample because the cyclotron motion is bounced back at the edge leading to the one-way 1D traffic of the electrons. One can consider this as a representative example of fractionalization in the sense that the directional degrees of freedoms are separated into right and left at each edge of the sample. The electrons in the chiral edge channels do not suffer the backward scattering since it requires the shift of the electrons from one to the other edge, which is prohibited by the gapped bulk states. In general, the fractionalized electrons are more robust against the external perturbations or electron-electron interactions because they are topologically protected by the bulk states.

This quantum Hall effect is a rather special phenomenon at low temperature and strong magnetic field. However, it is now recognized that it has many cousins in condensed matter systems. One direction is to raise the temperature, which has been realized by graphene [4]. There, the Dirac spectrum gives the peculiar Landau level formation with large energy splittings, and enables the quantum Hall effect at room temperature. Another direction is to get rid of the magnetic field. The magnetic field is replaced by the spontaneous magnetization combined with the relativistic spin-orbit interaction (SOI) in the anomalous Hall effect in metallic ferromagnet [5]. In this case, the Berry phase curvature in the momentum space is shown to be responsible for the intrinsic part of the Hall conductivity similar to the quantum Hall effect. Even the quantized anomalous Hall effect is expected without the Landau level formation [5]. Without the time-reversal (T) symmetry breaking, the Hall current or the Hall conductance $\sigma_{xy} - \sigma_{yx}$ is not allowed. However, when one considers the spin current instead of the charge current, the Hall effect is expected, and is called spin Hall effect [6,7]. Namely, the usual semiconductors with SOI are expected to show the spin Hall effect due to the Berry curvature in momentum space, which is represented by the SU(2) gauge field.

A recent breakthrough in the field is the discovery of the quantum spin Hall system characterized by the Z_2 topological invariant in two-dimensional spin-orbit coupled system [8] and its extension to three dimensions [9]. These are now called topological insulators. Furthermore, the global classification scheme including both the insulators and superconductors has been developed based on the three symmetries, i.e., time-reversal (Θ), particle-hole (Ξ), and the chiral (Π) symmetries [10,11]. Here Ξ symmetry is due to the superconductivity in the usual situation. These three symmetries are fundamental in the sense that they survive even when the disorder potential exists in sharp contrast to the translational or other point group symmetries. There are 10 classes and depending on the dimensionality d of the system, the topological invariant is specified to characterize the nontrivial topological states. This offers a topological periodic table. Later, this "10-fold way" has been extended to the classification including the textures such as domain walls and dislocations [12].

It is remarkable that most of the recent developments are triggered by the theoretical predictions which are followed by the experimental confirmations. Therefore, the theoretical design of the topological phenomena now becomes a realistic and tractable goal with the background described above, and the state-of-art first-principles band structure calculations, which is the issue discussed in this chapter.

For the theoretical design, we have (i) the methods, (ii) issues to be considered, and (iii) phenomena and effects to be designed. As for (i), we should start with the choice of the elements from the periodic table of the atoms, and consider the orbital character and strength of SOI. Then, we look at the basic symmetry and dimensionality of the system to be compatible with the topological periodic table. Even though the topological periodic table cares about only three symmetries, Θ, Ξ, and Π, we often need to consider the crystal structures and symmetry to design the band structure, including those in the artificial structures such as the superstructures and interfaces. On top of the band structure, we can add the disorder or put external fields such as the pressure, magnetic field, and electric field. Usually, the combinations of these methods enable the desired electronic states to be realized and to control the following issues, i.e., (ii). For the topological insulators, the band gap driven by the SOI is essential since the band structure without SOI is always spin degenerate and Z_2 number is trivial. This is not always the case for topological superconductors where the pairing potential often gives rise to the topological nature even without the SOI. In addition to this single-particle Hamiltonian, the role of electron correlation is an important issue. For the representative topological insulators such as HgTe and Bi-based compounds, there seems no sign of electron correlation experimentally, probably because the electrons are mostly in s- and p-orbitals. However, once the electrons in d-orbitals are considered, correlations will become very important and might lead to new many-body effects in combination with topology. Another direction is the magnetic ordering driven by the correlation, which can be also topologically nontrivial. With the disorder, the localization problem is crucial and is closely related to the quantization of the Hall conductance in the quantum Hall system. The basic symmetry, topology, and Berry phase effects affect the localization properties, and hence determine the physical phenomena. Last but not least, the surface or the interface is most important for the topological materials since the low energy states are located and the electron fractionalization occurs only there. In the case of topological superconductors, the real fermion called Majorana fermion [13] is expected to appear at the edge of the sample, and is a promising candidate for the qubit in quantum information technology.

The last item (iii), the phenomena and effects to be designed, includes quantum spin Hall effect, quantized (integer and fractional) anomalous Hall effect, quantized thermal Hall effect, topological magneto-electric (ME) effect, topological quantum phase transition and associated critical phenomena, and Andreev reflection and Josephson effect of topological origin. Many others are waiting to be explored and we will discuss below only a few examples studied up to now.

2 MAGNETIC AND MAGNETO-ELECTRIC PHENOMENA RELATED TO TOPOLOGICAL INSULATORS

The interplay between the magnetism and the topological insulators is an important issue for the novel physics and possible applications. If one is interested in the low energy phenomena, the surface or edge states are the only relevant degrees of freedom because the gap opens for the bulk states. As discussed in the Introduction, the electrons are fractionalized in the surface states of the topologically nontrivial matter. In the case of topological insulators, the spin degrees of freedom are halved as follows [1]. The time-reversal symmetry operation Θ for spin 1/2 system is given by

$$\Theta = -i\sigma_y K, \tag{1}$$

where K is the complex conjugation and σ_y is the y-component of Pauli matrices. Also an important feature of Θ is $\Theta^2 = -1$, which corresponds to the fact that the rotation of the spin by 2π results in the minus sign of the spinor wavefuntion. When the Hamiltonian H commutes with Θ, the eigenstate $|n\rangle$ of the Hamiltonian is always accompanied by the states $\Theta|n\rangle$ and these two states are orthogonal to and linearly independent of each other. This is because

$$\langle n|\Theta|n\rangle = \langle \Theta^2 n|\Theta n\rangle = -\langle n|\Theta|n\rangle = 0, \tag{2}$$

where the anti-unitary nature of Θ, i.e., $\langle \Theta a|\Theta b\rangle = \langle b|a\rangle$, has been used. Therefore, the time-reversal symmetry guarantees at least double degeneracy, Kramers doublet, for spin 1/2 (in general half-odd integer spin) particles. For the electrons in solids, Θ relates $|k\uparrow\rangle$ and $|-k\downarrow\rangle$ because the complex conjugation changes k to $-k$. When the spatial inversion symmetry I is also present, $|k\sigma\rangle$ and $|-k\sigma\rangle$ are degenerate, and the combination of Θ and I results in the degeneracy of $|k\uparrow\rangle$ and $|k\downarrow\rangle$ at each k-point.

From this viewpoint, the inversion symmetry I plays an important role in the presence of the SOI. Actually, the representative SOI in noncentrosymmetric system is the so-called Rashba interaction given by

$$H_R = \alpha (\mathbf{k} \times \boldsymbol{\sigma}) \cdot \mathbf{e}_z, \tag{3}$$

when the symmetry between z and $-z$ is broken. This results in the spin splitting between $|k \uparrow\rangle$ and $|k \downarrow\rangle$, and the energy dispersion is given by

$$E_\pm(k) = \frac{\hbar^2 k^2}{2m} \pm \alpha k, \tag{4}$$

and the momentum and spin direction is locked in each of the spin-split bands. This spin-momentum locking plays an important role in spintronics since it produces the coupling between the charge current and spin leading to the several interesting phenomena such as the anomalous Hall effect [5], spin Hall effect [6,7], and spin Galvanic effect [14–16].

This spin splitting is maximized at the edge or the surface of a topological insulator. In the case of 2D topological insulator, i.e., the quantum spin Hall system, the situation is clearly understood as follows [1]. There are two counterpropagating modes at the edge of the sample with opposite spin directions, e.g., the up spins are propagating to the right and the down spin to the left, and they are related as $|k \uparrow\rangle$ and $|-k \downarrow\rangle = \Theta|k \uparrow\rangle$. The backward scattering is forbidden as long as the perturbation H' respects the time-reversal symmetry, i.e., $[H', \Theta] = 0$. This is because

$$\langle -k \downarrow |H'|k \uparrow\rangle = \langle \Theta k \uparrow |H'|k \uparrow\rangle = \langle \Theta H' k \uparrow |\Theta^2 k \uparrow\rangle$$
$$= -\langle H' \Theta k \uparrow |k \uparrow\rangle = -\langle -k \downarrow |H'|k \uparrow\rangle = 0. \tag{5}$$

The crossing of the two dispersions occurs at the time-reversal symmetric points, i.e., $k = 0$ or $k = \pm \pi$, which is protected by the Kramers theorem.

For the case of 3D topological insulators, the surface state is described by [1]

$$H_D = \pm v_F (\mathbf{p} \times \boldsymbol{\sigma}) \cdot \mathbf{e}_z, \tag{6}$$

when the $\pm z$-direction is normal to surface. This is the limit of large Rashba splitting, i.e., the outer Fermi surface is missing. Therefore, the spin-momentum locking is perfect in the surface states of topological insulators, which is expected to be useful for the control of current by magnetism [17] in spintronics [18].

One consequence of this spin-momentum locking is that the magneto-transport phenomena on the surface of topological insulators are unique. Based on this expectation, we have considered the charge transport in 2D ferromagnet/ferromagnet junction on a topological insulator [19]. Usually, the conductance is large when the magnetizations of two sides are parallel to each other, while it is minimum for an anti-parallel configuration because

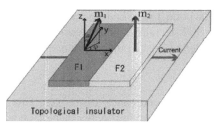

FIGURE 1 The ferromagnets attached to the surface of 3D topological insulator. Reproduced from Ref. [19].

the spin is conserved in the tunneling and hence is allowed only for the parallel configuration [17]. This is modified considerably for the surface Dirac fermions because of the strong spin anisotropy and the spin-momentum locking as described below [19].

We consider the 2D Dirac Hamiltonian coupled to the magnetization

$$H = \begin{pmatrix} m_z & k_x + m_x - i(k_y + m_y) \\ k_x + m_x + i(k_y + m_y) & -m_z \end{pmatrix}, \quad (7)$$

where m_x, m_y, and m_z are the components of the exchange field. (Here we follow the original notation in Ref. [19]. To be consistent with Eq. (6), one should rotate the momentum by 90°). We consider the two-dimensional ferromagnet/ferromagnet junctions, i.e., F1/F2, with different directions of the magnetization $\mathbf{m}_1$ and $\mathbf{m}_2$ as shown in Fig. 1. The interface is along the y-axis at $x = 0$. We define the exchange field in the F1 side as $\mathbf{m}_L = (m_x, m_y, m_z) = m_1(\sin\theta\cos\varphi, \sin\theta\sin\varphi, \cos\theta)$ while in the F2 side, we fixed it to be $m_x = m_y = 0$ and $m_z = m_R$. Then, the wavefunctions for each region 1 and 2 below F1 and F2 are given as follows. On the F1 side,

$$\psi(x \le 0) = \frac{1}{\sqrt{2E(E - m_z)}} e^{i(k_x + m_x)x} \begin{pmatrix} k_x + m_x - i(k_y + m_y) \\ E - m_z \end{pmatrix}$$

$$+ \frac{r}{\sqrt{2E(E - m_z)}} e^{-i(k_x + m_x)x} \begin{pmatrix} -k_x - m_x - i(k_y + m_y) \\ E - m_z \end{pmatrix}, \quad (8)$$

while in the F2 side,

$$\psi(x \ge 0) = \frac{t}{\sqrt{2E(E - m)}} e^{ik_x'x} \begin{pmatrix} k_x' - ik_y \\ E - m \end{pmatrix}, \quad (9)$$

with $E = \sqrt{m_z^2 + (k_x + m_x)^2 + (k_y + m_y)^2} = \sqrt{m^2 + k_x'^2 + k_y^2}$. This means that the Fermi surface in the F1 region is shifted by $(-m_x, -m_y)$ from the origin. Also, r and t are reflection and transmission coefficients, respectively. The momentum k_y is conserved, and hence the common factor $e^{ik_y y}$ is omitted above.

By matching the wavefunctions at the interface $x = 0$, we obtain

$$t = \frac{2(k_x + m_x)}{\left[k_x + m_x + i(k_y + m_y)\right]\sqrt{\frac{E-m}{E-m_z}} + (k'_x - ik_y)\sqrt{\frac{E-m_z}{E-m}}}.$$ (10)

We parametrize $k_x + m_x = k_F \cos\phi, k_y + m_y = k_F \sin\phi$. Then, we have $E = \sqrt{m_z^2 + (k_x + m_x)^2 + (k_y + m_y)^2} = \sqrt{m_z^2 + k_F^2} = \sqrt{m^2 + k_x'^2 + k_y^2}$.

Using these results, the normalized tunneling conductance is obtained as

$$\sigma = \frac{1}{2}\int_{-\frac{\pi}{2}}^{\frac{\pi}{2}} d\phi \, |t|^2 \, \text{Re}\left[\frac{k'_x}{E}\right].$$ (11)

In the following, we set $m_1 = \sqrt{0.9}E$.

Figure 2 shows the contour plot of the normalized tunneling conductance as a function of the polar coordinates θ and φ specifying the direction of $\mathbf{m}_1$. For (a), (b), the potential drop $V = 0$ between the regions 1 and 2, and the Fermi energy E measured from the Dirac point is set to be positive corresponding to the $n - n$ junction. For (c), (d), we take $V = 2E$ to make the $p - n$ junction. As for the magnetization in F_2, we set $m_2 = 0$ for (a), (c) and $m_2 = \sqrt{0.9}E$ for (b), (d). Namely, there is no effect from F2 in Fig. 2(a) and (c). The red color indicates the large conductance while the blue color small conductance. The periodic oscillation as a function of φ comes from the shift of the Fermi surface in region 1 by $(-m_x, -m_y)$ while that of region 2 does not change. Therefore, the discrepancy of k_y becomes maximum and the overlap of the two Fermi surfaces is minimized at $\varphi = \pi/2, 3\pi/2$ where the conductance is minimum, while the opposite occurs at $\varphi = 0, \pi$. As for the θ-dependence, it is noted that m_z opens a gap at the band crossing point, and the conductance is suppressed when the Fermi energy is within the gap of one or both of the two regions 1 and 2. This is the reason why the conductance is small at $\theta = 0, \pi$ in Fig. 2(a) and (c). On the other hand, the conductance is large at $\theta = 0$ (parallel) for (b) and at $\theta = \pi$ (anti-parallel) for (d). The latter is in sharp contrast to the conventional magneto-resistance where the parallel spin configuration is advantageous for the conductance. This can be understood when one considers the connection of the wavefunctions between two regions; for $p - n$ junction, the wavefunctions for electron and hole are more similar when the magnetization is the opposite direction. These are unique features of the magneto-transport phenomena to the surface of topological insulators.

Up to now, we have considered the magnetization coupled to the surface states as the given quantity. Now let us ask how the dynamics of the magnetization is influenced when coupled to the surface Dirac fermion on a topological insulator. To approach this problem, we have derived the Landau-Lifshitz-Gilbert (LLG) equation through the microscopic calculation [20]. Here the spin-momentum current naturally leads to the inverse spin-Galvanic effect

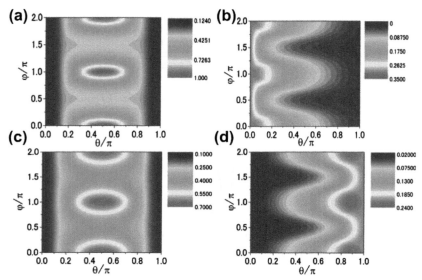

FIGURE 2 The contour map of the conductance as a function of the polar coordinates for the direction of the magnetization in F1 as defined in Fig. 1. Reproduced from Ref. [19].

in a TI/ferromagnet interface, predicting the current-induced magnetization reversal due to the Hall current on the TI [21]. Therefore, the LLG equation in the presence of the current is an intriguing problem.

We start with the same Hamiltonian equation (7) for the electrons, but now we consider the slowly varying $\mathbf{m}(r,t)$ as a function of space r and time t. By integrating over the electrons, we obtain the effective action for the magnetization $\mathbf{m}(r,t)$ and consequent equation of motion [20,22–24].

The obtained LLG equation reads:

$$\dot{n} - 2\sigma_{xy}\left(\frac{M}{ev_F}\right)^2 \dot{n}/(s_0 N) = \gamma_0 \boldsymbol{H}_{\text{eff}} \times \boldsymbol{n} + \left(\frac{M}{ev_F s_0 N}\right)\left(-\mathbf{j} + (\mathbf{n}\cdot\mathbf{j})\hat{z}\right)$$

$$+ (\alpha_0 + \alpha/N)\dot{\boldsymbol{n}} \times \boldsymbol{n} + t_{\text{el}}^{\beta}/(s_0 N), \qquad (12)$$

where $\mathbf{n} = \mathbf{m}/|\mathbf{m}|$, $\mathbf{j}$ is the charge current density, M is the exchange energy, and N is the number of ferromagnetic layers. Here we assume that $\mathbf{n}$ is near the $\mathbf{e}_z$-direction. Note that α-term, t_{el}^{β}, and Berry phase terms (σ_{xy}) originate from the interplay between Dirac fermions and local magnetization. The Gilbert damping constant α is given by

$$\alpha = \frac{1}{2}\left(\frac{Mv_F k_F}{\varepsilon_F}\right)^2 v_F \tau \frac{a^2}{\hbar S}, \qquad (13)$$

$$t_{\text{el}}^{\beta} = -\beta \frac{1}{2e} \left[\mathbf{n} \times (\mathbf{j} \cdot \nabla)\mathbf{n} - (\mathbf{j} - (\mathbf{j} \cdot \mathbf{n})\hat{z})\nabla \cdot (\mathbf{n} \times \hat{z}) \right.$$
$$\left. + (\nabla - (\mathbf{n} \cdot \nabla)\hat{z})\mathbf{n} \cdot (\mathbf{j} \times \hat{z}) \right], \tag{14}$$

where the charge current $\mathbf{j} = \sigma_C \mathbf{E}$, the conductivity $\sigma_C = \frac{e^2}{4\pi} \left(\frac{v_F k_F}{\varepsilon_F} \right)^2 \varepsilon_F \tau$, and

$$\beta = \frac{5\pi}{4\varepsilon_F \tau} \left(\frac{M}{v_F k_F} \right)^2. \tag{15}$$

Here ε_F is the Fermi energy measured from the Dirac point, v_F the Fermi velocity, k_F the Fermi momentum, and τ is the transport lifetime due to the impurities. Note here that in deriving Eq. (14), only the intraband contributions are considered.

From these equations, the following predictions are derived: (i) The strong coupling between the current and the spin is represented by the second term on the right-hand side of Eq. (12), which is active even without the spatial dependence of $\mathbf{n}$. (ii) The spin-transfer torque associated with the charge current is absent because the spin of an electron does not rotate when it goes through the spatially varying spin configuration due to the one-to-one correspondence between the momentum and spin. (iii) The Gilbert damping α, which is proportional to the diagonal conductivity σ_C of the Dirac fermion, can be very large. (iv) Considering the situation where $\varepsilon_F \tau \gg 1$, β is much smaller than α and t_{el}^{β} is negligible. (v) The Berry phase term is modified by the transverse conductivity σ_{xy} of the Dirac fermion, which will change the resonance frequency as $\Re\omega = \omega_0 + 2\sigma_{xy}(\frac{M}{ev_F})^2 \omega_0/(s_0 N)$.

These features can be tested experimentally by the ferromagnetic resonance or the inverse spin-Galvanic effect, or the domain wall motion under charge current on top of the topological insulator.

In the studies described above, we assumed that the Fermi energy is outside of the gap opening at the Dirac point, and the surface state is metallic. When the Fermi energy is within the gap, the surface Dirac system is a quantum Hall insulator with the quantized Hall conductance $\sigma_{xy} = e^2/(2h)$ [25]. The Hamiltonian is still the same, i.e., Eq. (7), and we consider the magnetic textures, such as domain walls and vortices, in a deposited ferromagnetic thin film. Because the total system is now insulating, one can discuss its charging and polarization properties. This is closely related to the multiferroics which attracted recent intensive interest [26]. Namely, in a certain class of insulating magnets, the coupling between the electric polarization and the magnetism is enhanced.

Now we consider the coupling of the surface Dirac fermion to the external electromagnetic field A_μ and the magnetization $\mathbf{n}$ [27]. The Hamiltonian density reads

$$\mathcal{H} = -i\hbar v_F \boldsymbol{\sigma} \times \mathbf{e}_z \cdot (\nabla - ie\mathbf{A}) + eA_0 - \Delta \mathbf{n} \cdot \boldsymbol{\sigma}, \tag{16}$$

where **n** is the direction of the magnetization (unit vector) and Δ is the exchange energy. From Eq. (16), one can read the correspondence as

$$
\begin{aligned}
n_x &\rightarrow A_y, \\
n_y &\rightarrow -A_x,
\end{aligned}
\tag{17}
$$

and hence

$$
\nabla \cdot \mathbf{n} \rightarrow \nabla \times \mathbf{A} = B_z.
\tag{18}
$$

This gives the dictionary to translate the electromagnetic properties of the surface Dirac fermion to the responses to the magnetization **n**. For example, the response of the external magnetic field B_z as $\rho \propto \sigma_H B_z$ is translated to

$$
\rho = -\left(\frac{\sigma_H \Delta}{e v_F}\right) \nabla \cdot \mathbf{n},
\tag{19}
$$

and the Hall response $\mathbf{j} \propto \sigma_H \mathbf{e}_z \times \mathbf{E}$ to

$$
\mathbf{j} = -\left(\frac{\sigma_H \Delta}{e v_F}\right)\left(\frac{\partial \mathbf{n}}{\partial t}\right).
\tag{20}
$$

Equations (19) and (20) indicate that **n** is equivalent to the polarization **P** of the surface electrons.

This strong coupling between the magnetism and polarization is a unique feature of the topological insulator, and hence should have many applications. As an example, we consider the domain (Neel) wall of the ferromagnet placed on the topological insulator as shown in Fig. 3. Because $\nabla \cdot \mathbf{n}$ is nonzero for this configuration, it becomes charged. Integrating this charge density along the x-direction, one can obtain the total charge which is estimated to be of the order of $\sim 3e$ [27]. This enables the acceleration of the magnetic domain wall by the external electric field E instead of the magnetic field. Assuming the Gilbert damping $\alpha \sim 10^{-3}$, the velocity v is estimated as $V \cong E[V/m] \times 10^{-2}$ [m/sec], which is reasonably high at the reasonable strength of E. Therefore, the coupled system of the magnets and the topological insulator can be regarded as an ideal multiferroics where the magnetization and polarization are identical.

Another multiferroics important effect in the field of spintronics and multiferroics is the magneto-electric (ME) effect, where the electric field E produces the magnetization M while the magnetic field B produces the electric polarization P. Concerning this effect, topological ME effect of a three-dimensional (3D) topological insulator has been proposed theoretically [1,28–30]. More concretely, the electromagnetic response is described by the Lagrangian for the axion electrodynamics,

$$
\mathcal{L} = \frac{1}{8\pi}\left(\epsilon \mathbf{E}^2 - \frac{1}{\mu}\mathbf{B}^2\right) + \left(\frac{\alpha}{4\pi^2}\right)\theta \mathbf{E} \cdot \mathbf{B},
\tag{21}
$$

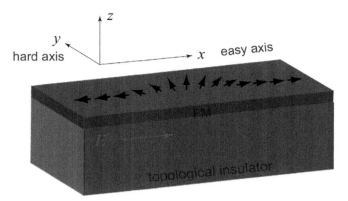

FIGURE 3 The Neel domain wall of the ferromagnet put on the topological insulator. Reproduced from Ref. [27].

where $\mathbf{E}$ and $\mathbf{B}$ are the electromagnetic fields, ϵ and μ are the dielectric constant and magnetic permeability, and $\alpha = e^2/\hbar c$ is the fine structure constant. The new ingredient is contained in the second term, called the θ-term [31]. The angle θ has the periodicity with respect to 2π and hence only two values are allowed in the presence of the time-reversal symmetry Θ, i.e., $\theta = 0$ or $\theta = \pm\pi$, because Θ transforms θ to $-\theta$. It is noted that $\mathbf{E} \cdot \mathbf{B}$ is proportional to $\varepsilon^{\mu\nu\lambda\eta}\partial_\mu A_\nu \partial_\lambda A_\eta = \partial_\mu[\varepsilon^{\mu\nu\lambda\eta}A_\nu \partial_\lambda A_\eta]$. This means that the θ-term is reduced to the "surface term" in the $(3 + 1)$-dimensional space-time. In other words, it is more appropriate to make θ a function of space $\theta(r)$, and the charge and current densities are given by

$$\rho \propto \nabla\theta \times \mathbf{E},$$
$$\mathbf{j} \propto \nabla\theta \cdot \mathbf{B}. \tag{22}$$

This means that the charge and current densities appear only at the surface of the topological insulator where θ has a jump from $\pm\pi$ (inside) to 0 (outside: vacuum).

Physically, these responses to the electromagnetic field originate from the quantum Hall system of the gapped Dirac fermions at the surface. Namely, when the time-reversal symmetry is broken by the magnets attached to the surface with the magnetization normal to the surface, it opens the gap and hence the system is a quantized Hall insulator with $\sigma_H = \pm e^2/(2h)$ when the Fermi energy is within the gap as discussed earlier [25]. The sign of θ and σ_H is determined by the direction of the magnetization. The requirement here is that the magnetization should always be pointing in the inward or outward direction over the whole surface. Otherwise, the reversal of the sign of the mass term for the Dirac fermion results in the 1D chiral channel and becomes conducting. When we apply the magnetic field $\mathbf{B}$ along the z-direction, the upper and lower surfaces will induce the charge density of opposite sign, which results in the *bulk* polarization P also along the z-axis because they are separated by the macroscopic distance. Similarly, when the electric field along the z-direction induces the Hall current on the side surface circulating around the sample. This surface current produces the *bulk* magnetization **m** along the z-axis because the moment of the surface current is macroscopic.

This topological ME effect is protected by the topological property of the bulk states [1,28–30], but experimentally there are several difficulties to realize this effect. (a) It is required to suppress the carriers in the bulk states. (b) One needs to attach the insulating ferromagnetic layer with the magnetization normal to the surface all pointing out or in. (c) The Fermi energy must be tuned accurately within the small gap of the surface Dirac fermion opened by the exchange interaction. These conditions are not easy to realize at present. (a) Is progressively realized [32,33], while (b) and (c) seem still very difficult even though recent experimental work shows that the gapless surface Dirac states of Bi_2Se_3 become gapped with magnetic impurities (Mn and Fe) doped into the crystal [34]. Therefore, the topological ME does not appear to be accessible even though several theoretical proposals have been made [21,27,28,35–40].

However, the situation is much better than expected when we consider the localization of the surface wavefunctions [41]. The surface Dirac fermion is known to be robust against the localization when random potential respects the time-reversal symmetry [42]. This is because the backward scattering is forbidden by the Kramers theorem as described above for the helical edge channel in 2D topological insulator. On the surface of a 3D topological insulator, the scattering process is allowed if it is not strictly backward (from **k** to $-$**k**), this suppression reduces the quantum interference of the scattered waves leading to the absence of localization. This results in the β-function [43] which remains positive for any value of the dimensionless conductance g [42]. This is in sharp contrast to the conventional spin-orbit system which belongs to the symplectic class and has a localization-delocalization transition at the critical $g = g_c$ where the sign change of the β-function occurs [43]. Intuitively, the surface states must connect the valence and conduction bands, and this connectivity does not allow the localization. Therefore, the surface of a 3D topological insulator is always metallic as long as the time-reversal symmetry is there and the bulk gap is not destroyed. On the other hand, when the magnetic impurities are present, the situation changes dramatically. First, the Hall conductivity becomes finite, and second all the states are localized irrespective of the strength of the disorder [41]. This is described by the two-parameter scaling similar to the case of quantized Hall system. In Fig. 4, we plot $(\sigma_{xx}, \sigma_{xy})$ by changing the system size L. Although the system size cannot be widely changed, it indicates that, as increasing system size, $(\sigma_{xx}, \sigma_{xy})$ flows and approaches two fixed points $(0, \pm 1/2)$ in units of e^2/h. The scaling law is reminiscent of that for the integer quantum Hall effect [44–46]. However, there is a crucial difference. The fixed points in the latter case are $(0, n)$ in units of e^2/h, n being an integer, while σ_{xy} is shifted by 1/2 in the present case. The 1/2-shift results in an important consequence, that is, $\sigma_{xy} = 0$ is unstable. Namely, even if the surface mass gap

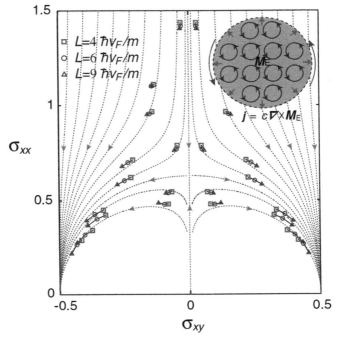

FIGURE 4 The scaling trajectories of the conductivities for the surface state of the 3D topological insulator with magnetic impurities. Reproduced from Ref. [41].

is infinitely small, $|\sigma_{xy}|$ increases and approaches $e^2/2h$. This is a generalization of parity anomaly to the case with disorder. Another crucial difference is that the extended states survive which carry the quantized Hall current in the integer Hall system, while the surface states are all localized and the bulk states carry the Hall current in the present case of topological insulator. Even when the surface is damaged by the disorder potential, the surface is "shifted inward" and is not destroyed as long as the bulk gap remains intact.

Once all the surface states are localized by magnetic disorder, the effective action including the θ-term in Eq. (21) is applicable to describe the topological ME effect [28,29,31,39]: This localization effect relaxes the condition (c) described above. The position of the Fermi energy can be anywhere in the bulk gap since all the surface states are localized, and the Hall conductance is quantized. Then the next issue is how to satisfy the condition (b). Usually, it is not easy to control the magnetic anisotropy. Especially the easy-axis anisotropy normal to the surface is a difficult issue. In addition, the sign of the Hall conductance σ_H should not change sign, which means that the magnetization on average should always point out or in over the whole surface. This seems to be difficult to realize, but as discussed below one can arrange the surface magnetization using the external electric and magnetic fields. The basic idea is that the θ-value at the surface determines the *bulk* energy. In other words, the direction of the surface magnetization is controlled by the *bulk* energy, which always dominates over the surface magnetic anisotropy energy in the limit of large sample size. An estimate for this bulk energy is

$$U_\theta = -\int d^3 x \left(\frac{\alpha}{4\pi}\right) \mathbf{E} \cdot \mathbf{B}$$
$$\cong -10^{10}(E[\text{V/cm}])(B[\text{T}])(L[\text{cm}])^3 [\text{eV}], \qquad (23)$$

where L is the linear size of the sample. On the other hand, the anisotropic energy [37] is estimated as $U_{\text{aniso}}/L^2 \sim 5 \times 10^{10}$ [eV/cm^2] and the Zeeman energy $U_{\text{Zeeman}}/L^2 \sim 10^9 (B[\text{T}])$ [eV/cm^2]. Therefore, with the typical strengths of the electric field $E \sim 10^3$[V/cm], magnetic fields $B \sim 1$ [T] are enough to flip and rearrange the surface magnetization so that the sign of θ is constant over the whole surface to reduce the bulk ME energy.

3 CORRELATED TOPOLOGICAL INSULATORS

The interplay between the topology and electron correlation is one of the most important directions for the developments. As described above, the only low energy states are localized at the edge or surface of the topological insulator. Therefore, the correlation effects are most relevant to these edge and surface channels as long as the electron-electron interaction is smaller than the gap size of the bulk states. This assumption is usually justified for the semiconductors such as HgTe, Bi$_2$Se$_3$, and Bi$_2$Te$_3$. Up to now, there is no experimental signature of the correlation effects on the surface Dirac fermions in 3D topological insulators. On the other hand, the correlation effect is expected to be important for the 1D helical edge channels in 2D topological insulators which is discussed below.

The electron fractionalization is again the key concept when the correlation effect is considered. Table 1 gives the list of various 1D channels realized

TABLE 1 The categories of fractionalized electrons in 1D. The fractionalization proceeds as one goes from right to left, and the system becomes more and more robust against the perturbations.

Chiral Majorana	Chiral Fermion	Helical Majorana	Spinless Fermion	Helical Fermion	Spinful Fermion	2-Spinful Fermion
$p+ip$SC 5/2 FQH STI + SC	1/3 FQH	Helical SC	Ferro wire	QSHS	Q-wire	Ladder

in the edge of the systems. One-dimensional electrons have three degrees of freedom, i.e., (i) spin, (ii) chirality (left-going and right-going), and (iii) positive and negative energy. (i) is reduced by the spin polarization, while (ii) by the cyclotron motion in the quantum Hall system, and (iii) by the Cooper pairing. When all the degrees of freedom are fractionalized, the electrons are separated into $2^3 = 8$ pieces. In the case of helical edge channel, (i) spin and (ii) chirality are entangled and are reduced to half at each of the edges, i.e., the spin and chirality are locked [1]. Several nontrivial features are expected for this 1D system such as the half-e charge near the ferromagnetic domain wall [47], and the robustness against the umklapp scattering [48,49].

In the Majorana edge channel of the spinless chiral superconductor, the electrons are completely fractionalized as will be discussed later in Section 4. The general principle is that the particle becomes more and more robust against the disorder, interactions, etc., as the fractionalization proceeds. This is because the scattering channels are reduced as the degrees of freedom are reduced.

On the other hand, one can "recombine" the fractionalized pieces of the electrons to construct the composite particle by the electron-electron interaction. It is noted that the system becomes more and more susceptible to the external stimuli as the degrees of freedom increase. The composite particle can be quite different from the original electrons before the fractionalization. Interactions between two helical edge channels have been studied from this viewpoint [50]. Compared with the single pair of helical edge channels, which has the same degrees of freedom as the interacting spinless fermion model and the Tomonaga-Luttinger liquid effect is induced by the interaction, the gap opening due to the backward scattering becomes possible for half of the degrees of freedom analogous to the Luther-Emery liquid [51]. However, the gap is not the simple spin gap but an entangled one of spin and helicity in the present case. Unusual physical properties are expected in this gapped liquid state [50].

Next we consider the strongly correlated electronic systems where the bulk states are influenced by the electron-electron interaction and the interaction energy is comparable to or larger than the bulk energy gap. It has been known for the d-electrons that the spin-orbit interaction is stronger for heavier elements, while the correlation effect becomes weaker because the d-orbitals are more extended. These two tendencies cross at 5d compounds, where both the SOI and correlation energy are of the order of 0.5 eV. This fact has been recognized in the studies of Ir oxides. In Sr_2IrO_4, the three t_{2g} orbitals (xy, yz, zx) are coupled by the SOI to form the $J_{\text{eff}} = 3/2, 1/2$ states [52]. This is because the correspondence like $xy \sim p_z$ exists between the t_{2g} orbitals and p-orbitals concerning the SOI except for the sign change in the matrix elements. One of the $5d$-electrons of Ir^{4+} ion occupies the $J_{\text{eff}} = 1/2$ orbitals while the other four electrons completely occupy the $J_{\text{eff}} = 3/2$ orbitals. Therefore, the system is effectively described by the one-band Hubbard model, and shows the Mott insulating state with the canted antiferromagnetism at low temperature [52].

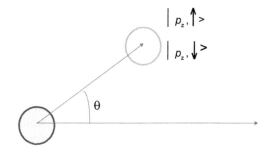

$$|1/2\rangle = (|xy \uparrow\rangle + |yz \downarrow\rangle + i|zx \downarrow\rangle)/\sqrt{3}$$
$$|-1/2\rangle = (-|xy \downarrow\rangle + |yz \uparrow\rangle - i|zx \uparrow\rangle)/\sqrt{3}$$

FIGURE 5   The transfer integrals between the d-orbitals and p-orbitals, which depend on the angle θ and on the spin. See the text for details.

Based on this fact, one can design the topological insulator in the following way.

First, the two wavefunctions for $J_{\text{eff}} = 1/2$ are given by

$$|+1/2\rangle = \frac{1}{\sqrt{3}}(|xy \uparrow\rangle + |yz \downarrow\rangle + i|zx \downarrow\rangle),$$
$$|-1/2\rangle = \frac{1}{\sqrt{3}}(-|xy \downarrow\rangle + |yz \uparrow\rangle + i|zx \uparrow\rangle). \quad (24)$$

An important aspect is that both wavefunctions are complex, and hence the transfer integrals between the d-orbitals and p-orbitals of oxygen are complex as shown in Fig. 5. Furthermore, those for $|-1/2\rangle$ are complex conjugate to those for $|+1/2\rangle$. This is a promising situation where the topological nature of the Bloch wave states arises with time-reversal symmetry, but the lattice structure needs to be designed in addition to this. Actually, in the case of the square lattice as in Sr_2IrO_4, the system remains topologically trivial. On the other hand, the honeycomb lattice is an interesting structure as originally proposed by Kane-Mele [8]. The crystal structure of Na_2IrO_3 can be regarded as the stacking of honeycomb layers, and hence a candidate for the oxide topological insulator. We have done a first-principles band structure calculation and the tight-binding fitting to conclude that it is a weak-topological insulator, i.e., the stacking of the 2D quantum spin Hall system very similar to the Kane-Mele model before the magnetic order sets in [53]. A model from the strong correlation limit concludes that Na_2IrO_3 is a model system for Hubbard-Kitaev model [54–57]. Furthermore, Kim et al. [58] study the phase diagram of Na_2IrO_3 where the transition is driven by the change in the long-range hopping and trigonal crystal field terms. Experimentally, the neutron and X-ray scattering are employed to reveal the crystal and magnetic structures [59–62].

A natural extension of this idea is the pyrochlore lattice where the three-dimensional strong topological insulator and topological spin liquid states are proposed [63]. However, there is no experimental report confirming the topological insulator states in transition metal oxides.

Recently, we have considered the possibility of the topological states in the superlattice system made of transition metal oxides [64]. Artificial structures such as the interfaces or superlattices are the powerful method to design the electronic states [65]. In these systems, one can tune various parameters such as the lattice constant, carrier densities, symmetry of the lattice, etc., by the choice of the substrate, choice of the surfaces, the external electric/magnetic fields, and so on, which offers an ideal system to design the appropriate electronic wavefunctions for any purposes. Especially, the perovskite transition metal oxides are the most actively studied systems, and the spin/orbital/charge orders, meta-insulator transition, colossal magneto-resistance (CMR), multiferroics, and superconductivity have been realized there [65]. Here we focus on the bilayers of perovskite-type transition metal oxides grown along the [1 1 1] crystallographic axis [64]. This is because it can be regarded as the honeycomb lattice seen from [1 1 1] direction, with the upper and lower layers of triangular lattices. In addition to the lattice structure, the symmetry of the environment surrounding the d-orbitals changes from the octahedral to trigonal. This results in lifting of the degeneracies as shown in Fig. 6. By the crystal field Δ of trigonal symmetry, the threefold degenerate t_{2g} orbitals with octahedral symmetry are split into doubly degenerate e'_g orbitals and nondegenerate a_{1g} orbital, while the e_g orbitals remain doubly degenerate. When the spin-orbit interaction λ is taken into account with the octahedral symmetry, the e_g orbitals remain doubly degenerate (fourfold degenerate when spin degrees of freedom are considered) because there is no matrix element of the SOI within the e_g manifold, while the t_{2g} orbitals are rearranged into $J_{\text{eff}} = 1/2$ and $J_{\text{eff}} = 3/2$ as described above. In the presence of both Δ and λ, which corresponds to the bilayer system, the t_{2g} orbitals split into three doubly degenerate Kramers pairs, while the e_g orbitals also split into two Kramers pairs. Namely, the combination of Δ and λ makes the SOI active also within the e_g orbitals, which leads to remarkable consequences as shown below.

The design principle is to start with a band structure that possesses "Dirac points" at the K and K' in the Brillouin zone without the SOI, and then to see if an energy gap opens or not when the SOI is switched on. If the gap opens, the system has a good chance to be a topological insulator when the Fermi energy is within this gap. We have analyzed the phenomenological tight-binding Hamiltonian, and found that the system shows the nontrivial Z_2 topological invariants for the sets of reasonable values of parameters for both e_g and t_{2g} orbitals, and hence the realistic first-principles band structure calculations are in order. Figure 7 shows two examples from t_{2g} and e_g systems, respectively; (a) $SrTiO_3 / SrIrO_3 / SrTiO_3$

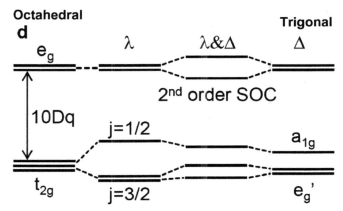

FIGURE 6 The d-orbitals under the octahedral (left) and trigonal (right) crystal field. With the spin-orbit interaction, the orbitals are further split. Reproduced from Ref. [64].

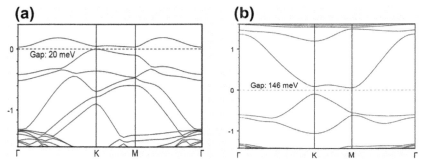

FIGURE 7 The energy band structure of the (1 1 1) bilayer systems of (a) $SrTiO_3/SrIrO_3$ and (b) $LaAlO_3/LaAUO_3/LaAlO_3$. Reproduced from Ref. [64].

and (b) $LaAlO_3/LaAuO_3/LaAlO_3$. Both systems are predicted to be the 2D topological insulators. Especially, the gap size for $LaAlO_3/LaAuO_3/LaAlO_3$ is of the order of $0.15eV$, i.e., much larger than the room temperature. We performed the first-principles density functional theory (DFT) calculations for the bilayers of variety of materials and found that $LaOsO_3$ (t_{2g}), and $LaAgO_3$ (e_g) are also the candidates for topological insulators.

Another interesting feature of the band structures in (1 1 1)-bilayer systems is that the reasonably flat bands appear in the cases of e_g electrons. This means that the electron correlation effects are enhanced, and intriguing phenomena are expected to occur. Actually, the most probable order is the magnetism due to the electron repulsion. We have studied the ferromagnetic order is the e_g systems, and found that the Chern numbers are produced for each of the spin-split bands, and the quantized anomalous Hall effect is realized when the Fermi energy is within one of the band gaps. Furthermore, when the flat bands are partially filled, e.g., 1/3 filled, there is a possibility of fractional anomalous quantum Hall state. By projecting the Hamiltonian into one of the flat bands to construct the single-band Hubbard-like model, we have studied the ground state degeneracy, Chern number for the many-body wavefunctions, and the gap in the excitation spectrum. All of these quantities are consistent with the fractional anomalous quantum Hall effect to be realized in the 1/3-filled flat band [66]. An important feature here is that the energy scale of this phenomenon is typically the effective Hubbard U or the intersite Coulomb interaction V, which is much larger than the room temperature. Therefore, the (1 1 1) bilayer perovskite transition metal oxides offer an interesting possibility of room temperature fractional anomalous quantum Hall effect [64].

4 TOPOLOGICAL SUPERCONDUCTORS AND MAJORANA FERMIONS

The concept of topology can be applied also to the superconductors by regarding the Bogoliubov-de Gennes Hamiltonian as that of the Bloch electrons. In the topological classification of the electronic states in solids [10,11], the Cooper pairing adds the particle-hole symmetry Ξ. In addition to the early pioneering theoretical works on ^{3}He [67], recently the p-wave superconductivity with broken T-symmetry in Sr_2RuO_4 has been discussed from this perspective [68]. Helical topological superconductors with time-reversal Θ-symmetry have also been studied [69]. From the viewpoint of electron fractionalization, the topological superconductivity or superfluidity split the electrons into positive and negative energy parts, which enables the Majorana fermion [70–74]. Majorana fermion is a half of the fermion or the real fermion. In the language of the operators, the creation and annihilation operators are identical to each other for Majorana fermions, i.e.,

$$\psi^\dagger = \psi. \tag{25}$$

One can express the Majorana fermion operator ψ in terms of the usual fermion operators f and $f^\dagger$ as

$$\psi = \frac{1}{\sqrt{2}}(f + f^\dagger). \tag{26}$$

Therefore, Eq. (25) is satisfied and $\psi^2 = 1/2$. Although the particle number is not a good quantum number any more, the parity of the particle number still remains a conserved quantity. This is because the pairing or the off-diagonal order separates the Hilbert space into two sectors, i.e., odd and even numbers of electrons, and ψ connects these two sectors.

The simplest model showing the topological superconductivity and Majorana fermion is given by Kitaev [75] and is described by the following

Hamiltonian in 1D [76]:

$$H = \frac{1}{2} \sum_{k:half} C_k^\dagger H_k C_k, \tag{27}$$

with $C_k^\dagger = (c_k^\dagger, c_{-k})$ and $H_k = \mathbf{h}(k) \cdot \boldsymbol{\sigma}$ and $\mathbf{h}(k) = (\mathrm{Re}\Delta_k, \mathrm{Im}\Delta_k, \xi_k)$ with $\Delta_k = i\Delta \sin k$ is the p-wave pairing amplitude and $\xi_k = -2t \cos k - \mu$ is the normal state dispersion. Because $(C_{-k}^\dagger)^T = \sigma^x C_k$, the $\mathbf{h}$'s at two momenta k and $-k$ are related as

$$\begin{aligned} h_{x,y}(k) &= -h_{x,y}(-k), \\ h_z(k) &= h_z(-k). \end{aligned} \tag{28}$$

Therefore, at the time-reversal invariant momentum (TRIM), i.e., the momentum k that is equivalent to $-k$, only h_z can be nonzero. In 1D, $k = 0$ and $k = \pm\pi$ are the TRIM and hence one can consider the two classes of the trajectories of $\mathbf{n}(k) = \mathbf{h}(k)/|\mathbf{h}(k)|$ on the unit sphere. One class comprises those paths which connect the north and south poles as k increases from 0 to π, while those in the other class make a loop starting and ending at the north or south pole. The former one is topologically nontrivial, while the latter is trivial. As a consequence, the Majorana bound state at zero energy appears at the two ends of the sample in the former case. This situation is schematically depicted in Fig. 8. The spinless fermion at each site is expressed in terms of two Majorana fermions, and the pairing of the two Majorana fermions occurs between the nearest neighbor sites. This results in an unpaired Majorana fermion at each end of the sample. In the case of 2D, one can define the Chern number for the Bogoliubov-de Gennes Hamiltonian to classify the topologically distinct systems. The spinless $p + ip$ pairing state is the representative showing the nonzero Chern number, and has the chiral Majorana edge channel [77].

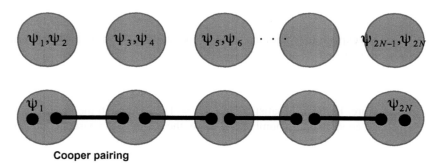

Cooper pairing

FIGURE 8   The schematic picture for the Kitaev model. The bond represents the Cooper pairing, which forms the "pair" of neighboring Majorana fermions, and results in the isolated Majorana bound state at each end of the sample.

Now the issue is how to implement/fabricate the topological superconductors/superfluid and Majorana fermions in real materials. The proximity-induced superconductivity of the surface state in the 3D topological insulator has been proposed to be a promising laboratory to realize these goals [78]. Furthermore, the superconductivity in a doped topological insulator $Cu_x Bi_2 Se_3$ has been also discussed from this viewpoint [79–83]. Here, we consider the Rashba systems as the promising candidates for this purpose since the spin splitting occurs at generic k-points which is required for the electron fractionalization.

With the time-reversal symmetry Θ, the Kramers pair exists even though the momenta are separated. Especially, the propagating Majorana edge channels become helical edge modes related by Θ, and the zero energy Majorana bound states are doubly degenerate. This possibility in Rashba superconductors has been discussed in [84,85]. This is rather realistic since the superconductivity has been found at the interface of $LaAlO_3$ and $SrTiO_3$ [86], and the role of Rashba interaction has been identified there [87]. With the Rashba splitting, the spin direction is locked to and rotates with the momentum, leading to the $p \pm ip$ pairing form when the superconducting order parameter is projected onto each of the bands. It is noted here, however, that the interference between the two spin-split bands is crucial, and the relative sign of the order parameters determines the topological properties of the superconductor. Namely, the spin triplet p-wave pairing should be dominant over spin singlet s-wave pairing for the realization of helical topological superconductivity [84,88,89]. Once the helical superconductivity is realized, there appear the helical Majorana edge channels which control the low energy transport phenomena. In particular the spin-dependent Andreev reflection occurs because of the spin-momentum locking of the helical edge modes. For example, a tiny external magnetic field produces the large spin conductance of the Andreev reflection, which might be useful for the spintronics applications [84]. However, the realization of the helical superconductivity depends on the details of the electron-electron interaction, which is not so easy to control.

One way to resolve this difficulty is to open a gap and delete the inner Fermi surface by the external Zeeman magnetic field or by the exchange splitting due to the proximity to a ferromagnet, leaving alone the $p + ip$ (or $p - ip$) pairing on the outer Fermi surface [85,90–93]. This results in the chiral Majorana edge channels around the sample. The behavior of these Majorana edge modes and their role in the Andreev reflection has been studied [94]. There is a quantum phase transition between the topologically trivial superconducting state and the chiral superconducting state as the chemical potential is increased. The behavior of the chiral Majorana mode is found to be classified into two classes: (i) The Majorana mode survives even on the trivial superconductor side while the gap opens in the spectrum, (ii) The Majorana mode disappears on the trivial side. According to this, the phase transition is classified into two types, and the behavior of the conductance is quite different between these two, which can be detected experimentally.

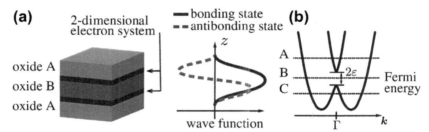

FIGURE 9 (a) The proposed structure of the bilayer Rashba system, and the corresponding bonding and anti-bonding bands. (b) The energy dispersion of the bands with the hybridization gap at the band crossing point. Note that each states are doubly degenerate. Reproduced from Ref. [96].

When one further reduces the dimension to 1, the system is reduced to Kitaev model with p-wave pairing, and the Majorana bound states are expected at the edge of the sample. A recent experiment using InSb has observed the zero energy state by the tunneling spectroscopy at the edge of the sample, which is stimulating further research of this possibility [95].

One possible method to make a helical superconductivity in a bilayer Rashba system has been proposed [96]. It is based on the idea to introduce the hybridization gap at the band crossing points. As shown in Fig. 9, the Rashba 2D-electrons at the two interfaces with the opposite direction of the electric field in the 3-layer structure is the candidate to realize this model. Actually, this structure can be fabricated both in the oxides [65] and semiconductors [97,98].

The single-particle Hamiltonian H_0 is given by

$$\mathcal{H}_0 = \frac{k^2}{2m} \mathbb{I}_s \otimes \mathbb{I}_\sigma - \varepsilon \mathbb{I}_s \otimes \sigma_x + \alpha \left(k_x s_y - k_y s_x \right) \sigma_z, \tag{29}$$

where s and σ denote the Pauli matrices for spin and layers, respectively, ε is the hybridization energy between the layers, and α is the strength of Rashba spin-orbit interaction. The eigenvalues of this Hamiltonian are given by

$$E_k = k^2/2m \pm \sqrt{\varepsilon^2 + \alpha^2 k^2} = k^2/2m \pm \lambda_k. \tag{30}$$

Note that the band dispersions are doubly degenerate at each k-point because this system has both the time-reversal symmetry, $\Theta = -i\sigma_y K$, where K is the complex conjugation operator, and space inversion (I-) symmetry, $I = \sigma_x$. The Hamiltonian for the electron-electron interaction is given by

$$\mathcal{H}_{\text{int}}\left(x\right) = -U \left(n_1^2 \left(x\right) + n_2^2 \left(x\right) \right) - 2V n_1\left(x\right) n_2(x), \tag{31}$$

where $n_{\sigma=1,2} = \sum_{s=\uparrow,\downarrow} c_{\sigma s}^\dagger c_{\sigma s}$ is the electron density in layer 1 and 2, and U and V are intra-layer interaction and inter-layer interaction, respectively.

TABLE 2 The classification of the pairing states in D_{4h} symmetry. Reproduced from Ref. [96].

Pairing Potential	Explicit Representation	Matrix	Parity
Δ_1:	$c_{1\uparrow}c_{1\downarrow}+c_{2\uparrow}c_{2\downarrow}, c_{1\uparrow}c_{2\downarrow}-c_{1\downarrow}c_{2\uparrow}$	I, σ_x	$+$
Δ_2:	$c_{1\uparrow}c_{2\downarrow}+c_{1\downarrow}c_{2\uparrow}$	$s_z\sigma_y$	$-$
Δ_3:	$c_{1\uparrow}c_{1\downarrow}-c_{2\uparrow}c_{2\downarrow}$	σ_z	$-$
Δ_4:	$(c_{1\uparrow}c_{2\uparrow}+c_{1\downarrow}c_{2\downarrow}, c_{1\uparrow}c_{2\uparrow}-c_{1\downarrow}c_{2\downarrow})$	$(s_x\sigma_y, s_y\sigma_y)$	$-$

The superconducting state can be described by the mean field theory with the pairing order parameters. Considering the interface between $SrTiO_3$ and $LaAlO_3$, the lattice symmetry is assumed to be D_{4h}. Then the pairing states are classified by the irreducible representation of this point group as shown in Table 2. (i) Δ_1: the most conventional even-parity pairing state with combination of the intra-layer singlet pairings ($c_{1\uparrow}c_{1\downarrow}+c_{2\uparrow}c_{2\downarrow}$) and the inter-layer singlet pairing ($c_{1\uparrow}c_{2\downarrow}-c_{1\downarrow}c_{2\uparrow}$). (ii) Δ_2: the inter-layer triplet pairing with odd parity, (iii) Δ_3: the intra-layer singlet with odd parity, and (iv) Δ_4: the inter-layer triplet pairing with odd parity. It is noted that the parity operation here is the mapping $r \rightarrow -r$, and is different from the even-odd symmetry with respect to the exchange of the two coordinates $(r_1, r_2) \rightarrow (r_2, r_1)$ of the pairing electrons.

From the analysis of the energy dispersion, only Δ_4 pairing state has point nodes while the other three are fully gapped states. By looking at the largest pairing susceptibility, one can draw the phase diagram as shown in Fig. 10. There are several important conclusions here. One is that the multiband effect combined with the Rashba SOI enables the unconventional pairing states (Δ_2, Δ_3). Without α, all the phase diagram is occupied by the conventional Δ_1 state or the non-superconducting state. In the absence of the SOI, the k-dependence of the pairing amplitude needs to be considered to realize the unconventional states, which is beyond the scope of the present study. Second, the phase diagram depends on the position of the Fermi energy. With the small Rashba spin-orbit interaction α, the unconventional pairing states appear when the Fermi energy is lower where α is more relevant to the electronic state. Lastly, the pairing state Δ_4 with nodes cannot be the dominant instability anywhere in the phase diagram. A remark for the real systems is in order here. The electrons at the interface $LaAlO_3/SrTiO_3$ are superconducting [86], which suggests that U is considered to be attractive assuming the conventional pairing state. Most probably, this attractive interaction U is given by the short-range electron-phonon coupling reduced by the Coulomb repulsive interaction, the repulsive inter-layer $|V|$ is expected to be larger than U. This indicates that the bilayer Rashba system realized at $LaAlO_3/SrTiO_3/LaAlO_3$ could belong to the situation where Δ_3 is stabilized [10,11].

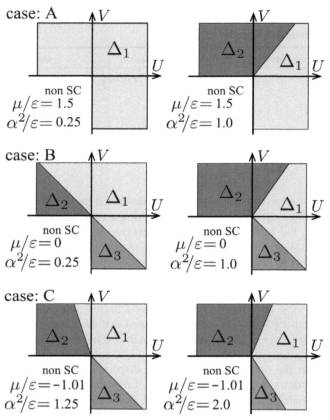

FIGURE 10 The phase diagrams of the bilayer Rashba model. A, B, and C correspond to the positions of the Fermi energy in Fig. 9. Reproduced from Ref. [96].

Now the topological nature of the pairing states is explored. According to the topological periodic table [10], the present system belongs to DIII class in 2D. Therefore, Z_2 number characterizes the topology of the pairing states in contrast to the Z in 3D. This model resembles the model of the three-dimensional doped topological insulator $Cu_x Bi_2 Se_3$ studied by Fu and Berg [80]. The criterion employed in Ref. [80] can be used also here to judge the topological nature of the pairing states although the present system is 2D. For the helical superconductivity, (i) it must have odd-parity pairing symmetry with full superconducting gap (Δ_2 or Δ_3) and (ii) odd number of Kramers pairs at time-reversal invariant momenta Γ_α's are below the Fermi surface. To satisfy these conditions in the present system, the Fermi surface must lie in the hybridization gap, i.e., case B, and the pairing symmetry must be Δ_2 or Δ_3. The existence of the helical Majorana edge channels has been confirmed numerically when these two conditions are satisfied [96].

Compared with the chiral superconductors and chiral Majorana fermion, the helical superconductors show several interesting effects and phenomena. First, the spin degrees of freedom are still alive for the Majorana edge channels, and Majorana bound state exists in the core of the vortex. Therefore, the magnetic response of the Majorana fermions is an interesting issue. The coupling of the magnetic impurity to the helical Majorana edge channels has been studied from this viewpoint [99]. In Ref. [99], the two types of 2D spin-triplet topological superconductors have been considered as described by the order parameter $\hat{\Delta}_k = id_k \cdot \sigma \sigma_2$ with the d-vector [100]

$$d_k = \begin{cases} (\hat{x}\sin\theta + \hat{y}\cos\theta)(k_x + ik_y), & \text{(chiral)}, \\ (\hat{x}\sin\theta + \hat{y}\cos\theta)k_x + (\hat{x}\cos\theta - \hat{y}\sin\theta)k_y, & \text{(helical)}. \end{cases} \quad (32)$$

Here $\hat{x}$ and $\hat{y}$ are unit vectors in the spin space. The direction of the d-vector d_k is independent of k in the chiral case, while it constitutes a coplanar spin texture in the k-space for the helical case.

The above two types of 2D spin-triplet topological superconductors support gapless Majorana edge modes which satisfy; (a) $\psi_\sigma(r) = ie^{i\theta_\sigma + i\phi}\psi_\sigma^\dagger(r)$ ($\sigma = \uparrow$ and $\downarrow$) for the chiral case, and (b) $\psi_\uparrow(r) = ie^{i\theta_\uparrow + i\phi}\psi_\uparrow^\dagger(r)$ and $\psi_\downarrow(r) = ie^{i\theta_\downarrow - i\phi}\psi_\downarrow^\dagger(r)$ for the helical case. These Majorana conditions depend on the phase of the pairing order parameter, and lead to the operator identities for the edge mode's spin density,

$$2\hat{s}_z(r) = \psi_\uparrow^\dagger\psi_\uparrow - \psi_\downarrow^\dagger\psi_\downarrow = 0, \quad (33a)$$

$$\hat{s}_+(r) = \psi_\uparrow^\dagger\psi_\downarrow = \begin{cases} -e^{-2i\theta}\hat{s}_-(r) & \text{(chiral)}, \\ -e^{-2i(\theta+\phi)}\hat{s}_-(r) & \text{(helical)}. \end{cases} \quad (33b)$$

These equations imply that the spin density is always Ising-like: $(\hat{s}_x,\hat{s}_y) \propto (\sin\theta, \cos\theta)$ for the chiral case and $(\hat{s}_x,\hat{s}_y) \propto (\sin(\theta+\phi), \cos(\theta+\phi))$ for the helical case [101,102]. When this Ising-like spin of Majorana edge modes is coupled to the impurity spin, it gives rise to an interesting anisotropic Kondo-like behavior, which can be detected by, for example, the ESR experiment.

Now we turn to the interaction effect on the Majorana edge modes. The interaction is given by the product of four field operators as

$$H_{int} = g \int dx\, \psi_i \psi_j \psi_k \psi_\ell, \quad (34)$$

where the indices i,j,k,ℓ specify the species of the fermions. Here we assume the local interaction with x being common for the four operators, since the derivative with respect to x increases the engineering dimension of the interaction and hence becomes more irrelevant in the sense of renormalization group. (Because of the Majorana condition $\psi^\dagger \propto \psi$, we did not put † in Eq. (34).) When we have only one species, the product of the four field operators is always a constant since ψ_i^2 is a constant. Therefore, the chiral Majorana edge channel is robust against the electron-electron interaction. That is also the case for helical Majorana edge channels where two species of fermions exist. We need at least four species of Majorana fermions to have the interaction effect. When the two helical Mojorana edge channels are interacting with each other, this condition is satisfied. The system is equivalent to the spinless 1D fermion or helical edge channel, where the forward scattering is allowed leading to the Tomonaga-Luttinger liquid behavior. This situation has been considered in Ref. [103], which studied a local tunnel junction between two helical Majorana edge channels. As discussed in Section 3, the system becomes more and more robust as the electron fractionalization proceeds and the chiral Majorana fermion is the most tolerant system, while the system gets more susceptible as the internal degrees of freedom increase. Helical Majorana edge channels enjoy both the robustness and responses to the external stimuli, and are expected to be an interesting laboratory to explore various novel phenomena in the future.

5 DISCUSSION AND CONCLUSIONS

In this chapter, we have discussed the theoretical design of the topological materials and their functions. The basic idea is the "twist" of the Bloch wavefunctions in the Brillouin zone and the consequent edge/surface states, which are classified by the topological invariants. The electron fractionalization at the edge/surface states leads to the robustness and spin-momentum locking, which will have a wide range of applications in spintronics. Various physical effects including the spin-current coupling, topological magneto-electric effect, spin-dependent reflection, Majorana fermions, and Andreev/Josephson effects in superconductors are designed and partly realized experimentally. There is still much to be studied, and we mention some of the future directions in this concluding section.

One is the role of topological defects. Teo and Kane [12] generalized the topological periodic table to the case of textures in crystals. Remarkably, the conclusion is simple; spatial dimension d is shifted by the dimension D of the manifold which encloses the texture, and $\delta = d - D$ plays the same role of d for the pure case. This also deepens our understanding of the Bott periodicity which relates the homotopy classes of different symmetries and dimensions. Practically, this opens up many possible designs of the topological phenomena with defects, dislocations, surfaces, and composite systems [104]. For example, we have studied the two-dimensional spinless $p + ip$ superconductor with the focus on the Majorana fermions appearing at its dislocation and vortex [105]. In this case, the "weak" topological invariants play the crucial role to determine the presence of the zero energy Majorana state.

Another direction is the interaction. There appear several theoretical works on the topological correlated states. It is an important and intriguing issue to look for the topological spin liquid state and the topological invariants to characterize it. However, it is also important to look for the topologically nontrivial states with the magnetic ordering. The search for the quantized anomalous Hall effect is one of the examples which still is waiting to be realized. Weyl metal in a pyrochlore crystal with the robust surface states is another example [106].

To conclude, the concept of topology has enriched the content of condensed matter physics tremendously, and opened the new direction to applications also. This will be a major topic of physics research in the years to come.

ACKNOWLEDGMENTS

The author acknowledges D.Asahi, Y. Asano, A.V. Balatsky, A. Furusaki, H. Katsura, J. Kune, S. Murakami, K. Nomura, S. Okamoto, X.L. Qi, Y. Ran, M. Sato, R. Shindou, A. Shitade, Y. Tanaka, D. Xiao, A. Yamakage, T. Yokoyama, S.C. Zhang, and W. Zhu for fruitful collaborations reviewed in this article. This work was supported by Grant-in-Aid for Scientific Research (S) (Grant No. 24224009) from the Ministry of Education, Culture, Sports, Science and Technology of Japan; Strategic International Cooperative Program (Joint Research Type) from Japan Science and Technology Agency; the Funding Program for World-Leading Innovative RD on Science and Technology (FIRST Program).

REFERENCES

[1] Hasan MZ, Kane CL. Rev Mod Phys 2010;82:3045; Qi X-L, Zhang S-C. Rev Mod Phys 2011;83:1057.
[2] Prange RE, Girvin SM, editors. The quantum Hall effect. New York: Springer-Verlag; 1987.
[3] Thouless DJ, Kohmoto M, Nightingale MP, den Nijs M. Phys Rev Lett 1982;49:405.
[4] Castro Neto AH, Guinea F, Peres NMR, Novoselov KS, Geim AK. Rev Mod Phys 2009;81:109.
[5] Nagaosa Naoto, Sinova Jairo, Onoda Shigeki, MacDonald AH, Ong NP. Rev Mod Phys 2010;82:1539.
[6] Murakami S, Nagaosa N, Zhang SC. Science 2003;301:1348; Sinova J et al. Phys Rev Lett 2004;92:126603.
[7] Murakami S, Nagaosa N. Comprehensive semiconductor science and technology (SEST). In: Bhattacharya Pallab, Fornari Roberto, Kamimura Hiroshi, editors, vol. 1. Elsevier; 2011.
[8] Kane CL, Mele EJ. Phys Rev Lett 2005;95:226801.
[9] Moore JE, Balents L. Phys Rev B 2007;75:121306; Fu L, Kane CL. Phys Rev B 2007;76:045302; Roy R. Phys Rev B 2009;79:195321.
[10] Schnyder AP, Ryu S, Furusaki A, Ludwig AWW. Phys Rev B 2008;78:195125.
[11] Kitaev A. AIP Conf Proc 2009;1134:22.
[12] Teo JC, Kane CL. Phys Rev B 2010;82:115120.
[13] Majorana E. Nuovo Cimento 1937;5:171; Wilczek F. Nat Phys 2009;5:614.
[14] Manchon A, Zhang S. Phys Rev B 2008;78:212405.
[15] Chernyshov A et al. Nat Phys 2009;5:656.
[16] Miron IM et al. Nat Mater 2010;9:230.
[17] Zutic I, Fabian J, Das Sarma S. Rev Mod Phys 2004;76:323.
[18] Wolf SA, Awschalom DD, Buhrman RA, Daughton JM, von Molnar S, Roukes ML, et al. Science 2001;294:1488.
[19] Yokoyama T, Tanaka Y, Nagaosa N. Phys Rev B 2010;81:121401(R).
[20] Yokoyama Takehito, Zang Jiadong, Nagaosa Naoto. Phys Rev B 2010;81:241410.
[21] Garate I, Franz M. Phys Rev Lett 2010;104:146802.
[22] Kohno H, Tatara G, Shibata J. J Phys Soc Jpn 2006;75:113706.
[23] Tserkovnyak Y, Skadsem HJ, Brataas A, Bauer GEW. Phys Rev B 2006;74:144405; Tserkovnyak Y, Brataas A, Bauer GE. J Magn Magn Mater 2008;320:1282.

[24] Tserkovnyak Y, Fiete GA, Halperin BI. Appl Phys Lett 2004;84:5234.
[25] Jackiw R. Phys Rev D 1984;29:2375.
[26] Nagaosa N, Tokura Y. Physica Scripta 2012;T146:014020.
[27] Nomura K, Nagaosa N. Phys Rev B 2010;82:161401.
[28] Qi X-L, Hughes TL, Zhang S-C. Phys Rev B 2008;78:195424.
[29] Essin AM, Moore JE, Vanderbilt D. Phys Rev Lett 2009;102:146805.
[30] Chen K-T, Lee PA. Phys Rev B 2011;83:125119.
[31] Wilczek F. Phys Rev Lett 1987;58:1799.
[32] Ren Z et al. Phys Rev B 2010;82:241306.
[33] Shuang Jia et al. Phys Rev B 2011;84:235206.
[34] Chen YL et al. Science 2010;329:659.
[35] Liu Q et al. Phys Rev Lett 2009;102:156603.
[36] Guo H-M, Franz M. Phys Rev B 2010;81:041102.
[37] Nunez AS, Fernandez-Rossier J. arxiv:1003.5931.
[38] Abanin DA, Pesin DA. arxiv:1010.0668.
[39] Tse W-K, MacDonald AH. Phys Rev Lett 2010;105:057401; Tse W-K, MacDonald AH. Phys Rev B 2010;82:161104.
[40] Maciejko J, Qi X-L, Drew HD, Zhang S-C. Phys Rev Lett 2010;105:166803.
[41] Nomura Kentaro, Nagaosa Naoto. Phys Rev Lett 2011;106:166802.
[42] Nomura K, Koshino M, Ryu S. Phys Rev Lett 2007;99:146806; Nomura K et al. Phys Rev Lett 2008;100:246806.
[43] Lee Patrick A, Ramakrishnan TV. Rev Mod Phys 1985;57:287.
[44] Pruisken AMM. in Ref. [2].
[45] Khmelnitsky D. JETP Lett 1983;38:552.
[46] Dolan BP. Nucl Phys B 1999;460[FS]:297.
[47] Qi X-L, Hughes T, Zhang S-C. Nat Phys 2008;4:273.
[48] Wu C, Bernevig BA, Zhang SC. Phys Rev Lett 2006;96:106401.
[49] Xu C, Moore JE. Phys Rev B 2006;73:045322.
[50] Tanaka Yukio, Nagaosa Naoto. Phys Rev Lett 2009;103:166403.
[51] Giamarchi T. Quantum physics in one-dimension. Oxford: Oxford Science Publications; 2004.
[52] Kim BJ et al. Phys Rev Lett 2008;101:076402; Kim BJ et al. Science 2009;323:1329.
[53] Shitade Atsuo, Katsura Hosho, Kune Jan, Qi Xiao-Liang, Zhang Shou-Cheng, Nagaosa Naoto. Phys Rev Lett 2009;102:256403.
[54] Chaloupka Jili, Jackeli George, Khaliullin Giniyat. Phys Rev Lett 2010;105:027204.
[55] Jiang Hong-Chen, Gu Zheng-Cheng, Qi Xiao-Liang, Trebst Simon. Phys Rev B 2011;83:245104.
[56] Reuther Johannes, Thomale Ronny, Trebst Simon. Phys Rev B 2011;84:100406.
[57] Singh Yogesh, Manni S, Reuther J, Berlijn T, Thomale R, Ku W. Phys Rev Lett 2012;108:127203.
[58] Kim Choong H, Kim Heung Sik, Jeong Hogyun, Jin Hosub, Yu Jaejun. Phys Rev Lett 2012;108:106401.
[59] Singh Yogesh, Gegenwart P. Phys Rev B 2010;82:064412.
[60] Liu X, Berlijn T, Yin W-G, Ku W, Tsvelik A, Kim Young-June, et al. Phys Rev B 2011;83:220403.
[61] Choi SK, Coldea R, Kolmogorov AN, Lancaster T, Mazin II, Blundell SJ, et al. Phys Rev Lett 2012;108:127204.
[62] Ye Feng, Chi Songxue, Cao Huibo, Chakoumakos Bryan C, Fernandez-Baca Jaime A, Custelcean Radu, et al. Phys Rev B 2012;85:180403.
[63] Pesin D, Balents L. Nat Phys 2010;6:376; Guo HM, Franz M. Phys Rev Lett 2009;103:206805; Bohm-Jung Yang, Yong Baek Kim. Phys Rev B 2010;82:085111; Kurita M, Yamaji Y, Imada M. J Phys Soc Jpn 2011;80:044708; Kargarian M, Wen J, Fiete GA. Phys Rev B 2011;83:165112.
[64] Xiao Di et al. Nat Commun 2011;2:596.
[65] Hwang HY, Iwasa Y, Kawasaki M, Keimer B, Nagaosa N, Tokura Y. Nat Mater 2012;11:103.
[66] Sheng DN et al. Nat Commun 2011;2:389; Neupert T et al. Phys Rev B 2011;84:165107; Neupert T et al. Phys. Rev Lett 2011;106:236804; Sun K et al. Phys Rev Lett 2011;106:236803; Tang E, Mei JW, Wen XG. Phys Rev Lett 2011;106:236802.

[67] Volovik GE. The Universe in a Helium Droplet. Oxford: Clarendon; 2003.
[68] Mackenzie Andrew Peter, Maeno Yoshiteru. Rev Mod. Phys 2003;75:657.
[69] Qi XL, Hughes TL, Raghu S, Zhang SC. Phys Rev Lett 2009;102:187001.
[70] Nayak C, Simon SH, Stern A, Freedman M, Das Sarma S. Rev Mod Phys 2008;80:1083.
[71] Das Sarma S, Nayak C, Tewari S. Phys Rev B 2006;73:220502.
[72] Read N, Green D. Phys Rev B 2000;61:10267.
[73] Ivanov DA. Phys Rev Lett 2001;86:268.
[74] Tanaka Y, Sato M, Nagaosa N. J Phys Soc Jpn 2012;81:011013.
[75] Kitaev AY. Physics-Uspekhi 2001;44:131.
[76] Alicea J. Rep Prog Phys 2012;75:076501.
[77] Furusaki A, Matsumoto M, Sigrist M. Phys Rev B 2001;64:054514.
[78] Fu L, Kane CL. Phys Rev Lett 2008;100:096407.
[79] Hor YS, Williams AJ, Checkelsky JG, Roushan P, Seo J, Xu Q, et al. Phys Rev Lett 2010;104:057001.
[80] Fu L, Berg E. Phys Rev Lett 2010;105:097001.
[81] Sasaki S, Kriener M, Segawa K, Yada K, Tanaka Y, Sato M, et al. Phys Rev Lett 2011;107:217001.
[82] Hao L, Lee TK. Phys Rev B 2011;83:134516.
[83] Hsieh TH, Fu L. Phys Rev Lett 2012;108:107005.
[84] Tanaka Y, Yokoyama T, Balatsky AV, Nagaosa N. Phys Rev B 2009;79:060505.
[85] Sato M, Fujimoto S. Phys Rev B 2009;79:094504.
[86] Reyren N et al. Science 2007;317:1196.
[87] Caviglia AD, Gabay M, Gariglio S, Reyren N, Cancellieri C, Triscone J-M. Phys Rev Lett 2010;104:126803.
[88] Iniotakis C, Hayashi N, Sawa Y, Yokoyama T, May U, Tanaka Y, et al. Phys Rev B 2007;76:012501.
[89] Vorontsov AB, Vekhter I, Eschrig M. Phys Rev Lett 2008;101:127003.
[90] Sau JD, Lutchyn RM, Tewari S, Das Sarma S. Phys Rev Lett 2010;104:040502.
[91] Qi X-L, Hughes TL, Zhang S-C. Phys Rev B 2010;82:184516.
[92] Lutchyn RM, Sau JD, Das Sarma S. Phys Rev Lett 2010;105:077001.
[93] Alicea J. Phys Rev B 2010;81:125318.
[94] Yamakage A, Tanaka Y, Nagaosa N. Phys Rev Lett 2012;108:087003.
[95] Mourik V et al. Science 2012;336:1003.
[96] Nakosai S, Tanaka Y, Nagaosa N. Phys Rev Lett 2012;108:147003.
[97] Bernardes E et al. Phys Rev Lett 2007;99:076603.
[98] The helical topological superconductivity has been proposed in the interacting two surface Dirac fermions in 3D topological insulator. See Liu C-X, Trauzettel B. Phys Rev B 2011;83:220510.
[99] Shindou R, Furusaki A, Nagaosa N. Phys Rev B 2010;82:180505.
[100] Leggett AJ. Rev Mod. Phys 1975;47:331.
[101] Stone M, Roy R. Phys Rev B 2004;69:184511.
[102] Chung SB, Zhang SC. Phys Rev Lett 2009;103:235301.
[103] Asano Y, Tanaka Y, Nagaosa N. Phys Rev Lett 2010;105:056402.
[104] Ran Y, Zhang Y, Vishwanath A. Nat Phys 2009;5:298.
[105] Asahi D, Nagaosa N. Phys Rev B 2012;86:100504.
[106] Wan X et al. Phys Rev B 2011;83:205101.

Chapter 11

Beyond Band Insulators: Topology of Semimetals and Interacting Phases

Ari M. Turner* and **Ashvin Vishwanath**[†]

*Institute for Theoretical Physics, University of Amsterdam, Science Park 904, P.O. Box 94485, 1090 GL Amsterdam, The Netherlands
[†]Department of Physics, University of California, Berkeley, CA 94720, USA

Chapter Outline Head

The recent developments in topological insulators have highlighted the important role of topology in defining phases of matter (for a concise summary, see Chapter 1 in this volume). Topological insulators are classified according to the qualitative properties of their bulk band structure; the physical manifestations of this band topology are special surface states [1] and quantized response functions [2]. A complete classification of free fermion topological insulators that are well defined in the presence of disorder has been achieved

Topological Insulators. http://dx.doi.org/10.1016/B978-0-444-63314-9.00011-1

[3,4], and their basic properties are well understood. However, if we loosen these restrictions to look beyond insulating states of free fermions, do new phases appear? For example, are there topological phases of free fermions in the absence of an energy gap? Are there new gapped topological phases of interacting particles, for which a band structure based description does not exist? In this chapter we will discuss recent progress on both generalizations of topological insulators.

In **Part 1** we focus on topological semimetals, which are essentially free fermion phases where band topology can be defined even though the energy gap closes at certain points in the Brillouin zone. We will focus mainly on the three-dimensional Weyl semimetal and its unusual surface states that take the form of "Fermi arcs" [5] as well as its topological response, which is closely related to the Adler-Bell-Jackiw (ABJ) [6] anomaly. The low-energy excitations of this state are modeled by the Weyl equation of particle physics. The Weyl equation was originally believed to describe neutrinos, but later discarded as the low-energy description of neutrinos following the discovery of neutrino mass. If realized in solids, this would provide the first physical instance of a Weyl fermion. We discuss candidate materials, including the pyrochlore iridates [5] and heterostructures of topological insulators and ferromagnets [7]. We also discuss the generalization of the idea of topological semimetals to include other instances including nodal superconductors.

In **Part 2**, we will discuss the effect of interactions. First, we discuss the possibility that two phases deemed different from the free fermion perspective can merge in the presence of interactions. This reduces the possible set of topological phases. Next, we discuss physical properties of gapped topological phases in 1D and their classification [8–11]. These include new topological phases of bosons (or spins), which are only possible in the presence of interactions. Finally we discuss topological phases in two and higher dimensions. A surprising recent development is the identification of topological phases of bosons that have no topological order [12,13]. These truly generalize the concept of the free fermion topological insulators to the interacting regime. We describe properties of some of these phases in 2D, including a phase where the thermal Hall conductivity is quantized to multiples of eight times the thermal quantum [12], and integer quantum Hall phases of bosons where only the even integer Hall plateaus are realized [14–16]. Possible physical realizations of these states are also discussed. We end with a brief discussion of topological phases with topological order, including fractional topological insulators.

1 PART 1: TOPOLOGICAL SEMIMETALS

A perturbation does not significantly change the properties of a fully gapped insulator, hence readily allowing us to define robust topological properties. Is it possible for a gapless system to have a defining topological feature? How can one distinguish phases of a system with gapless degrees of freedom that can be rearranged in many ways?

We will consider a particular example of a gapless system, a Weyl semimetal, as an example. There are three answers to the question of how this state can be topologically characterized despite the absence of a complete gap: in momentum space, the gapless points are "topologically protected" by the behavior of the band structure on a surface enclosing each point [17,18]. There are topologically protected surface states, as in 3D topological insulators, which can never be realized in a purely two-dimensional system [5]. If a field is applied, there is a response whose value depends only on the location of the gapless points, but on no other details of the band structure (and so is topological).

In Weyl semimetals, the valence and conduction bands touch at points in the Brillouin zone. A key requirement is that the bands are individually non-degenerate. This requires that either time reversal symmetry or inversion symmetry (parity) is broken. For simplicity, we assume time reversal symmetry is broken, but inversion is preserved. In practice this is achieved in solids with some kind of magnetic order. Consider the following Hamiltonian [19,20], a simple two-band model where the 2×2 Pauli matrices refer to the two bands:

$$H = - \sum_{k} \left[2t_x (\cos k_x - \cos k_0) + m(2 - \cos k_y - \cos k_z) \right] \sigma_x$$
$$+ 2t_y \sin k_y \sigma_y + 2t_z \sin k_z \sigma_z. \tag{1}$$

We assume the σ transforms like angular momentum under inversion and time reversal, which implies the above Hamiltonian is invariant under the former, but not the latter symmetry. It is readily seen to possess nodes. At $\mathbf{k} = \pm k_0 \hat{\mathbf{x}}$, and only there (if m is sufficiently large), the gap closes. This will turn out not to be due to symmetry, but to topology. Even if all symmetries are broken (time reversal already is) a gap will not open. The gapless points can move around but we will see that they can only disappear if they meet one another, at which point they can annihilate. This is a key difference from the analogous dispersion in 2D (e.g., graphene) which may be gapped by breaking appropriate symmetries.

If the bands are filled with fermions right up to the nodes, then this is a semimetal with vanishing density of states at the Fermi energy. We expand the Hamiltonian around the gapless points setting $\mathbf{p}_\pm = (\pm k_x \mp k_0, k_y, k_z)$, and assuming $|\mathbf{p}_\pm| \ll k_0$:

$$H_\pm = v_x [p_\pm]_x \sigma_x + v_y [p_\pm]_y \sigma_y + v_z [p_\pm]_z \sigma_z, \tag{2}$$
$$E^2 = v_x^2 [p_\pm]_x^2 + v_y^2 [p_\pm]_y^2 + v_z^2 [p_\pm]_z^2, \tag{3}$$

where $v_x = 2t_x \sin k_0, v_{y,z} = -2t_{y,z}$. This is closely related to Weyl's equation from particle physics:

$$H_{\text{Weyl}}^\pm = \pm c \mathbf{p} \cdot \sigma,$$

which reduced the 4×4 matrices of the Dirac equation to 2×2 Pauli matrices. Since there is no "fourth" Pauli matrix that anticommutes with the other three, one cannot add a mass term to the Weyl equation. Thus, it describes a massless fermion with a single helicity, either right- or left-handed, which corresponds to the sign $\kappa = \pm 1$ before the equation. The velocity, $\mathbf{v} = \pm c\boldsymbol{\sigma}$, is either along or opposite to the spin depending on the helicity. In a 3D lattice model, Weyl points always come in pairs of opposite helicity; this is the fermion doubling theorem [21], which is justified below.

The low-energy theory of a Weyl point is in general an anisotropic version of the Weyl equation,

$$H = \sum_i v_i (\hat{\boldsymbol{n}}_i \cdot \mathbf{p}) \sigma_i + v_0 (\hat{\boldsymbol{n}}_0 \cdot \mathbf{p}) \mathbb{1}. \tag{4}$$

The three velocities v_i describe the anisotropy of the Weyl point, and the $\hat{\boldsymbol{n}}_i$'s are the principal directions, which are linearly independent but not necessarily orthogonal. (We have chosen the basis for the Pauli matrices such that they match the principal directions.) The helicity is given by $\kappa = \text{sign}[\hat{\boldsymbol{n}}_1 \cdot (\hat{\boldsymbol{n}}_2 \times \hat{\boldsymbol{n}}_3)]$ (the sign in front of the Weyl equation after transforming to coordinates where the velocity is isotropic).

1.1 Topological Properties of Weyl Semimetals

1.1.1 Stability

A well-known argument about level crossing in quantum mechanics [22] can be applied to argue that Weyl nodes are stable to small perturbations regardless of symmetry [23]. Consider a pair of energy levels (say two bands) that approach each other. We would like to understand when the crossing of these energy levels (band touching) is allowed. We can write a general Hamiltonian for these two levels as: $H = \begin{pmatrix} \delta E & \psi_1 + i\psi_2 \\ \psi_1 - i\psi_2 & -\delta E \end{pmatrix}$, ignoring the overall zero of energy. The energy splitting then is $\Delta E = \pm \sqrt{\delta E^2 + \psi_1^2 + \psi_2^2}$. For $\Delta E = 0$ we need to individually tune each of the three real numbers to zero. This gives three equations, which in general need three variables for a solution. The three-dimensional Brillouin zone provides three parameters which in principle can be tuned to find a node. While this does not guarantee a node, once such a solution is found, a perturbation which changes the Hamiltonian slightly only shifts the location of the solution, i.e., moves the Weyl node in crystal momentum. Note, this does not tell us the energy of the Weyl node—in some cases it is fixed at the chemical potential from other considerations. This is a convenient situation to discuss, but several of the results we mention below do not strictly require this condition. Later we will discuss how Dirac points can be stabilized in lower dimensions as well, as in graphene, provided that some symmetries are present.

1.1.2 Magnetic Monopoles in Momentum Space

We will now show that the Weyl points' stability is connected to an integer-valued topological index. That is the first sense in which a Weyl semimetal is a topological state. There is a conservation law of the net charge of the Weyl points. That is, a single Weyl point cannot disappear on its own. However, on changing parameters in the Hamiltonian the Weyl points can move around and eventually pairs with opposite signs can annihilate, meaning that they first merge and then the bands separate. In order for Weyl points to have well-defined coordinates, one needs crystal momentum to be a well-defined quantum number. Therefore, unlike topological insulators which are well defined even in the presence of disorder, here we will assume crystalline translation symmetry to sharply define this phase.

The Weyl nodes are readily shown to be associated with a quantized Berry flux of $2\pi\kappa$ (where $\kappa = \pm 1$ is the chirality). The Berry flux $\mathbf{B}$ is defined as:

$$\mathcal{A}(\mathbf{k}) = -i \sum_{n,\text{occupied}} \langle u_{n,\mathbf{k}} | \nabla_{\mathbf{k}} | u_{n,\mathbf{k}} \rangle, \tag{5}$$

$$\mathbf{B}(\mathbf{k}) = \nabla_{\mathbf{k}} \times \mathcal{A}. \tag{6}$$

The Berry flux of a Weyl point is readily found from the following simple expression for the wave function of filled bands near the Weyl point:

$$\psi(\theta,\phi) = \begin{pmatrix} \sin\frac{\theta}{2} \\ -\cos\frac{\theta}{2} e^{i\phi} \end{pmatrix}, \tag{7}$$

where the momentum coordinates relative to the Weyl point are expressed in terms of the angle: θ and ϕ. The Berry phase of this spinor is the same as the Berry phase of a spin-1/2 object in a field; one can then determine the flux through a small sphere about the Weyl point; it is $\pm 2\pi$. The Berry flux through any surface containing it is the same, thanks to Gauss's law. Thus a Weyl point may be regarded as a magnetic monopole, since its flux corresponds to the quantized magnetic charge of monopoles.

Now the stability of Weyl points can be connected to Gauss's law: a Gaussian surface surrounding the Weyl point detects its charge, preventing it from disappearing surreptitiously. It can only disappear after an oppositely charged monopole goes through the surface, and later annihilates with it. Also, the net charge of all the Weyl points in the Brillouin zone has to be zero (which is seen by taking a Gauss-law surface that goes around the whole Brillouin zone). Thus, the minimum number of Weyl points (if there are some) is two, and they have to have the opposite chirality, as in the model above. This is the proof of the fermion doubling theorem.

Ideally, to isolate effects of the Weyl points, it is best that they lie right at the Fermi energy, and that there are no Fermi surfaces (whose contributions to response properties would drown out signals from the Weyl points). Now it might seem that it would be necessary to have a very special material and doping to ensure that the Weyl points are at the same energy and that this coincides with the Fermi energy. However, a symmetry can ensure the former, and stoichiometric filling, via Luttinger's theorem, the latter. Later examples of materials that have been proposed are given.

Let us compare this approach to identifying band topology with the situation in classifying topological insulators. The filled bands of a regular topological insulator give a continuous map from the Brillouin zone to the relevant target space, and one classifies topologically distinct maps. In a semimetal, the map isn't continuous since the occupied states in the vicinity of a gapless point change discontinuously. However, a topological property can still be identified by considering a *subspace* of the Brillouin zone that avoids the gapless point, in this case a two-dimensional closed surface surrounding the Weyl point.

1.1.3 Topological Surface States

Now we will look at the surface states of this model. How can surface states be protected in a gapless system? What keeps the states from hybridizing with the bulk states? The answer is translational symmetry: if the surface momentum is conserved, surface states at the Fermi energy can be stable at any momenta where there are no bulk states (at the same energy). For example, if the Fermi energy is tuned to the Weyl point in the bulk, the bulk states at this energy are only at the projections of the Weyl points to the surface Brillouin zone. Everywhere else, surface states are well defined because they cannot decay into bulk states at the same energy and momentum. As the Weyl point momenta are approached, the surface states penetrate deeper into the bulk.

It turns out (see Fig. 1) that surface states form an *arc* at the Fermi energy, instead of a closed curve. Our usual intuition is that Fermi surfaces form a closed loop since they are defined by the intersection of the dispersion surface $\epsilon = \epsilon(k_x, k_y)$ with the Fermi level $\epsilon = 0$. If the Fermi surface is an arc, which "side" corresponds to occupied states?

The key to answering these questions is that surface states are not well defined across the entire Brillouin zone, due to the presence of bulk states as can be seen from Fig. 1(b). The bulk states intervene and allow for the non-intuitive behavior, that is impossible to obtain in a two-dimensional system, and in this particular case, also impossible on the surface of a fully gapped insulator. For example, the Fermi arcs end at the gap nodes where the surface mode is no longer well defined. Furthermore, the surface representing the dispersion of the surface states joins up with the bulk states (the blue cones) at positive and negative energies.

Intuitively, we may imagine a complete Fermi surface being present when we begin with a thin sample of Weyl semimetal. On increasing the thickness, complementary parts of the Fermi surface get attached to the top and bottom surfaces, resulting in arcs. Thus the Fermi surface can be completed by

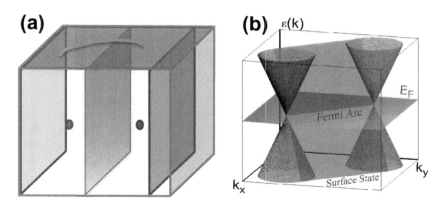

FIGURE 1 Surface states of a Weyl semimetal. (a) The surface states of a Weyl semimetal form an arc connecting the projections of the two Weyl points to one another. Points on the Fermi arc can be understood as the edge state of a two-dimensional insulator, and the Berry flux of the Weyl points ensures that the insulators represented by either the red and blue planes are integer Hall states, which have edge modes. Here we depict the former. (b) A graph of the dispersion of the surface modes (the pink plane) and how this joins up with the bulk states (represented by solid red and blue cones) [5]. (For interpretation of the references to color in this figure legend, the reader is referred to the web version of this book.)

combining the arcs living on the opposite sides of the sample. The surface states are obtained by mapping three-dimensional Weyl semimetals to Chern insulators in two dimensions as we now explain.

Semimetals and Topological Insulators—Relations between dimensions Topological semimetals inherit the topology of an insulator of a lower dimension. The spherical surface enclosing a Weyl node used above to detect the Berry monopole has another interpretation: it describes the band structure of an insulator with no symmetries (class A) in two dimensions. That is because a 2D insulator is described by a matrix-function $H(k_x, k_y)$ that has a gap everywhere; we only need to trade the Brillouin zone torus for a sphere in this case. The Hamiltonian matrices become gapped because the surface avoids the node. These maps are classified by the Chern number, which is identical to the net Berry flux of the magnetic monopole, given by the integral $\int \frac{d^2\mathbf{k}}{(2\pi)^2} \mathbf{B}(\mathbf{k})$ where $\mathbf{B}$ is defined in Eq. (6).

This observation explains why there are surface states in the Weyl semimetal. We will show that the Fermi arcs connect the projections of Weyl points onto the surface. The starting point is the fact that any closed surface in momentum space surrounding one of the Weyl points has a Berry flux through it. Such a surface therefore defines a two-dimensional insulator with a Hall conductance, which must have edge states, and for the right geometry these will correspond to real edge states of the three-dimensional crystal.

Let us consider a crystal that is cut parallel to the xy-plane. For a clean surface, momentum is conserved, and the wave functions can be taken to have a definite crystal momentum $k_\perp = (k_x, k_y)$ in the direction parallel to the surface. Let us in particular fix k_x, so that $\psi(x,y,z) = e^{ik_x x} f_{k_x}(y,z)$. The Hamiltonian then reduces to a two-dimensional problem, different for each given momentum,

$$H_{k_x} f_{k_x}(y,z) = \epsilon f_{k_x}(y,z). \tag{8}$$

This describes a 2D system with an edge, corresponding to the top surface of the original semimetal. If two cross-sections contain a monopole between them, then there is a non-zero net flux through the two of them. Thus the Chern numbers differ by 1, and so one of the two systems at least is a 2D quantum Hall state. For example, in Fig. 1(a) the planes in between the two monopoles all have Chern number 1. Thus there is an edge state for each fixed value of k_x. Putting all these edge states together, we find that there is a Fermi *arc* connecting the two Weyl points.

That completes the argument for the surface states. There is another connection between Weyl semimetals and topological insulators in a different dimension: this one appears on *increasing* the dimension and considering $4 + 1$D topological insulators (which exist in the absence of any symmetry). The boundary states of these phases correspond to Weyl fermions, i.e., a *single* Weyl point. Since this is the boundary of a four-dimensional bulk, the fermion doubling theorem does not apply. This immediately allows us to draw certain physical conclusions. Disorder on the surface of a 4D topological insulator will not localize the surface states. Hence, if we have a 3D Weyl semimetal with disorder that scatters particles only *within* a node, we may conclude that this will *not* lead to localization.

1.1.4 The Chiral Anomaly

We now consider "topological responses" of Weyl semimetals.

At first, it may seem that the number of particles at a given Weyl point has to be conserved, as long as there is translational symmetry. Yet there is an "anomaly" that breaks this conservation law. The conservation of momentum implies that the particles cannot scatter between the two Weyl points, and hence they can be described by separate equations. The effective description of either one in an electromagnetic field is

$$H_{R/L} = \mp i\hbar v \psi_{L/R}^\dagger \sigma \cdot \left(\nabla - \frac{ie\mathbf{A}}{\hbar}\right)\psi_{L/R}, \tag{9}$$

where we have assumed the velocity to be isotropic. Response to a low-frequency electromagnetic field can be calculated from these relativistic equations, using methods from field theory. The first step is to add in a cut-off to give a finite answer. This cut-off does not respect the conservation law, and so it is found that the charge is not conserved:

$$\frac{\partial}{\partial t}(n_{R/L}(\mathbf{r})) = \pm\frac{e^2}{h^2}\mathbf{E}\cdot\mathbf{B}. \tag{10}$$

This is known as the Adler-Bell-Jackiw anomaly in field theory.

Condensed matter derivation of the anomaly—Consider a Weyl semimetal with cross-sectional area A and length L_z in an electric and a magnetic field applied along z. The magnetic field causes the electron states to split into a set of Landau levels. As opposed to the two-dimensional case, each Landau level is a one-dimensional mode dispersing along z. The zeroth levels are the most interesting. They provide one-dimensional chiral modes with the linear dispersion $\epsilon_{R/L} = \pm v p_z$ for the right- and left-handed species. These modes can be pictured as chiral wires parallel to the magnetic field, with a density per unit area of B/Φ_0 where $\Phi_0 = \frac{h}{e}$ is the flux quantum. The derivation uses the fact that lowest Landau levels in a 2D Dirac Hamiltonian correspond to an eigenstate of σ_z, which sets the velocity of z direction propagation. The other modes do not cross through zero, so they cannot respond much to external fields. Thus, the three-dimensional anomaly reduces essentially to the one-dimensional one.

Now, on applying an electric field the electrons accelerate $\dot{p}_z = e E_z$ and so the Fermi momenta increase as well:

$$\dot{p}_{F,L/R} = e E_z, \tag{11}$$

creating a larger Fermi sea at the right Weyl point, and a smaller one at the left Weyl point. So effectively the electrons move from the left Weyl point to the right. The rate of increase gives the exactly correct form for the anomaly

$$\frac{dN_R}{dt} = (e E_z) \left(\frac{L_z}{h} \right) \left(\frac{AB}{\Phi_0} \right), \tag{12}$$

where $\frac{L_z}{h}$ is the density of states in a chiral mode and $\frac{AB}{\Phi_0}$ is the density of modes. The number density is obtained from $N_{R/L}$ by dividing by the volume $A L_z$, and the result agrees with Eq. (10) for the anomaly.

There is a close connection between the topological surface states and the chiral anomaly. Consider the following paradox: In a magnetic field, there are two sets of chiral modes, one traveling along and the other opposite to the field, at different transverse momenta. When a particle in the mode traveling upward reaches the surface, where does it go? It must get to the opposite node and return back through the bulk, but that is at a different momentum. In fact, it travels along the Fermi arc which connects the projection of the right-handed Weyl point to the left-handed one (the Lorentz force pushes it along the arc).

From the anomaly, it is clear that a single Weyl point is disallowed in a 3D band structure since that would violate particle number conservation. Note, the $\mathbf{E} \cdot \mathbf{B}$ term appears here in an equation of motion, while the same term appears in the effective action for topological insulators [2]. This connection arises from the fact that the transition between trivial and topological insulators can be understood in terms of Weyl fermions [24,25].

Anomalous Hall Effect: Although this anomaly is a quantized response, it is hard to measure separately the charge at individual Weyl points. Instead, consider the anomalous Hall effect associated with this simple Weyl semimetal with two Weyl points. The region of the Brillouin zone between the Weyl points can be considered as planes of 2D band structures each with a quantized Hall effect. Thus, the net Hall effect is proportional to the separation between the nodes. Consider creating a pair of Weyl nodes at the center of the Brillouin zone. As the nodes separate, the Hall conductance increases, until they meet at the face of the Brillouin zone and annihilate. At this point the Hall conductance is quantized, in the sense that the system can be considered a layered integer quantum Hall state (Chern insulator) with a layering in the direction $\mathbf{G}$, the reciprocal lattice vector where the nodes annihilated. Thus the Weyl semimetal is an intermediate state between a trivial insulator and a layered Chern insulator. Likewise, the Fermi arc surface states are understood as the evolution of the surface to ultimately produce the chiral edge state of the layered Chern insulator. In a general setting [19], the Hall conductivity is given by the vector:

$$\mathbf{G}_H = \frac{e^2}{2\pi h} \sum_i \kappa_i \mathbf{k}_i, \tag{13}$$

where $\mathbf{k}_i$ is the location of the Weyl points and κ_i is their charge. In an electric field, current flows at the rate $\mathbf{J} = \mathbf{E} \times \mathbf{G}_H$. This phenomenon is a type of quantized response. The Hall conductivity just depends on the topological properties of the Weyl points as well as their locations [71]. If the latter are determined from a separate experiment, such as ARPES, then a truly quantized response can be extracted. Since this is a gapless state, the accuracy of quantization needs to be theoretically investigated, and is a topic for future work. Certainly, the temperature-dependence of deviations from the quantized value vanishes, at best, as a power law of temperature given the absence of a finite gap.

Remarkably, this is directly related to the chiral anomaly. Apply both an electric $\mathbf{E}$ and a magnetic $\mathbf{B}$ field to a topological insulator. Charge will be transferred from the left-handed to the right-handed Weyl points. The points are at different momenta, and hence momentum is not conserved. One can compute the rate of change of momentum per unit volume using Eq. (10):

$$\frac{d\mathbf{p}}{dt} = \frac{e^2}{h^2} \sum_i \kappa_i \hbar \mathbf{k}_i \mathbf{E} \cdot \mathbf{B}. \tag{14}$$

This is exactly the Hall coefficient, i.e., $\frac{d\mathbf{p}}{dt} = \mathbf{G}_H \mathbf{E} \cdot \mathbf{B}$. This is understood from the fact that electric and magnetic fields acting on a system with an anomalous Hall conductance induce a charge $\rho_H = \mathbf{B} \cdot \mathbf{G}_H$ and current density $\mathbf{J}_H = \mathbf{E} \times \mathbf{G}_H$, which in turn respond with a Lorentz force $\frac{d\mathbf{p}}{dt} = \rho_H \mathbf{E} + \mathbf{J}_H \times \mathbf{B}$, which is equal to $\mathbf{G}_H \mathbf{E} \cdot \mathbf{B}$. This argument gives an alternative derivation of the Hall conductivity from the anomaly.

1.2 Topological Semimetals: Generalizations

As for topological insulators, topological semimetals can exist in various numbers of dimensions and with different symmetries, and they all have surface states. In particular, superconducting systems with nodes can have flat *bands* on their surface [26–29], which could potentially form interesting phases because interactions would be extremely important.

A topological semimetal is a material whose metallic behavior has a topological origin; that is, it is not possible to eliminate gapless points without annihilating them with one another. Such a material could also have nodes (i.e., band touchings) along loops that are stable, unless the loops are shrunk down to a point. (We will ignore the energy dispersion along these loops since they can be smoothly deformed to have the same energy). In general, we define a topological semimetal in d spatial dimensions as one that has a band touching along a q-dimensional space in the Brillouin zone (where $q \leq d - 2$) which is protected by non-trivial band topology of a topological insulator derived from a submanifold of the Brillouin zone surrounding the nodal space. For a q-dimensional node, this topological insulator will be $(d - q - 1)$ dimensional, e.g., in $d = 3$, a point node will be protected by a sphere surrounding it and a line node will be protected by a circle linked with it. Here we will restrict attention to those topological states whose symmetries leave the crystal momentum invariant—or example, we will not consider time reversal symmetry, since time reversal changes momentum, and thus it does not protect nodes at generic points [72]. However, the classes A and AIII are both allowed; in the former there are no symmetries, while the latter is based on sublattice symmetry, which leaves the momentum untouched. Invariance of momentum under a defining symmetry allows us to ignore the fact that the submanifold can be rather different than a Brillouin zone. (The topology of the manifold is more important when symmetry is present; e.g., on a sphere, no point is invariant under reversing momenta while on a torus there are four invariant points).

Is Graphene a Topological Semimetal?: Graphene has two point nodes in its 2D bulk band structure. However, there is no 1D topological invariant in class A that could be used to protect the degeneracy. Therefore we conclude that real graphene is a non-topological semimetal. However, if we consider a slightly idealized version of graphene, where the Hamiltonian only involves hopping between opposite sublattices of the honeycomb lattice (such as only nearest neighbor hopping), this turns out to satisfy the criteria for topological semimetal. This version of graphene has a sublattice symmetry, in which changing the sign of amplitudes on one sublattice reverses the sign of the Hamiltonian. This puts it in class AIII, for which there is a 1D topological invariant [3]. Does this topological protection have other consequences? Indeed the edge nodes in graphene can be understood from the hidden topological 1D insulators embedded in the band structure. Each line across the Brillouin zone gives a one-dimensional Hamiltonian. If a line is in a non-trivial class, it must have edge states, and in class AIII, these edge states have strictly zero energy. Lines on opposite sides of a node are in different one-dimensional classes, so if one side has no edge states, the other must have zero-energy states. Indeed, the zero-energy, state expected, e.g., at the zig-zag edge [30] of graphene conforms to a topological surface state. However, breaking the sublattice symmetry, for example by changing the potential near the edge, displaces these zero modes.

The same idea protects *line* nodes in three dimensions in models with sublattice symmetry. A line node can be linked with a loop, and if the Hamiltonian on the loop is non-trivial, then the line node is topological.

Nodal Superconductors: As pointed out in Refs. [17,26–29], this is particularly interesting in the context of superconductors. In particular, consider a superconductor with time reversal symmetry T. The superconductor also has a particle-hole symmetry Ξ (resulting from the redundancy in the Bogoliubov-de Gennes equations), and the product, ΞT, is a unitary symmetry that anticommutes with the Hamiltonian, and leaves the crystal momentum **k** invariant. Thus, the loop is again classified by AIII symmetry. If it is non-trivial, then the line node cannot be removed. This is the most natural way for the AIII symmetry to arise—the chiral symmetry could just happen to be an exact symmetry, as for ideal graphene, but any small farther-neighbor hopping will generically break it.

The definition of a topological semimetal as we formulated it generalizes the "topological protection" of Weyl points by their magnetic charge to other systems. Do the other properties, surface states and topological responses, come as consequences?

The argument for surface states given in the graphene case and for the Weyl semimetal certainly generalizes; they are the edge states of $(d - q - 1)$-dimensional cross-sections of the Brillouin zone that have a non-trivial band topology on account of the non-trivial topology of the nodes. For example, there are exactly zero-energy states on the surface of a superconductor with topological node-loops: For any facet direction that the superconductor is cut along, the loop of nodes casts a "shadow" onto the surface Brillouin zone. On either the inside or the outside there has to be a state with energy that is strictly equal to zero.

Indeed, the well-known zero bias peaks in d-wave cuprate superconductors are also an example of edge states of a topological semimetal (the d-wave singlet superconductor) [26]. Applications of this idea [26,27] could help determine pairing symmetry in a nodal superconductor.

Do quantized responses, which we define as ones that depend only on the locations and topological properties of nodes, generalize as well? So far there is only one example, the Hall effect in a Weyl semimetal. It is "topological" in the sense that it is determined by the location of the Weyl nodes and their indices, but does not depend on the wave functions as, e.g., electrical polarizability does. There is a second example if we allow an ordinary metal to count as topological: Luttinger's theorem. Luttinger's theorem says that the density of electrons (which plays the role of the response) is proportional to the volume of the Fermi surface. (The Fermi surface in an ordinary metal can be regarded as a topologically protected node with $q = d - 1$—however, because it is so ordinary we ruled it out explicitly above by assuming $q \le d - 2$.)

1.3 Candidate Materials

While there is no experimental confirmation of a Weyl semimetal phase, several promising proposals exist:

(i) *Pyrochlore iridates:* $A_2 Ir_2 O_7$, form a sequence of materials (as the element A is changed) with increasingly strong interactions. A can be taken to be a Lanthanide element or Yttrium. While some members such as $A = Pr$ are excellent metals, others such as $A = Nd, Sm, Eu$, and Y evolve into poor conductors below a magnetic ordering transition. These may be consistent with an insulating or semimetallic magnetic ground state [31].

Recent ab initio calculation [5] using $LSDA + SO + U$ method predicts a magnetic ground state with all-in all-out structure. On varying the on-site Coulomb interactions from strong to weak (see Fig. 2), the electronic structure is found to vary. At very strong interactions, an ordinary magnetically ordered Mott insulator is expected. (This would have no topological properties since strong interactions overwhelm hopping, thus ruling out interesting charge-carrying surface states.) As the interactions become weaker, the system possibly realizes an "axion insulator," which

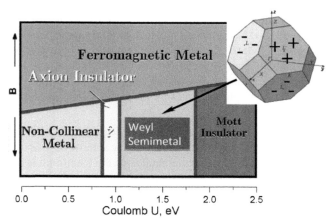

FIGURE 2 A sketch of the calculated phase diagram of pyrochlore iridates with interaction strength (U) on the horizontal axis. The Weyl semimetal appears in a regime of intermediate correlations that is believed to be realized in some members of the pyrochlore iridates. It appears as a transition phase between a trivial (Mott) and a topological (Axion) insulator. The inset shows the location of the 24 Weyl nodes, labeled by their ± chirality, as predicted for this cubic system [5].

is analogous to a strong topological insulator but with inversion rather than time reversal symmetry. A competing state is a magnetic metal. Intermediate between these is the Weyl semimetal phase. This sequence of phase transitions is natural because an axion insulator requires two band inversions while a Weyl semimetal requires only one. There were 24 Weyl points (12 of each chirality) predicted that are related to one another by the cubic point group of this crystal. This implies all 24 Weyl nodes are at the same energy, and at stoichiometry, the chemical potential passes through the nodes. The anomalous Hall effect cancels out because of the symmetry, but if the material is strained, there will be a Hall coefficient that is proportional to the amount of the strain [19]. Other effective theories find broadly similar results [32,33] and related phases have been predicted in spinel osmates [34]. Currently experiments are underway to further investigate these materials.

(ii) *TI-Based Heterostructures:* Ref. [7] showed that layering a topological insulator with a ferromagnet produces a Weyl semimetal with just two Weyl points. Ref. [35] showed that it is also possible to make a Weyl semimetal without breaking time reversal symmetry: it proposes layering $HgTe$ and $CdTe$ and then applying an electric field perpendicular to the layers to break *inversion* symmetry. Magnetic doping near the quantum phase transition between a topological and trivial insulator is also expected to open a region of Weyl semimetal [36].

(iii) *Spinel $HgCr_2Se_4$:* Ab initio calculations of Xu et al. [37] on $HgCr_2Se_4$ show that it is likely to realize a pair of Weyl nodes. It is known to be ferromagnetic, thus breaking time reversal symmetry. The Weyl nodes are "doubled" in that they carry Chern number ± 2, and the low temperature metallic nature is explained by the presence of other bands crossing the Fermi energy. Nevertheless unusual surface states should be visible.

(iv) *Photonic Crystals:* Lu et al. [38] showed that light can be given a Weyl dispersion by passing it through a suitable photonic crystal. This scenario includes Weyl line nodes, which imply flat dispersions on the surface of the crystal, i.e., photon states on the surface with zero velocity.

The symmetry properties of crystals that could realize Weyl semimetals are discussed in [39] and a related massless Dirac semimetal was discussed in [40]. Given how natural the Weyl semimetal is as the electronic state of a 3D solid, these materials candidates may be just the tip of the iceberg.

1.4 Other Directions

In Weyl semimetals, both the surface and the bulk entertain low-energy excitations. The role of disorder and interactions on these and their interplay needs much more study. Initial studies have focused on transport. Simple power counting arguments reveal that disorder is irrelevant, unlike in graphene where it is marginally relevant. Hence weak disorder will not induce a finite density of states in a Weyl semimetal. The conductivity displays a more intricate behavior than this simple analysis suggests [7,29,41]. Turning to the effect of interactions, long-range Coulomb interactions are marginally irrelevant, and the possibility that they dominate transport was considered in [41]. In real samples, charged impurities are likely to move the chemical potential away from the Weyl nodes [29]. Anomalous magnetotransport signatures of Weyl nodes were discussed in [42–45], but other smoking gun signatures of the ABJ anomaly would be of great interest. The question of analogous responses in nodal superconductors is also an open problem. Instabilities of the Weyl semimetal state to density wave [46,47] and superconducting [48–50] instabilities, as well as the nature of topological defects [51] were also recently studied.

2 PART 2: TOPOLOGY OF INTERACTING PHASES

A major theme that saturates the theory of topological insulators is "What are the possible phases of systems that do not break symmetries?" More specifically, what are some qualities that distinguish them? We will focus on states characterized by unusual surface modes, but which are otherwise conventional. Although this ignores interesting topologically ordered states such as fractional Quantum Hall phases in two dimensions, it also captures many interesting phases, and is a very natural generalization of topological insulators, whose surface states (such as Dirac modes and Majorana fermions) have been measured with photoemission, tunneling, and so on.

Interactions can change the classification of topological insulators. There are two possible effects they can have—they can cause phases to merge, or they can allow for new phases to exist which could not exist without interactions. A surprising example of the former type is that the classification of a certain type of one-dimensional topological insulator (class BDI) changes [52]: it is classified by integers when there are no interactions, so there are infinitely many phases; when interactions are added, the classification changes to $\mathbb{Z}_8$, see Fig. 3.

When one asks how interactions change the classification of topological insulators, one means "how does the connectivity of the regions in the phase diagram change due to interactions," not, for example, more involved questions like how the dynamics of excitations is affected by scattering or even about the specific form of the phase diagram. If there is some path joining one state to another without passing through a discontinuity, e.g., in the ground state energy, then these states are defined to be in the same phase. Imagine adding an axis with a parameter that represents the strength of interactions, as in Fig. 3. When the interaction is turned on, some distinct phases in the non-interacting case may merge, similar to how the liquid and vapor phases of water merge at high pressure. This is exactly what happens in the BDI class of superconductors: When interactions are turned on, the two phases that are labeled by 0 and 8 join,

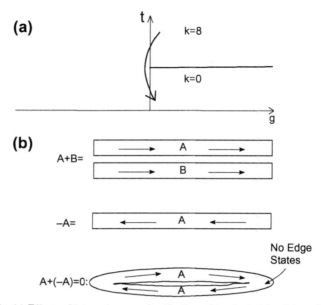

FIGURE 3   (a) Effects of interactions on classification of topological insulators. A schematic phase diagram (based on two of the phases of Majorana fermion chains) with a parameter (t) related to hopping and an interaction parameter (g). Along the non-interacting vertical axis, this diagram contains two distinct phases. When a negative interaction is turned on, the two phases connect to each other. (b) The group operations on topological phases: addition is represented by layering states, the negative of a state is its mirror image. The last figure shows why this is. The arrows are supposed to indicate that the states may break reflection symmetry.

as in the phase diagram shown, although in the non-interacting limit these states remain separated by a phase transition no matter how many axes are added.

2.0.1 Phases with Short-Range Entanglement

We will restrict ourselves to a certain type of phase with Short-Range Entanglement (**SRE**). Physically, this means that we look at phases with (i) An energy gap in the bulk and (ii) No topological order, i.e., a unique ground state when defined on a closed manifold. At the same time these requirements impose a certain locality condition on the quantum entanglement. More precisely, we require that the conditional mutual information of the variables in any three disjoint regions A,B,C tends to zero exponentially as the distance between the two closest points in A and B tends to infinity. The conditional mutual information between three sets of variables A,B,C is defined as $S(A,B|C) := S(AC) + S(BC) - S(C) - S(ABC)$ where S is the entanglement entropy. It roughly captures the correlations between variables A and B after variable C is measured. In gapped phases with topological order, this yields the topological entanglement entropy [53]. This differs slightly from the definition by Chen et al. [13] of SRE phases, which are further required to be able to be continuously changed into a product state without a phase transition when symmetries are allowed to be broken.

A characteristic of topological states with only SRE is unusual edge excitations, namely edge states that cannot exist in a true system of $d - 1$ dimensions, but only on the boundary of the d-dimensional topological state. For example, the doubling theorem says that a chiral mode cannot exist on a discrete one-dimensional lattice, yet it can be simulated at the edge of a two-dimensional quantum Hall state. Another example is a chain made up of integer spin particles. As we will discuss, they can have half-integer spins at the ends, even though half-integer spins cannot be made out of integer spin particles in a zero-dimensional system.

When long-range-entanglement is allowed, an even more fascinating phenomenon can occur: anyon excitations in the bulk, like in the fractional quantum Hall effect. These states however are much more complicated to classify, and so we stick to the SRE states.

Part of the interest of SRE phases is that they have an orderly and interesting structure: they form a group; thus classifying states (such as the integer quantum Hall states) is tractable, much simpler than classifying phases with anyons in the bulk. We can give a simple argument for this based on the assumption that SRE phases can be classified entirely by their edge states, whereas Kitaev has given a more general argument [12]. (The assumption is correct in one dimension as long as none of the symmetries imposed are spatial symmetries like translation or inversion symmetry, and there are no counterexamples known in two dimensions.)

The operations for the group are shown in Fig. 3(b). The addition operation is obvious: take two states and put them side by side. But it is not so obvious that a state has an inverse: how is it possible to *cancel out* edge states? Two copies of a topological insulator cancel one another, because the Dirac points can be coupled by a scattering term that makes a gap. A bigger surprise is that eight copies of the one-dimensional Majorana chain can be coupled by quartic interactions in a way that makes a gap.

The inverse of a phase is its mirror image. To see this, we must show that the state and its mirror image cancel; the argument is illustrated in the last frame of the figure. In one dimension, for example, take a closed loop of the state, and flatten it. The "ends" are really part of the bulk of the loop before it was squashed, so they are gapped. Therefore this state has no edge states. Because topological SRE phases are classified by their edge states, it must be the trivial state.

Phases with long-range-entanglement on the other hand do not form a group because they are not classified by their boundary excitations. When two layers are combined as above, the edges can be gapped, but now there are even more species of anyons in the bulk (one for each layer), and the state is even farther from being equal to the identity. Hence long-range-entangled states do not have as much structure to organize them.

2.1 One-Dimensional Topological Insulators

Classifying topological insulators becomes difficult with interactions because there is no single-particle band structure that allows us to reduce the problem to studying the topology of matrix-valued functions on a torus. Since it is very complicated to solve a many-body Hamiltonian, we start with the ground state directly. In one dimension (with the exception of Majorana phases of fermions), a symmetry must be imposed in order to produce distinct phases.

Since we cannot take advantage of band structure in the interacting case, all we have is (1) the many-body low-energy eigenstates which are localized around the edge, and (2) the symmetry operations. How can these be combined together to distinguish different phases?

A model of an interacting topological phase (that is analogous to a topological insulator) is the Heisenberg spin chain of spin 1 atoms, which has the Hamiltonian

$$H = \sum_i \mathbf{S}_i \cdot \mathbf{S}_{i+1}. \tag{15}$$

This Hamiltonian has an SO_3 symmetry. The bulk of the ground state has an SO_3 symmetry as well (thanks to the Mermin-Wagner theorem), while the edge states are spin $\frac{1}{2}$ excitations.

These excitations have been seen in a compound called NINAZ (among others, see also Ref. [54]) that consists of chains of spin 1's (carried by nickel atoms). Cooling the material through a structural transition breaks the chains up, creating ends. Then the signal from electron spin resonance and the temperature dependence of the magnetic susceptibility [55] behave as if there are free spin $\frac{1}{2}$'s making up the system, even though the atoms all have spin one.

What does it mean to say that the spin one chain has spin $\frac{1}{2}$'s at the end? First of all, this Hamiltonian has four nearly degenerate ground states, the same as one would expect from a pair of spin-$\frac{1}{2}$ particles that are far enough apart that they cannot interact. In order to justify this interpretation, note that the four ground states are not symmetric under rotations of the full chain, $\Sigma(\hat{\mathbf{n}}, \theta) = \prod_j e^{-i\mathbf{S}_j \cdot \hat{\mathbf{n}}\theta}$. Since the order parameter in the bulk vanishes, it seems (although this may be classical intuition at work) that the asymmetry must be associated with asymmetric operators at the ends [73]. This can be tested by applying a magnetic field to just one end. This lifts the degeneracy partly: afterward the levels form two doublets, which indicates that there is one spin $\frac{1}{2}$ at the end the field is applied to, and another at the other end that remains unpolarized.

The bulk is symmetric under rotations, and only the ends change [74]. Therefore, we can write Σ as

$$\Sigma(\theta, \hat{\mathbf{n}})|\alpha\rangle = L(\theta, \hat{\mathbf{n}})R(\theta, \hat{\mathbf{n}})|\alpha\rangle, \tag{16}$$

for all the ground states $|\alpha\rangle$, where $L(\theta, \hat{n}), R(\theta, \hat{n})$ are some operators on sites within the correlation length of the ends [75]. These operators are called the *effective edge symmetries* [76].

We can now introduce effective spin operators by defining $L(\theta,\hat{n}) := e^{-i\mathbf{s}_L\cdot\hat{n}\theta}$ and similarly for $R(\theta,\hat{n})$. We then have

$$\mathbf{S}_{total} = \mathbf{s}_L + \mathbf{s}_R. \tag{17}$$

This is true *within the space of ground states*. When we say that the edges have half-integer spins, we mean that they transform as half-integer spin multiplets under $\mathbf{s}_L$ and $\mathbf{s}_R$. What is the actual form of these operators? They are not just the sum of a sequence of N spins at one of the ends (for some N) because any eigenvalue of this would have to be an integer. Such a sum is inappropriate for defining the spin of the edge because it does not commute with the density matrix of the bulk; measuring this operator breaks bonds from the Nth atom, the last whose spin is summed. This alters the answer because it creates non-zero angular momentum around the Nth atom. The correct operator must be modified to keep the bonds intact.

Now spin chains can be divided into two classes. The eigenstates of $\mathbf{s}_L$ and $\mathbf{s}_R$ come in multiplets. Either all multiplets are integer or all multiplets are half-integer spins. Otherwise, one could pair a half-integer state on the left end with an integer state on the right end (since the two ends are independent). Then the net spin is a half-integer which is impossible, since it must match the physical spin $\mathbf{S}_{total}$ and that is an integer.

The key distinction among spin chains is whether the edge spin is an integer or a half-integer spin; the actual value of the spin is not important. For example, a state with an integer spin at the end (e.g., a spin 2 Heisenberg chain, which has a spin 1 at the end) can be adiabatically transformed into a state without a spin at the end; there will be a level crossing at some point, but this only has to occur at the ends. When the edge spin changes by a *half*-integer there must be a gapless mode in the bulk at some point.

General Classification of Spin Chains and Bosonic Systems—The idea of fractional edge states can be generalized beyond this special example. For each symmetry group, there is a definite list of phases associated with the group. The classification depends only on the abstract type of the symmetry group, and not on what it describes. For example, the classification applies to systems of mobile bosons, as well as fixed spins. Conservation of particle number produces a $U(1)$ symmetry $\psi \to e^{i\theta}\psi$. If there are two species of particle, then there are two $U(1)$ symmetries. Alternatively the symmetries can be spatial symmetry with respect to a cross-section: for example, four chains with a square cross-section can have the symmetry group D_4.

For an ordinary symmetry-breaking transition, a phase is specified by *the symmetry group* and the *list of broken symmetries*.

In a topological state no symmetries are broken in the bulk, so the rule is more subtle. One first observes that the edges might spontaneously break symmetry, even though the bulk does not. Thus, the topological phases are defined by how the edge states transform under the symmetries. Generalizing Eq. (16) to other symmetry groups gives a factorization

$$g_i|\alpha\rangle = L_{g_i} R_{g_i}|\alpha\rangle, \tag{18}$$

for each element g_i of the group.

We now need a way to distinguish "fractional" and "integer" symmetries L_g for discrete symmetry groups. We classified the SO_3-symmetric phases above (i.e., spin chains) by whether the spins of the ends were integer or half-integer valued—that is, by checking whether the eigenvalues of s_{Lz} were integers or half-integers. In the more general case we cannot simply look at the eigenvalues of L_{g_i} because the factorization Eq. (18) into L_{g_i} and R_{g_i} is not unique, thanks to the transformations $L_{g_i} \rightarrow e^{i\alpha} L_{g_i}, R_{g_i} \rightarrow e^{-i\alpha} R_{g_i}$. This ambiguity makes the eigenvalues of individual L_{g_i}'s meaningless, but it turns out also to be the key to a different approach to defining fractionalization.

We must consider several symmetries simultaneously. The action of the symmetries on the left end of the chain defines a "projective representation" of the actual symmetry group. For each g_i, make an arbitrary choice for the phase of L_{g_i}. The effective symmetries then multiply in a way that differs from the original group of symmetries by phase factors, called "factor sets,"

$$L_{g_i} L_{g_j} = e^{i\rho(g_i, g_j)} L_{g_i g_j}, \tag{19}$$

on account of the ambiguity in the phase. *Topological phases can then be classified by the values of the factor sets for the edge states* [8–10].

Understanding how this works requires some more elaboration. Although the phase of L_g is ambiguous, there is a certain part of the phase $\rho(g_1, g_2)$ that cannot be attributed to this ambiguity. This is the part that can be used to classify the phases. In fact, the ρ's can be divided into a discrete number of inequivalent classes, and it is not possible to go from one class to another without either a discontinuous transition or a point where the gap closes (so that the edge state operators temporarily lose their meaning and thus the ρ phases can change). The following two examples illustrate how discrete phases come about.

Classifying states with $\mathbb{Z}_2$ and $\mathbb{Z}_2 \times \mathbb{Z}_2$ symmetry. For a single order two symmetry g, it is possible to eliminate the ρ-phase (there is only one) by making a different choice for the phase of L_g in the first place. If $L_g^2 = e^{i\rho}\mathbb{1}$, then redefine $\bar{L}_g = e^{-i\rho/2}\mathbb{1}$, and the phase has been eliminated.

For two order two symmetries that commute the phases cannot necessarily be completely absorbed by redefining the phases of the L_g's. Let us call the symmetries Σ_x and Σ_z and call the effective edge operators at the left end L_x and L_z. As in the above, L_x^2 must be a phase, but this phase can be absorbed into the definition of L_x. More usefully,

$$\Sigma_x \Sigma_z \Sigma_z^{-1} \Sigma_x^{-1} = \mathbb{1}, \tag{20}$$

which implies

$$L_x L_z L_x^{-1} L_z^{-1} = e^{i\phi}\mathbb{1}, \tag{21}$$

because of the phase ambiguity. *This* phase cannot be eliminated no matter how the phases of the L's are redefined, and so the value of ϕ can be used to discriminate between topological states. In addition, from the fact that $L_x^2 \propto \mathbb{1}$ it follows that $\phi = 0$ or $\phi = \pi$, giving two distinct phases of systems with this $\mathbb{Z}_2 \times \mathbb{Z}_2$ symmetry. The $\phi = \pi$ case is identified as a fractional phase: because the symmetries anticommute, the ends must be twofold degenerate, like spin-$\frac{1}{2}$s.

One application of this example is to spin chains as discussed above, but with crystal anisotropy. Suppose that the spin chain has an anisotropy such that the x, y, and z components of the spin have different coupling strengths (i.e., $H = \sum_i J_x S_{i,x} S_{i+1,x} + J_y S_{i,y} S_{i+1,y} + J_z S_{i,z} S_{i+1,z}$). Then the only remaining symmetries are 180° rotations around the three axes: we will focus on the rotations around x and z, Σ_x and Σ_z (since the third rotation is just their product).

We want to be able to distinguish between integer and half-integer edge-states. One idea is to characterize a half-integer spin by the fact that it picks up a minus sign under a 360° rotation. The phase of the rotation $L_x^2 = \pm \mathbb{1}$ acting on the edge is not well defined however. The characterization above does distinguish two phases, though. The two symmetries commute, $\Sigma_x \Sigma_z = \Sigma_z \Sigma_x$ (most pairs of rotations do not commute, but two 180° rotations around orthogonal axes do, as one can check by rotating a book around two axes of symmetry). Thus there are two phases possible.

One can check that 180° rotations around x and z of a spin either commute or anticommute depending on whether the spin is an integer or half-integer; thus the two possible phases are the vestiges of the phases of SO_3-symmetric spin chains with integer and half-integer spins at the ends, respectively; the full SO_3 symmetry is not necessary to distinguish the states.

General Groups—Two states are in the same phase if the factor set $\rho(g_i, g_j)$ is the same; indeed there is always a path connecting them continuously in this case, as shown by Refs. [9,10]. Thus, the phases for a general group can be analyzed by solving for all the possible factor sets. The ρ's must satisfy a set of simultaneous equations,

$$\rho(g_i g_j, g_k) + \rho(g_i, g_j) = \rho(g_i, g_j g_k) + \rho(g_j, g_k) \bmod 2\pi, \qquad (22)$$

which follow from associativity of the product $L_{g_i} L_{g_j} L_{g_k}$. Now, two states are in the same phase if their factor sets are equivalent. That is, they are in the same phase whenever their factor sets can be transformed into one another by making a "gauge transformation" on the effective symmetry operators of one of the states, $L_{g_i} \to e^{i\alpha(g_i)} L_{g_i}$. Thus the two sets of phases

$$\rho'(g_i, g_j) = \rho(g_i, g_j) + \alpha(g_i) + \alpha(g_j) - \alpha(g_i g_j), \qquad (23)$$

describe the same phase. When Eq. (22) is solved and the solutions are broken into distinct classes according to Eq. (23), then there are a finite number of distinct classes. The group structure of the phases is easy to see in this result;

a pair of side-by-side chains whose factor sets are ρ_1 and ρ_2 has a combined factor set of $\rho_1 + \rho_2$. With this operation, the set of classes for a given symmetry group G also becomes a group, which is called the second cohomology group $H^2(G)$, or the "Schur multiplier of G" [56].

Note that *any* state with a non-trivial ρ that cannot be reduced to zero must have edge states. Indeed, if the ground state is unique, it is an eigenfunction of all the L_g's, with eigenvalue $e^{-i\alpha(g)}$ say. Eq. (19) then says that $\rho(g_i, g_j)$ can be gauged to zero.

The minimal symmetry that can be used to define topological phases is time reversal, where we assume that $T^2 = 1$ (i.e., the particles are spinless). T is antiunitary, so L_T is as well. It squares to a pure phase; however, the phase cannot be absorbed for an antiunitary operator, and has to be $+1$ or -1, giving two phases. In the latter case, the edge state forms a doublet by Kramers's theorem. Thus even particles which individually have no "spin degrees of freedom," i.e., ones whose energy levels are all singlets, can form a chain with edge states!

Topological phases might occur in insulating systems of hopping bosons as well as in spin chains; a state like this would be a topological variant on a Mott insulator. An example of this type was proposed by [57], but this state requires inversion symmetry, so there are no edge states. Further work is needed to find realistic topological phases besides the antiferromagnetic spin 1 chain.

Topological Phases of Fermions: Interacting Version of Topological Insulators—Table 1 gives some examples of symmetry groups and the topological phases for them. The last two lines of the table are examples of classifications for fermions (see below for the explanation). These examples generalize the classification of topological insulators to interacting systems, although the symmetry groups are different from the ones that usually appear in the classification of topological insulators. One has to assign the symmetry groups in a different way to each of the Altland-Zirnbauer classes when there are interactions, because one is not just studying the single-particle band structure. For example, the class A (the quantum Hall case, in two dimensions) is usually said not to have any symmetry, because the matrices describing the one-body equation of motion do not have any symmetry. But this system does have particle number conservation which is an important symmetry for classifying this state (the Hall conductivity is the amount of charge transported). Thus the symmetry group is $U(1)$. Superconducting phases have a particle-hole symmetry in the standard classification, but it is not present when there are interactions. Superconducting phases are characterized by the fact that they do not have a full $U(1)$ symmetry, but only $\mathbb{Z}_2^{charge}$ symmetry: the term $\Delta \psi_\uparrow \psi_\downarrow$ that describes pairing (at least in mean-field theory) breaks $U(1)$ symmetry. Thus, the class DIII is described [77] by $\mathbb{Z}_2^T \times \mathbb{Z}_2^{charge}$, where $\mathbb{Z}_2^T$ refers to time reversal. Note that we model a superconductor by explicitly inserting a pairing term into the Hamiltonian to ensure that the chain is gapped. We do not consider the effects of phase fluctuations.

TABLE 1 Classification of some gapped $D = 1 + 1$ dimensional phases of bosons and fermions.

Symmetry	Topological Classification	Comments
No symmetry	0	Topological states require symmetry in 1D
SO_3	$\mathbb{Z}_2$	Spin chain with *integer* spins
Z_2^T such that $T^2 = 1$.	$\mathbb{Z}_2$	Time reversal symmetry
$U(1)$	0	Charge conserved
$\mathbb{Z}_n$	0	
$\mathbb{Z}_n \times \mathbb{Z}_m$	$\mathbb{Z}_{(n,m)}$	$(a,b) \equiv$ greatest common divisor of a and b
$U(1) \rtimes \mathbb{Z}_2^{particle\text{-}hole}$	$\mathbb{Z}_2$	Particle-hole symmetry
Fermions; $\mathbb{Z}_2^{charge}$	$\mathbb{Z}_2$	Superconductors
Fermions; $\mathbb{Z}_2^{charge} \times \mathbb{Z}_2^T$; $T^2 = 1$	$\mathbb{Z}_8$	Superconductor with time reversal of spinless fermions

Besides these 10 Altland-Zirnbauer symmetry groups, the system can have internal symmetries. Unlike in the interacting case, each possible symmetry group has to be treated separately. For the non-interacting case, only the symmetries that are non-unitary (or flip the sign of the energy) need to be classified, explaining why the Altland-Zirnbauer classes are so important. Any ordinary internal symmetry commutes with the momentum, so it can be used to divide the band structure up into blocks according to the states' quantum numbers. Then each block may be classified according to the Altland-Zirnbauer classification [78].

On the other hand, in an interacting system, interactions exist between the different blocks, and hence all symmetries (not just the antiunitary ones) can become fractionalized by interactions; there are more (infinitely many more) than 10 classes in the presence of interactions.

The methods of the previous section can be applied to fermions as well with a small modification; this gives a methodical explanation of the $\mathbb{Z}_8$ classification of superconductors of the type BDI [11]. Two special features of fermions have to be taken into account. These features already protect two phases even when no symmetry is imposed. The first feature is that there is no way to break the conservation of fermion parity Q (defined as ± 1 depending on whether there is an odd or an even number of fermions) because unpaired fermions cannot form a condensate. This is just a single $\mathbb{Z}_2$ symmetry, and for bosons that would not be enough to protect distinct phases. However, the second key feature is that the factors L_Q and R_Q of Q may either *commute* or *anticommute* because they can contain fermionic creation and annihilation operators. These correspond to distinct phases.

When L_Q and R_Q anticommute, the consequence is that there is a fermion state that is split between the two ends of the chain. The commutation properties of L_Q and R_Q show that there is a twofold degenerate ground state associated with the system as a whole, not with the separate ends. This represents the two states (occupied and not occupied) associated with a fermion. If there is an observer at each end, then either one of them can switch the system between these two states, by applying L_Q or R_Q. On the other hand *neither* observer acting alone can measure which of the two states the system is in (as L_Q, R_Q are fermionic, they are not observable). This two-level system can therefore act as a robust q-bit because it cannot decohere through "observation" by the environment [59].

This classification into two phases agrees with the appropriate non-interacting classification, i.e., the type D topological insulators. In fact, these are classified into two phases, depending on whether there is an even or an odd number of Majorana operators $a_{R/Li}$ (satisfying $a_{R/Li}^2 = 1$) at an end. The total parity of a state below the gap in such a non-interacting insulator is given by the product over all Majorana and ordinary fermion edge modes: $\prod_i a_{Li} \prod_i a_{Ri} (-1)^{\sum_i f_{Li}^\dagger f_{Li} + \sum f_{Ri}^\dagger f_{Ri}}$ (up to a phase). The L_Q and R_Q operators are the parts corresponding to the two ends. Thus L_Q is fermionic in a phase with an odd number of Majorana modes at each end. The arguments above show that a twofold degeneracy survives in this case even when the Majorana modes interact.

When an additional symmetry is added, the interacting and non-interacting cases behave differently. If we consider "spinless time reversal symmetry," i.e., $T^2 = 1$, then the classification table for class BDI says that the phases are classified by $\mathbb{Z}$. However, it seems unlikely to be possible to get a classification into an infinite number of states by studying the symmetries of the edge states, since each symmetry or set of symmetries gives rise to only a few discrete phases like ϕ in Eq. (21). In fact, analyzing the symmetries shows that there are just eight interacting phases. They are classified by the binary choice just described (whether L_Q and R_Q are bosonic or fermionic) together with one that describes whether $L_T^2 = \pm 1$ and another that describes whether L_T and Q_T commute or anticommute, giving eight phases in all. One can also see this starting from the non-interacting limit where any integer number of Majorana zero modes at the edge may be realized. Suppose k identical chains each with a Majorana mode at the end are placed side by side. Say the modes are all even under time reversal. Then these cannot couple to one another in quadratic order because of time reversal symmetry (a coupling such as $a_1 a_2$ isn't Hermitian, while $i a_1 a_2$ violates time reversal symmetry) so a degeneracy of 2^k for the entire segment is protected. However, if there are interactions, there can be a coupling of 4 Majoranas ($a_1 a_2 a_3 a_4$). This reduces the degeneracy *per end* from fourfold to twofold. That suggests that when there are enough Majoranas at the ends of the chain, it should be possible to couple them in a way that removes all the degeneracy. And in fact that is possible once there are 8 Majoranas. The twofold degeneracy of an end of the $k = 4$ state is protected by time reversal. The doublet is a Kramers doublet (although $T^2 = 1$ in the microscopic degrees of freedom, $T^2 = -1$ in the fractionalized edge states). It can be described as a spin-$\frac{1}{2}$ (though not with rotational symmetry). When two sets of four chains are coupled together, the spin-$\frac{1}{2}$ in each can form a singlet, and the edge degeneracy is removed.

Summary—In one dimension, distinct phases exist only when the Hamiltonian is symmetric, with the exception of Majorana fermions (these are also related to a symmetry, Q, but *this* symmetry is automatic). These phases have fractionalized edge states, and the states can be characterized based on the factor set for the projective transformations of the edge states. These phases form a group when addition is defined as placing two states side by side.

2.2 Topological SRE Phases in Higher Dimensions

Now we turn to two dimensions. In 2D topological order, as exemplified by the fractional quantum Hall state, becomes possible for the first time. In contrast with topological insulators, as exemplified by the integer quantum Hall and quantum spin Hall states, these systems have non-trivial bulk excitations with anyon statistics. Many instances of topologically ordered states have been studied, in connection with the fractional quantum Hall effect and quantum spin liquids. The surprise in recent times has been the discovery of interacting topological phases with SRE [12, 13] in $d > 1$, the analog of the *integer* quantum Hall effect for interacting systems. Hence we will focus on these phases. The notable feature of topological insulators is their surface states, and the SRE phases share this property with them: there are no non-trivial excitations in their bulk. In contrast, topologically ordered states do not necessarily have edge states (although some do). We will specialize to gapped phases built out of *bosons*. This ensures that interactions are essential to the description (since non-interacting bosons can only form superfluids), so that we will not mistakenly reproduce a documented free fermion topological insulator.

The most basic distinction between phases arises in the absence of any symmetry. For one-dimensional gapped phases, there are no non-trivial distinctions in the absence of symmetry. However, in two dimensions, this is no longer true, as pointed out by Kitaev. A convenient way to distinguish topological phases is via quantized response. While the best known quantized response is Hall conductance, this implicitly assumes the presence of a conserved electric charge that couples to the applied electric field. In the absence of symmetry, there is no conserved charge. However, one can still define a quantized thermal Hall conductance κ_{xy}. Its ratio to the temperature $\frac{\kappa_{xy}}{T}$ is quantized in units of $L_0 = \frac{\pi^2 k_B^2}{3h}$. Since energy is always conserved, this phase remains distinct even when charge conservation is violated. It is well known that in the context of free fermion topological phases, the chiral superconductors that break time reversal symmetry (class D) may be distinguished by the number of Majorana modes at the edge, with thermal Hall conductance quantized, in units of $\kappa_{xy} = \frac{1}{2}L_0 T$ while the integer quantum Hall phases have Hall conductances quantized in units of L_0. Note, as usual this is sharply defined only in the limit of zero temperature ($T \to 0$), and the quantized response is actually the ratio of thermal Hall conductance to temperature. This quantized thermal conductance can also be defined for interacting phases. Kitaev showed that it is possible to construct a bosonic insulator (which is necessarily interacting) with chiral edge modes that has a quantized thermal Hall conductance. Moreover, assuming there is no bulk topological order, the thermal Hall conductance κ_{xy}/T of such an insulator must be quantized to multiples of $8L_0$ (16 times the minimum conductance of an SRE system of fermions). When the bulk has topological order, the thermal Hall conductance can be a fraction of this, similar to the ordinary Hall conductance in a fractional quantum Hall state.

One can also discuss topological phases protected by a symmetry—for example, U(1) charge conservation. This leads to an integer classification of SRE phases [13–15] of gapped bosons in 2D. Just as in the previous case, the topological phases are distinguished by a quantized response, which in this case is quantized Hall conductance. Interestingly, for bosons, the Hall conductances are quantized to *even* integers, i.e., $\sigma_{xy} = 2n\frac{q^2}{h}$ [14], assuming no topological order. While there are different ways to see these results, the easiest is via the Chern-Simons K-Matrix formulation which we briefly describe.

K-matrix formulation: In the K-matrix formulation of quantum Hall states, a symmetric integer matrix appears in the Chern-Simons action: ($\hbar = 1$, and summation is implied over repeated indices $\mu, \nu, \lambda = 0, 1, 2$):

$$4\pi \mathcal{S}_{CS} = \int d^2 x dt \sum_{I,J} \epsilon_{\mu\nu\lambda} a_\mu^I [\mathbf{K}]_{I,J} \partial_\nu a_\lambda^J. \tag{24}$$

While this has been utilized to discuss quantum Hall states with Abelian topological order, it is also a powerful tool to discuss topological phases in the *absence* of topological order. The latter requires $|\det \mathbf{K}| = 1$ (i.e., $\mathbf{K}$ is a *unimodular* matrix). Furthermore, in order that all excitations are bosonic, we demand that the diagonal entries of the matrix are *even* integers. The bulk action also determines topological properties of the edge states. For example, the signature of the K matrix (number of positive minus negative eigenvalues) is the chirality of edge states—the imbalance between the number of right and left moving edge modes. Maximally chiral states have all edge modes moving in the same direction. Physically, the fluxes $j_I^\lambda = \epsilon^{\mu\nu\lambda} \partial_\mu a_\nu^I$ are often related to densities and currents of bosons of different flavors.

We begin by looking for maximally chiral states of bosons without topological order. These are bosonic analogs of the integer quantum Hall effect or chiral superconductors of fermions; they have a *thermal* Hall coefficient of nL_0T where n is the dimension of the K matrix. The smallest dimension of bosonic unimodular K matrix that yields a maximally chiral state is 8 [14]. This is consistent with the prediction of Kitaev, derived from topological field theory. An example of such a K matrix is given by:

$$K^{E_8} = \begin{pmatrix} 2 & -1 & 0 & 0 & 0 & 0 & 0 & 0 \\ -1 & 2 & -1 & 0 & 0 & 0 & 0 & 0 \\ 0 & -1 & 2 & -1 & 0 & 0 & 0 & -1 \\ 0 & 0 & -1 & 2 & -1 & 0 & 0 & 0 \\ 0 & 0 & 0 & -1 & 2 & -1 & 0 & 0 \\ 0 & 0 & 0 & 0 & -1 & 2 & -1 & 0 \\ 0 & 0 & 0 & 0 & 0 & -1 & 2 & 0 \\ 0 & 0 & -1 & 0 & 0 & 0 & 0 & 2 \end{pmatrix}. \tag{25}$$

This K matrix is the Cartan matrix of the group E_8, and hence we call this phase the Kitaev E_8 phase; see Fig. 4(a). It has edge modes protected by the thermal Hall effect, which does not require any symmetry.

We now consider non-chiral states of bosons, with equal number of left and right moving edge modes, in order to investigate other routes to protected edge modes. In the absence of symmetry we believe there are no non-chiral topological phases with $|\det \mathbf{K}| = 1$. However, the presence of a symmetry can lead to new topological phases. The integer quantum Hall states of bosons are in fact modeled by the very simple K matrix: $K = \begin{pmatrix} 0 & 1 \\ 1 & 0 \end{pmatrix}$. Due to the presence of a conserved charge, the electromagnetic response of the system can be computed by coupling an external gauge field A_μ. The field theory then is:

$$\mathcal{L} = \frac{1}{2\pi}\epsilon^{\mu\nu\lambda}\left[a_{1\mu}\partial_\nu a_{2\lambda} - (a_{1\mu} + a_{2\mu})\partial_\nu A_\lambda\right], \tag{26}$$

where we have taken the coupling to have the simplest interesting form [79]. Integrating out the internal fields gives the effective action for the electromagnetic response:

$$\mathcal{L}_{em} = -\frac{1}{2\pi}\epsilon^{\mu\nu\lambda}A_\mu\partial_\nu A_\lambda, \tag{27}$$

which corresponds to a $\sigma_{xy} = 2$. A pictorial representation of this phase is given in Fig. 4(a). Intuitively, the non-chiral modes at the edge implied by Eq. (26) can be viewed as oppositely propagating modes, one of which is charged and the other neutral [14–16]. This prohibits backscattering between the modes. Another viewpoint is that it is a regular Luttinger liquid except in that it transforms under the internal $U(1)$ symmetry in a way that is impossible to realize

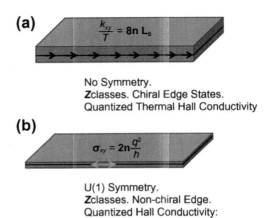

(a)
$$\frac{k_{xy}}{T} = 8n\,L_0$$

No Symmetry.
Zclasses. Chiral Edge States.
Quantized Thermal Hall Conductivity

(b)
$$\sigma_{xy} = 2n\frac{q^2}{h}$$

U(1) Symmetry.
Zclasses. Non-chiral Edge.
Quantized Hall Conductivity:

FIGURE 4 Summary of some simple "integer" bosonic topological phases. (a) A chiral phase of bosons (no symmetry required). An integer multiple of eight chiral bosons at the edge is needed to evade topological order, leading to a quantized thermal Hall conductance $\kappa_{xy}/T = 8nL_0$ in units of the universal thermal conductance $L_0 = \frac{\pi^2 k_B^2}{3h}$. These are bosonic analogs of chiral superconductors. (b) A non-chiral phase of bosons protected by $U(1)$ symmetry (e.g., charge conservation). Distinct phases can be labeled by the quantized Hall conductance $\sigma_{xy} = 2n\sigma_0$, which are even integer multiples of the universal conductance $\sigma_0 = q^2/h$ for particles with charge q. These are bosonic analogs of the integer quantum Hall phases.

in a 1D Luttinger liquid [14,16]. For example, in this particular case, both the conjugate fields θ, ϕ of the Luttinger liquid transform like charge phases under the $U(1)$ symmetry. Recent work has argued that the Integer Quantum Hall state of bosons could appear in interacting systems of bosons in the lowest Landau level [15], that could be realized in cold atom systems exposed to an artificial gauge field.

In the presence of more complicated symmetries, there are many more topological phases. These phases are at least partly classified by the third cohomology group of G, as shown by Ref. [13,60] The classification is still not entirely complete, especially when Majorana edge modes are present. Besides producing new phases in bosonic systems, interactions can also reduce the number of topological insulating phases in fermionic systems in higher dimensions (similar to the one-dimensional case) [14,61,62].

SRE Phases and Topological Order: There are interesting and deep relations between SRE topological phases and states with topological order. Levin and Gu [63] pointed out that the relation can be viewed as a duality. It is well known that the 2 + 1D quantum Ising model with Z_2 global symmetry is dual to a Z_2 gauge theory, with the disordered phase mapping to the deconfined phase of the gauge theory. However, there exists a SRE topological phase with Z_2 symmetry, distinct from the disordered phase of the Ising model. Under the same duality it maps to a twisted version of the Z_2 gauge theory, the doubled semion model. This approach can be used to give an alternative interpretation of the classification obtained by Refs. [13,60]. The idea is to study the properties of the gauge excitations, point-like excitations similar to vortices in a superconductor that are created by passing a flux line through the system for an element g_i of the symmetry group. If we take three such gauge-fluxes g_1, g_2, g_3 arrayed along a line [80] then they can be combined to make an excitation with flux $g_1 g_2 g_3$. This can be done in two orders, fusing $1, 2$ first and then 3 or $2, 3$ first and then 1. The difference between the two orders gives a phase factor $e^{i\kappa(g_1, g_2, g_3)}$. This phase factor is ambiguous because a phase can accumulate as two of the fluxes are brought together, depending on arbitrary choices one makes about how they are moved. Nevertheless there are still inequivalent choices of κ's, similar to the inequivalent classes of ρ's in the one-dimensional case. This gives a way to distinguish phases. The group of κ's is called the third cohomology group, which therefore classifies the bosonic states (at least partly) as Ref. [13,60] found. A similar result about discrete gauge theories has been known [64], that there can be extra phase factors in the mutual statistics and fusion of the anyons for exactly this reason.

2.3 Edge States and Topological Order

We have been characterizing SRE phases by their edge states. However edge states can exist in fractional quantum Hall states as well. Is there a connection between the fractional edge states and the existence of anyons? After all, anyons and edge states both have a handedness. In fact, the thermal Hall conductance of the edge of a system of bosons is *partly* determined by the collection of anyons through the following strange relation [65]:

$$e^{\frac{\pi i}{4}\left[\frac{\kappa}{TL_0}\right]} = \frac{\sum_{i=1}^{N_A} e^{2\pi i s_i}}{\sqrt{N_A}}, \tag{28}$$

where the sum is over the N_A species of anyons and $2\pi s_i$ are their statistical angles (and we have also assumed that the phase is abelian). Thus systems with fractional values of $\frac{\kappa}{8TL_0}$ must have anyons (explaining why the Kitaev E_8 state has $\kappa = 8TL_0$). Conversely, the set of anyons determines $\frac{\kappa}{TL_0}$ partly; it must have the form $\frac{p}{q} + 8n$ where $\frac{p}{q}$ is a certain fraction and n is an indeterminate integer. If this is not a multiple of 8, the edge states are protected without assuming any symmetry.

A related subject that has received much attention is the interplay between topological order and surface states in the presence of global symmetries. Perhaps most relevant to this article is the question of stability of edge states in fractional topological insulators, in the presence of time reversal symmetry. A general condition for abelian states was worked out in [66]. Similar considerations in the context of fractionalized three-dimensional phases with time reversal symmetry [67,68] were discussed. In insulators that are adiabatically connected to free fermions, the magnetoelectric polarizability, θ, is fixed to be multiples of π to preserve time reversal symmetry. In Refs. [68] it was argued that fractional values of θ/π are compatible with time reversal symmetry only if the bulk possesses topological order.

2.4 Outlook

There are several open questions in this rapidly evolving field. Perhaps most urgently, the physical nature of SRE topological phases in $d > 1$ needs to be more deeply understood. Some recent work has begun to discuss properties of 3D SRE topological phases [69,70] predicted in Ref. [13]. Another important open question concerns the physical realization of the bosonic topological phases in quantum magnets and lattice cold atom systems, and the types of interactions required to realize them in various dimensions. It also remains to be seen if interacting fermions can lead to new topological phases. So far, all the cases of fermionic topological states that have been examined [14,61] either reduce to bosonic phases formed from bound states of the fermions or to non-interacting topological insulators. The deep connections between topological order and SRE topological phases, as well as the interplay of symmetry and topological order, remain to be more deeply explored.

ACKNOWLEDGMENTS

We would like to acknowledge insightful discussions with Alexei Kitaev, Lukasz Fidkowski, and Leon Balents and acknowledge our collaborators on related projects Sergey Savrasov, Xiangang Wan, Yuan-Ming Lu, T. Senthil, Tarun Grover, Frank Pollman, Erez Berg, and Masaki Oshikawa. This work was supported in part by Grant Number NSF DMR-1206728.

REFERENCES

[1] Hasan MZ, Kane CL. Rev Mod Phys 2010;82:3045.
[2] Qi X-L, Zhang S-C. Rev Mod Phys 2011;83:1057.
[3] Schnyder Andreas P, Ryu Shinsei, Furusaki Akira, Ludwig Andreas WW. Phys Rev B 2008;78:195125.
[4] Kitaev, Alexei, arXiv:0901.2686.
[5] Wan Xiangang, Turner Ari M, Vishwanath Ashvin, Savrasov Sergey Y. Phys Rev B 2011;83:205101.
[6] Adler S. Phys Rev 1969;177:2426; Bell JS, Jackiw R. Nuovo Cimento 1969;60A:47.
[7] Burkov AA, Balents Leon. Phys Rev Lett 2011;107:127205.
[8] Pollmann Frank, Turner Ari M, Berg Erez, Oshikawa Masaki. Phys Rev B 2010;81:064439.
[9] Chen Xie, Gu Zheng-Cheng, Wen Xiao-Gang. Phys Rev B 2011;83:035107.
[10] Schuch Norbert, Pérez-Garcia David, Ignacio Cirac J. Phys Rev B 2011;84:165139.
[11] Fidkowski Lukasz, Kitaev Alexei. Phys Rev B 2011;83:075103; Turner Ari M, Pollmann Frank, Berg Erez. Phys Rev B 2011;83:075102.
[12] Kitaev A. unpublished. See <http://online.kitp.ucsb.edu/online/topomat11/kitaev/>.
[13] Chen Xie, Gu Zheng-Cheng, Liu Zheng-Xin, Wen Xiao-Gang. Phys Rev B 2013;87:155114. arXiv:1106.4772.

[14] Lu Yuan-Ming, Vishwanath Ashvin. Phys Rev B 2012;86:125119.
[15] Senthil T, Levin Michael. Phys Rev Lett 2013;110:046801. arXiv:1206.1604.
[16] Chen Xie, Wen Xiao-Gang. Phys Rev B 2012;86:235135. arXiv:1206.3117; Liu Zheng-Xin, Wen Xiao-Gang. Phys Rev Lett 2013;110:067205. arXiv:1205.7024
[17] Volovik GE. JETP Lett 2002;75:55; Volovik GE. The Universe in a Helium Droplet. Oxford: Clarendon Press; 2003.
[18] Horava Petr. Phys Rev Lett 2005;95:016405.
[19] Yang Kai-Yu, Lu Yuan-Ming, Ran Ying. Phys Rev B 2011;84:075129.
[20] Delplace Pierre, Li Jian, Carpentier David. EPL 2012;97:67004.
[21] Nielsen HB, Ninomiya Masao. Phys Lett B 1983;130:389.
[22] Von Neumann J, Wigner EP. Phys Z 1929;30:467.
[23] Herring C. Phys Rev 1937;52:365.
[24] Murakami S. New J Phys 2007;9:356.
[25] Hosur P, Ryu S, Vishwanath A. Phys Rev B 2010;81:045120.
[26] Sato M. Phys Rev B 2006;73:214502; Béri B. Phys Rev B 2010;81:134515.
[27] Sato M, Tanaka Y, Yada K, Yokoyama T. Phys Rev B 2011;83:224511; Schnyder AP, Ryu S. Phys Rev B 2011;84:060504R; Brydon PMR, Schnyder Andreas P, Timm Carsten. Phys Rev B 2011;84:020501(R); Wang Fa, Lee Dung-Hai. Phys Rev B 2012;86:094512; Schnyder AP, Brydon PMR, Timm C. Phys Rev B 2012;85:024522.
[28] Heikkilä T, Kopnin N, Volovik G. JETP Lett 2011;94:233; Silaev MA, Volovik GE. Phys Rev B 2012;86:214511. arXiv:1209.3368.
[29] Burkov AA, Hook MD, Balents Leon. Phys Rev B 2011;84:235126.
[30] Castro Neto AH, Guinea F, Peres NMR, Novoselov KS, Geim AK. Rev Mod Phys 2009;81:109.
[31] Yanagishima D, Maeno Y. J Phys Soc Jpn 2001;70:2880; Taira N, Wakeshima M, Hinatsu Y. J Phys: Condens Matter 2001;13:5527; Zhao Songrui, Mackie JM, MacLaughlin DE, Bernal OO, Ishikawa JJ, Ohta Y. Phys Rev B 2011;83:180402; Tafti FF, Ishikawa JJ, McCollam A, Nakatsuji S, Julian SR. Phys Rev B 2012;85:205104.
[32] Witczak-Krempa William, Kim Yong Baek. Phys Rev B 2012;85:045124.
[33] Chen Gang, Hermele Michael. Phys Rev B 2012;86:235129. arXiv:1208.4853.
[34] Wan Xiangang, Vishwanath Ashvin, Savrasov Sergey Y. Phys Rev Lett 2012;108:146601.
[35] Gábor Halász, Balents Leon. Phys Rev B 2012;85:035103.
[36] Cho Gil Young. arXiv:1110.1939.
[37] Xu Gang, Weng Hongming, Wang Zhijun, Dai Xi, Fang Zhong. Phys Rev Lett 2011;107:186806.
[38] Lu Ling, Fu Liang, Joannopoulos John D, Soljačić Marin. Nat Photonics 2013;7:294–9 arXiv:1207.0478.
[39] Mañes JL. Phys Rev B 2012;85:155118.
[40] Young SM, Zaheer S, Teo JCY, Kane CL, Mele EJ, Rappe AM. Phys Rev Lett 2012;108:140405.
[41] Hosur Pavan, Parameswaran SA, Vishwanath Ashvin. Phys Rev Lett 2012;108:046602.
[42] Aji Vivek. Phys Rev B 2012;85:241101.
[43] Zyuzin AA, Burkov AA. Phys Rev B 2012;86:115133.
[44] Son Dam Thanh, Yamamoto Naoki. Phys Rev Lett 2012;109:181602.
[45] Son DT, Spivak BZ. Phys Rev B 2013;88:104412. arXiv:1206.1627.
[46] Wang Zhong, Zhang Shou-Cheng. Phys Rev B 2013;87:161107(R). arXiv:1207.5234.
[47] Wei Huazhou, Chao Sung-Po, Aji Vivek. Phys Rev Lett 2012;109:196403. arXiv:1207.5065.
[48] Sau Jay D, Tewari Sumanta. Phys Rev B 2012;86:104509.
[49] Meng Tobias, Balents Leon. Phys Rev B 2012;86:054504.
[50] Cho Gil Young, Bardarson Jens H, Lu Yuan-Ming, Moore Joel E. Phys Rev B 2012;86:214514. arXiv:1209.2235.
[51] Liu Chao-Xing, Ye Peng, Qi Xiao-Liang. Phys Rev B 2013;87:235306. arXiv:1204.6551.
[52] Fidkowski Lukasz, Kitaev Alexei. Phys Rev B 2010;81:134509.
[53] Levin M, Wen X-G. Phys Rev Lett 2006;96:110405; Kitaev A, Preskill J. Phys Rev Lett 2006;96:110404.
[54] Renard JP, Verdaguer M, Regnault LP, Erkelens WAC, Rossat-Mignod J, Stirling WG. Europhys Lett 1987;3:945; Hagiwara M, Katsumata K, Affleck Ian, Halperin BI, Renard JP. Phys Rev Lett 1990;65:3181.

[55] Granroth GE et al. Phys Rev B 1998;58:9312.

[56] Schur JJ. Reine. Angew Math 1904;127:20.

[57] Dalla Torre Emanuele G, Berg Erez, Altman Ehud. Phys Rev Lett 2006;97:260401.

[58] Dyson Freeman J. J Math Phys 1962;3:1199.

[59] Kitaev A Yu. Phys Usp 2001;44(suppl.):131–6; see also Beenakker Carlo. Annu Rev Condens Matter Phys 2013;4:113. arXiv:1112.1950 and its references.

[60] Chen Xie, Liu Zheng-Xin, Xiao Gang Wen. Phys Rev B 2011;84:235141.

[61] Gu Zheng-Cheng, Wen Xiao-Gang. arXiv:1201.2648.

[62] Qi Xiao-Liang. New J Phys 2013;15:065002. arXiv:1202.3983; Ryu Shinsei, Zhang Shou-Cheng. Phys Rev B 85:245132; Yao Hong, Ryu Shinsei. Phys Rev B 2013;88:064507. arXiv:1202.5805.

[63] Levin Michael, Gu Zheng-Cheng. Phys Rev B 2012;86:115109.

[64] Dijkgraaf Robbert, Witten Edward. Comm Math Phys Volume 1990;129:393.

[65] Fröhlich J, Gabbiani F. Rev Math Phys 1990;2:251; Rehren KH. In: Kastler D editor. The Algebraic Theory of Superselection Sectors: Introduction and Recent Results, Proceedings of the Convegno Internationale Algebraic Theory of Superselection Sectors and Field Theory, Instituto Scientifico Internationale G.B. Guccia, Palermo 1989. Singapore: World Scientific Publishing; 1990; See also Kitaev A. Annals Phys 2006:321.

[66] Michael Levin, Stern Ady. Phys Rev Lett 2009;103:196803; Michael Levin, Stern Ady. Phys Rev 2012;86:115131.

[67] Pesin D, Balents L. Nat Phys 2010;6:376–81.

[68] Maciejko Joseph, Qi Xiao-Liang, Karch Andreas, Zhang Shou-Cheng. Phys Rev Lett 2010;105:246809; Swingle B, Barkeshli M, McGreevy J, Senthil T. Phys Rev B 2011;83:195139.

[69] Vishwanath A, Senthil T. Phys Rev X 2013;3:011016. arXiv:1209.3058.

[70] Swingle B. arXiv:1209.0776.

[71] Sometimes, lattice symmetries can cause this sum to vanish, in which case a different quantity is needed.

[72] Nodes at time reversal invariant momenta may be protected.

[73] An exotic alternative is that there is some asymmetric correlation among all the spins in the chain simultaneously, but it is not the case.

[74] The wave function cannot be divided into a bulk wave function and an edge wave function, on account of entanglement. The precise meaning of this statement is that the *reduced density matrix* of the interior of the chain is not affected by applying Σ.

[75] Intuitively, Σ is defined by cutting off several spins starting at the end, rotating them, and then correcting the bonds that have been broken.

[76] Since there are only four ground states, this condition does not uniquely define the edge spin operators. However, we can enlarge the space of edge states by supposing that the coupling between spins increases adiabatically over many spins to a very large value in the bulk; there will then be many sub-gap states and Eq. (17) can be required to hold for all of them.

[77] In fact there are really at least two ways to interpret each of the D and C symmetry classes when interactions start to play a role. In the non-interacting case, this can describe either a system with an actual particle-hole symmetry or a superconductor, where the particle-hole symmetry is a mathematical artifact. But in the interacting case, these two interpretations correspond to different symmetry groups.

[78] A block may be classified by a different Altland-Zirnbauer than the system as a whole, if, e.g., the action of the internal symmetry on the block is described by a complex representation. Ref. [58] explains how this works.

[79] The coupling between A_μ and $a_{j\mu}$ must have an integer coefficient; curl a_j represents a density of bosons of some type, which must have integer charges. We set both coefficients equal to one since this gives the smallest non-zero Hall conductivity.

[80] Arranging them on a line allows us to label fluxes by group elements rather than conjugacy classes.

Index